MODERN GEOMETRICAL OPTICS

MODERN GEOMETRICAL OPTICS

RICHARD DITTEON

A WILEY-INTERSCIENCE PUBLICATION

JOHN WILEY & SONS, INC.

New York • Chichester • Weinheim • Brisbane • Singapore • Toronto

This text is printed on acid-free paper.

Library of Congress Cataloging in Publication Data:

Ditteon, Richard, 1953–
Modern geometrical optics / by Richard Ditteon.
p. cm.
Includes index.
ISBN 0-471-16922-6 (cloth : alk. paper)
1. Geometrical optics. II. Title.
QC381.D48 1997
621.36—dc21 97-24252

Printed in the United States of America

10 9 8 7 6 5 4 3 2 1

To Margaret for her patience and support

CONTENTS

PREFACE

The ultimate purpose of the study of geometrical optics is the design of optical systems. Scientists and engineers who are asked to design optical systems frequently have very little background in geometrical optics. Typically, they have been exposed to geometrical optics only in their introductory physics courses. At the introductory physics level, important topics such as field of view, stops, pupils and windows, exact ray tracing, image quality, and optimization of the image are not covered. Advanced books on lens design and aberrations cover these topics, but do not present enough background information to be easily understood by the novice lens designer. This book is intended to bridge the gap between introductory physics texts and advanced texts on lens design.

The best way for students to learn a new subject is by doing. With this in mind, the text contains many example problems which range from very basic to more advanced. Each chapter also includes an extensive problem set. There is a set of conceptual problems which require little, if any, computation. Some problems are "made up" with ease of calculation in mind. These problems are designed so that the student can practice applying the underlying concept without using a computer. More realistic, and more computationally difficult, problems are also included. These problems can be solved practically only by using special lens design and analysis software. Many companies sell software for this purpose, but such software is generally expensive and not readily available. Therefore, a program which is capable of solving most of the problems in the book is available for downloading on the Wiley ftp server. Directions for downloading the software and a brief user's manual for the software is included in Appendix C.

Chapter 1 presents an overview of the field of optics and the relationship of the subfields of geometrical optics with physical optics and quantum optics. The basic properties of optical materials, index of refraction and dispersion, are discussed. The chapter concludes with several nonimaging applications of geometrical optics.

Chapter 2 discusses the basic properties and requirements of imaging systems. Much of the material in the first two chapters is the type of material that a scientist or engineer would encounter in college. An important reason for repeating this material is to introduce the notation and sign conventions which will be used throughout the text.

Optical engineers have developed *y-nu* ray tracing as a method of analyzing complex optical systems in order to determine basic first-order properties of the systems. This technique is fully developed in Chapter 3. Also, the cardinal points of an optical system are defined here.

Chapter 4 contains several additional paraxial optics topics. Here, the aperture stop and entrance and exit pupils are defined. Also, defined are the field stop and entrance and exit windows. Methods for finding each of these items are presented in detail. Vignetting and vignetting diagrams are discussed. Next, the only first-order aberrations, axial and lateral chromatic aberration, are presented. Finally, the concept of optical invariants is introduced and applied to simplify our calculations.

Many people are not familiar with *y-nu* ray tracing or simply prefer to use matrix techniques. Chapter 5 covers basically the same material as Chapters 3 and 4, but using matrix methods rather than the *y-nu* ray trace tables. The relationship between the two techniques is also discussed here.

Chapter 6 covers exact ray tracing and presents several different ways of evaluating image quality.

Chapter 7 presents third-order aberration theory as a means to understanding and correcting defects in image quality.

Chapters 3 through 7 discuss various methods for analyzing existing optical systems. The remaining chapters of the book serve as an introduction to the design of optical systems. Chapter 8 covers y–$\bar{y}$ diagrams, a technique for laying out the first-order design of a system. With this technique, a lot of trial and error can be avoided.

Chapter 9 discusses computer-aided optimization of optical systems. First, the error function is introduced as a measure of image quality. Next, the theory behind optimization is presented. Finally, optimization is applied to a few selected systems in Chapter 10.

MODERN GEOMETRICAL OPTICS

CHAPTER 1

THE NATURE OF LIGHT

Light is like an elephant.

An old story tells of three blind men who were asked to describe an elephant. One blind man touched the elephant's tail and said the elephant was long and thin like a rope. The second blind man touched the elephant's leg and described the elephant as round and hard like a tree trunk. The third man felt an ear and said that the elephant was thin and flat like a huge leaf. Each man's description was correct, but didn't give the complete picture.

Scientists who study the nature of light are like the blind men in the story. They try to describe light, but their descriptions depend very much on which aspects of light they study. It is important for students in any science or engineering discipline to understand that all of the laws, rules, and formulae that they learn are merely approximations to reality. Science progresses by making better and better approximations. For example, Newton's laws, which you learned in an introductory physics course, accurately describe the motion of objects whose velocity is small compared to the velocity of light. However, they fail to adequately explain what happens when the velocity is very large. Newton's laws are a low-velocity approximation to Einstein's equations of relativity.

Each description of light is merely an approximation to the reality that is light. In this chapter, each of the descriptions of light will be discussed briefly. The study of light is called *optics*, and each description of light makes up a separate branch of optics. The purpose of the current chapter is not to give a comprehensive description of the nature of light, but, rather, to put one aspect of the study of light, geometrical optics, into perspective. In addition, this chapter presents some background information on optical materials and a few applications of the laws of geometrical optics.

The analogy that light is like an elephant is appropriate since there are three different ways to describe light: geometrical optics, physical optics, and quantum optics.

1.1 GEOMETRICAL OPTICS

Geometrical optics is not really concerned with the nature of light. That is, geometrical optics can't answer the question, What is light? Rather, geometrical optics is the study of the propagation of light which can be described with a few simple geometrical relationships (hence the name geometrical optics). These relationships are the four laws of geometrical optics. The four laws are based on observation of a wide variety of phenomena. Since the observations require no special scientific apparatus, the observations were made long ago. Geometrical optics is by far the oldest branch of optics. Euclid postulated the laws of rectilinear propagation and reflection around 300 B.C. Ptolemy tabulated angles of incidence and refraction for several media around 130 A.D. In 1621, Snell proposed a mathematical relationship between the angles of incidence and refraction.

The first law of geometrical optics is the law of rectilinear propagation. This is a fancy way of saying that in an isotropic, homogeneous medium, light travels in straight lines. An isotropic homogeneous medium is one whose physical properties are constant and the same in every direction. Examples of such media are a vacuum, optical glass, and calm air. The first law may be demonstrated by placing two screens with small holes some distance apart in front of your eye. A certain part of an object beyond the screens can be seen (Figure 1.1). The two holes define a line along which light travels from the object to the eye. A third screen placed along this line will not affect the view. But if the third screen is moved away from the original line, the view is blocked. A modern demonstration of rectilinear propagation is to shine a laser beam through chalk dust suspended in air. The laser beam clearly travels in a straight line.

The line along which light travels can be represented graphically by drawing a line (a geometrical construction). The direction of propagation along the line is represented by an arrow, and the line becomes a ray.

When light encounters a surface between two media, its direction of propagation will be changed in two ways. The light will bounce off the surface, which we call *reflection*; and the light will be transmitted, which we call *refraction*. In other words, the incident ray is actually split into a reflected ray and a refracted ray. Normally, we treat these two rays independently. The law of rectilinear propagation, by itself, is not sufficient here because we no longer have a homogeneous medium.

As an example of reflection, let's look at a plane mirror. The mirror surface is a discontinuity in the medium. The angle that the incoming light makes with the normal to the surface is called the angle of incidence i. The incoming ray and the

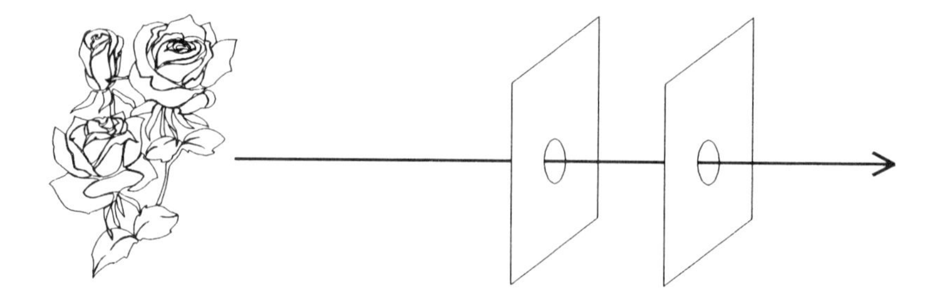

Figure 1.1 Demonstration that light travels in a straight line in a homogeneous medium.

normal lie on two lines which define a plane of incidence. The incoming ray is redirected by the mirror into an outgoing ray (Figure 1.2). The angle that the outgoing ray makes with the normal to the surface of the mirror is called the angle of reflection i'. The law of reflection states that the outgoing ray lies in the plane of incidence and that the angle of reflection is equal in magnitude to the angle of incidence. Since both angles are measured from the normal, but lie on opposite sides of the normal, they have different algebraic signs. Our sign convention is that all angles are positive when measured counterclockwise. Thus, the law of reflection may be stated as

$$i' = -i . \tag{1.1}$$

Of course the negative sign can be placed on the other side of the equation to make both angles positive if so desired. Also, it is very important to remember that both the incident and reflected rays and the normal to the surface all lie in the same plane.

As mentioned earlier, the law of refraction is based on observation and can be stated mathematically as

$$n \sin i = n' \sin i' . \tag{1.2}$$

In this equation i is once again the angle that the incident ray makes with the normal to the surface, but now i' indicates the angle that the refracted ray makes with the normal on the opposite side of the interface (Figure 1.3). Two new parameters n and n' are also introduced. These parameters represent an optical property of the material called the index of refraction. On the incident side of the interface the material has index of refraction n. On the refracted side, the index is n'. Table 1.1 shows numerical values for index of refraction for a range of materials. Note that the index is always greater than one. Indeed, the index of refraction of air is only slightly

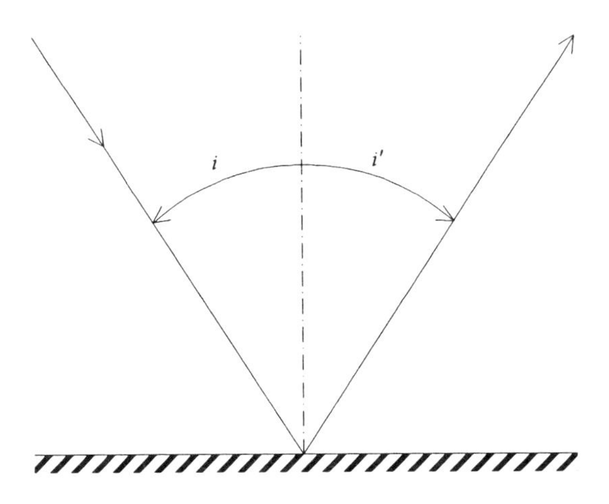

Figure 1.2 Reflection at a mirrored surface.

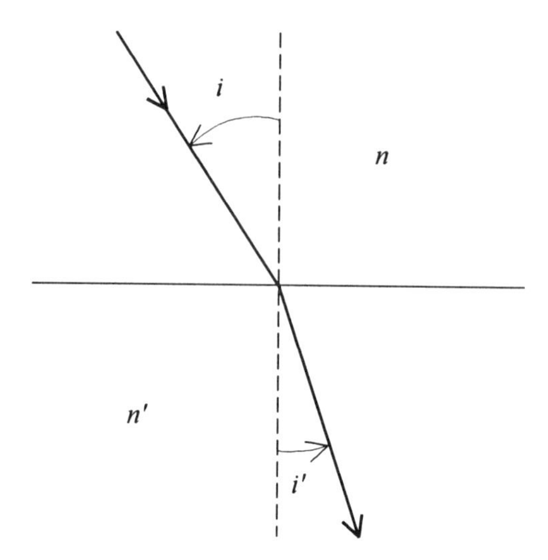

Figure 1.3 Refraction at an interface between two media.

larger than one. Since optical measurements are normally made in air, the indices used throughout this text are actually relative to air and not a vacuum. Equation 1.2 shows that the angle of refraction is smaller than the angle of incidence if n' is larger than n. Again, refraction always takes place in the plane of incidence which is defined by the incident ray and the normal to the surface.

The fourth law of geometrical optics is the law of reversibility. This law states that if the direction of a ray is reversed, it will trace exactly the same path backwards. The fourth law can be supported by examining the laws of reflection and refraction and Figures 1.2 and 1.3. In each case, the role of the incident and refracted or reflected angles could be reversed, but the equations would remain the same. The law of rectilinear propagation also agrees with the law of reversibility. After all, a straight line is a straight line in both directions. The law of reversibility may seem trivial, but its use will simplify many discussions later on.

Table 1.1 Index of refraction for selected materials.

Material	Index
air (1 atm, 0°C)	1.000292
water	1.333
plastic (methyl methacrylate)	1.49166
crown glass (BK7)	1.51680
flint glass (LaF21)	1.78831
diamond	2.426

Example 1.1

What is the angle of incidence required to send a ray of light into the block of glass shown in Figure 1.4 if we want the light to come out parallel to the base of the block? Assume that the index of refraction of the air surrounding the glass is 1.000 and that the index of the glass is 1.500.

One approach to this problem would be to derive an equation for the incidence angle as a function of the final angle of refraction. This approach will be taken with the applications presented in Section 1.5. Here we can use a more direct approach. Since we will be repeatedly using the same equations, we will number the surfaces in the order that the rays hit them. We know from the problem that we want the final refracted angle to be 45°. We can use the law of reversibility to determine what the incidence angle must be at the last surface. Solving

$$(1.500)\sin i_3 = (1.000)\sin(45^\circ)$$

gives $i_3 = 28.13^\circ$.

Next, apply a little trigonometry to find the reflected angle i_2'. The refracted ray, the mirror surface, and the sloped surface form a triangle with one 45° angle. Because the sum of the interior angles of a triangle must equal 180°, the angle between the ray and the mirror surface must be 16.87°. This makes the reflected angle 73.13°. Of course, the incidence angle at the mirror must have the same magnitude.

The refracted ray after the first surface and the two surface normals form a right triangle. Again, using the sum of the interior angles, we can find the incidence angle at the first surface to be 16.87°. One more application of the law of refraction gives

$$(1.000)\sin i_1 = (1.500)\sin(16.87^\circ)\ .$$

From which we find the answer to be $i_1 = 43.54^\circ$. ■

In the following section, the wave nature of light will be discussed. This discussion

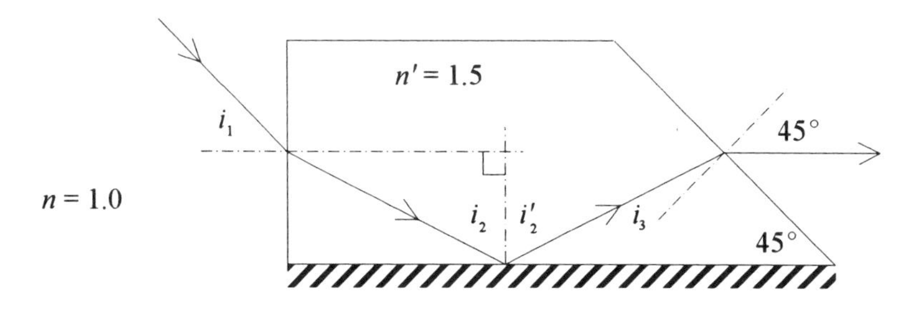

Figure 1.4 Example problem 1.1.

will provide a description of the nature of light and establish a theoretical basis for the four laws of geometrical optics.

1.2 PHYSICAL OPTICS

As mentioned in the previous section, geometrical optics has been known and studied for literally hundreds of years. But the nature of light was hotly debated until about 1860, when Maxwell showed that light was a form of electromagnetic radiation. Maxwell combined the laws of electricity and magnetism to show that electromagnetic waves existed. A simplified version of Maxwell's work is given below. The derivations given here are not meant to be rigorous and complete. A full treatment of physical optics requires its own text such as Hecht or Jenkins and White.

From introductory physics, you should be familiar with Gauss' law for electricity in the form

$$\oint \boldsymbol{E} \cdot d\boldsymbol{A} = \frac{q}{\epsilon} , \tag{1.3}$$

where q is the charge enclosed by the surface and ϵ is a property of the medium called the permittivity. Equation 1.3 leads to Coulomb's law and allows calculating electric field strength for simple charge distributions.

There is also a Gauss' law for magnetism:

$$\oint \boldsymbol{B} \cdot d\boldsymbol{A} = 0 . \tag{1.4}$$

The fact that the right-hand side of Eq. 1.4 is observed to be zero implies that there are no magnetic monopoles (i.e. single, isolated magnetic sources).

Faraday's law,

$$\oint \boldsymbol{E} \cdot d\boldsymbol{s} = -\frac{d\Phi_B}{dt} , \tag{1.5}$$

shows that electric fields can be created by a changing magnetic field. In this equation, Φ_B is the magnetic flux through the area defined by the integral.

Finally, the Ampere–Maxwell law is given by

$$\oint \boldsymbol{B} \cdot d\boldsymbol{s} = -\mu \epsilon \frac{d\Phi_E}{dt} + \mu I , \tag{1.6}$$

where μ is the magnetic permeability of the medium. This equation means that magnetic fields are created by currents or changing electric fields.

When discussing the propagation of light, we can simplify these equations because the material through which the light propagates is normally not a conductor. Therefore, there are no free charges ($q = 0$) or currents ($I = 0$) to contend with.

The equations given above are in the familiar integral form. For our purposes, the differential form is preferable. The equations can be converted from one form to another via Gauss' divergence theorem and Stokes' theorem from vector calculus. If you haven't had vector calculus, you will have to take my word that the following four equations are equivalent to the preceding four equations with $q = 0$ and $I = 0$.

$$\nabla \cdot \boldsymbol{E} = 0 \, , \tag{1.7}$$

$$\nabla \cdot \boldsymbol{B} = 0 \, , \tag{1.8}$$

$$\nabla \times \boldsymbol{E} = -\frac{\partial \boldsymbol{B}}{\partial t} \, , \tag{1.9}$$

$$\nabla \times \boldsymbol{B} = \mu \epsilon \frac{\partial \boldsymbol{E}}{\partial t} \, . \tag{1.10}$$

You have probably seen a version of these rather cryptic equations on a sweatshirt. Again, unless you have had vector calculus, these equations don't mean much. But we can write them out in Cartesian component form which should be more understandable. Equation 1.7 is really just

$$\frac{\partial E_x}{\partial x} + \frac{\partial E_y}{\partial y} + \frac{\partial E_z}{\partial z} = 0 \, , \tag{1.11}$$

and Eq. 1.8 looks like

$$\frac{\partial B_x}{\partial x} + \frac{\partial B_y}{\partial y} + \frac{\partial B_z}{\partial z} = 0 \, . \tag{1.12}$$

Equation 1.9 is really three equations:

$$\frac{\partial E_z}{\partial y} - \frac{\partial E_y}{\partial z} = -\frac{\partial B_x}{\partial t} ,$$
$$\frac{\partial E_x}{\partial z} - \frac{\partial E_z}{\partial x} = -\frac{\partial B_y}{\partial t} , \tag{1.13}$$
$$\frac{\partial E_y}{\partial x} - \frac{\partial E_x}{\partial y} = -\frac{\partial B_z}{\partial t} .$$

And finally, Eq. 1.10 really looks like the following three equations

$$\frac{\partial B_z}{\partial y} - \frac{\partial B_y}{\partial z} = \mu \epsilon \frac{\partial E_x}{\partial t} ,$$
$$\frac{\partial B_x}{\partial z} - \frac{\partial B_z}{\partial x} = \mu \epsilon \frac{\partial E_y}{\partial t} , \tag{1.14}$$
$$\frac{\partial B_y}{\partial x} - \frac{\partial B_x}{\partial y} = \mu \epsilon \frac{\partial E_z}{\partial t} .$$

As a special case, let's consider an electromagnetic disturbance which is uniform in the x and y directions (i.e., a plane wave). This assumption greatly simplifies the equations, but our results will be generally true nonetheless. Mathematically, the assumption means that

$$\frac{\partial E_x}{\partial x} = \frac{\partial E_y}{\partial y} = \frac{\partial B_x}{\partial x} = \frac{\partial B_y}{\partial y} = 0 . \tag{1.15}$$

Substituting Eq. 1.15 into Eqs. 1.11 and 1.12 shows that the partial derivative in z is also zero. This means that the fields in the z direction are either constant or zero. If they were constant, we would end up with only a static set of fields. We are interested in the case where E_z and B_z are zero. Equation 1.13 now looks like

$$-\frac{\partial E_y}{\partial z} = -\frac{\partial B_x}{\partial t} ,$$
$$\frac{\partial E_x}{\partial z} = -\frac{\partial B_y}{\partial t} , \tag{1.16}$$
$$\frac{\partial E_y}{\partial x} - \frac{\partial E_x}{\partial y} = 0 .$$

Similar equations come out of Eqs. 1.14 for the magnetic field. Let's temporarily concentrate on the first equation from the set of Eqs. 1.13. If we take the second partial derivative of this equation with respect to z and substitute for the magnetic field with the appropriate equation from the set of Eq. 1.14, we get

$$\frac{\partial^2 E_y}{\partial z^2} = \mu \epsilon \frac{\partial^2 E_y}{\partial t^2} . \tag{1.17}$$

Identical equations can be derived for E_x, B_x, and B_y. These equations are called *wave equations*. The solution of a wave equation like Eq. 1.17 can be any function in the form

$$E_y = f\left(t - \frac{z}{v} \right) . \tag{1.18}$$

This represents a disturbance traveling in the z direction with speed v. In particular, the function could be a cosine function of the form

$$E_y = E_m \cos \omega \left(t - \frac{z}{v} \right) , \tag{1.19}$$

where ω is the angular frequency of the wave and v is its speed. Equation 1.19 represents a wave of a particular frequency

$$f = \frac{\omega}{2 \pi} \tag{1.20}$$

and wavelength

$$\lambda = \frac{v}{f} . \tag{1.21}$$

It is easy to show that the wave speed is related to the electromagnetic properties of the medium by

$$v = \frac{1}{\sqrt{\mu \epsilon}} . \tag{1.22}$$

In a vacuum, we replace μ and ϵ with their free space values μ_0 and ϵ_0. When Maxwell worked this all out, the speed of light was known fairly well as were the values for μ_0 and ϵ_0. The agreement between the speed of an electromagnetic wave as determined by Eq. 1.22 and the experimentally determined speed of light was too unlikely an event to be merely a coincidence. Light must be an electromagnetic wave. Today, light is defined to have a speed c = 299,792,458 m/s in free space.

There are other forms of electromagnetic radiation besides what we perceive as light. These forms encompass an enormous range of wavelengths and frequencies. The boundaries between one type of radiation and the next are not fundamental in nature, but are based on the type of source which produces the radiation or the type of detector used to sense the radiation. Light is radiation which can be sensed by the eye. It is important to note that all of the laws of geometrical optics, as well as the laws of physical optics, apply equally well to all forms of electromagnetic radiation. This text restricts discussion to visible light because, historically, geometrical optics was developed for visible light.

Additional support for the idea that light is a wave phenomenon came from the observation of diffraction and interference effects with light. Diffraction and interference are distinct wave phenomena. Diffraction arises from the wave bending around the edge of a blocking object. Diffraction changes the direction of propagation of light without reflection or refraction. Therefore, diffraction is an affect outside the realm of geometrical optics. We must place a restriction on the use of geometrical optics, namely that geometrical optics will be applicable only when the dimensions of our optical system are large compared with the wavelength of light. Interference arises from the superposition of two separate waves. If the peaks of two waves coincide, then they add to give a larger amplitude wave. But, if the peak of one wave coincides with the trough of another wave, the two waves cancel out. The study of the wave nature of light is called *physical optics.*

Since both geometrical optics and physical optics are well established by experiment, the laws of both must agree with each other when the system dimensions are large.

The law of rectilinear propagation is demonstrated by Eq. 1.19. A wave traveling in the z direction continues to travel in the z direction in a homogeneous medium. Reversibility is nothing more than a sign change on the time in Eq. 1.19.

The laws of reflection and refraction take effect at an interface. Therefore, we need to investigate how the electric and magnetic fields behave at an interface. Assume that the interface lies in the x–y plane and that there is no free charge at the interface. We can apply Eq. 1.5 to see how the tangential components of the electric field behave across the interface. Let the path for the integral begin a distance a above the surface, and extend in the y direction a distance b along the interface. Next, the path goes a distance a below the interface, and finally back to the starting point as shown in Figure 1.5. We can rewrite Eq. 1.5 as

$$\oint \boldsymbol{E} \cdot d\boldsymbol{s} = -\int \frac{d\boldsymbol{B}}{dt} \cdot d\boldsymbol{A} \ . \tag{1.23}$$

If we make the distance a very small, then the area of the path goes to zero and so does the right-hand side of this equation. The only left-hand side contributions come from the long side b:

$$E_{1y} b - E_{2y} b = 0 \ , \tag{1.24}$$

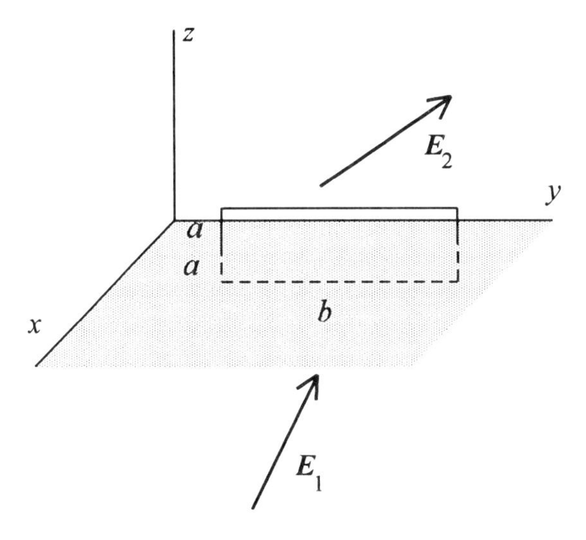

Figure 1.5 Rectangular path for the line integral. Half of the path is above the interface, while half is below.

from which we get

$$E_{1y} = E_{2y} , \tag{1.25}$$

which means that the tangential components of the electric fields are constant across the interface.

The next step is to write a more general form for a plane wave. Assume that the direction of propagation for the plane wave incident on the interface is given by the unit vector $\boldsymbol{a}$

$$\boldsymbol{a} = \sin\theta \, \cos\phi \, \boldsymbol{i} + \sin\theta \, \sin\phi \, \boldsymbol{j} + \cos\theta \, \boldsymbol{k} . \tag{1.26}$$

Similar equations hold for the refracted wave in the direction given by $\boldsymbol{a}''$ and for the reflected wave by $\boldsymbol{a}'$. The electric field for the incident wave can be written as

$$\boldsymbol{E} = \boldsymbol{E}_m \cos\omega\left(t - \frac{\boldsymbol{a}\cdot\boldsymbol{r}}{v} \right) . \tag{1.27}$$

The reflected wave is given by

$$\boldsymbol{E}' = \boldsymbol{E}_m' \cos\left[\omega'\left(t - \frac{\boldsymbol{a}'\cdot\boldsymbol{r}}{v} \right) - \delta'\right], \tag{1.28}$$

and the refracted ray is represented by

$$\boldsymbol{E}'' = \boldsymbol{E}''_m \cos\left[\omega''\left(t - \frac{\boldsymbol{a}''\cdot\boldsymbol{r}}{v''}\right) - \delta''\right]. \tag{1.29}$$

In these equations, $\boldsymbol{r}$ is the position vector for a location on the plane wave. Primes have been added to the angular frequencies and wave numbers to allow for a possible change in frequency. The δ's have been added to allow for possible phase shifts.

We can now apply the tangential boundary condition to get

$$\begin{aligned} &E_{my} \cos\omega\left(t - \frac{x\sin\theta}{v}\right) \\ &\quad + E'_{my} \cos\left[\omega'\left(t - \frac{x\sin\theta'\cos\phi' + y\sin\theta'\sin\phi'}{v}\right) - \delta'\right] \\ &= E''_{my} \cos\left[\omega''\left(t - \frac{x\sin\theta''\cos\phi' + y\sin\theta''\sin\phi''}{v''}\right) - \delta''\right]. \end{aligned} \tag{1.30}$$

For this equation to be satisfied at all times and for all points along the interface, the arguments of the cosine functions must be the same or differ by multiples of π. Therefore, the phase shifts may be 0 or π. The frequency must remain constant. In writing this equation, I assumed that the incident direction vector $\boldsymbol{a}$ was in the x–z plane as shown in Figure 1.6. Since the incident wave is not dependent on y, neither can the incident or refracted waves be dependent on y. Thus, the reflected and refracted waves stay in the same plane as the incident wave. Furthermore, if the x dependence of the reflected wave is to be the same as the incident wave, then

$$\theta = 180^\circ - \theta', \tag{1.31}$$

which is equivalent to the law of reflection. Finally, if the x dependence of the refracted wave is to be the same as the incident wave, then

$$\frac{\sin\theta}{v} = \frac{\sin\theta''}{v''}. \tag{1.32}$$

This equation will reduce to the law of refraction if the index of refraction is defined as

$$n = \frac{c}{v}. \tag{1.33}$$

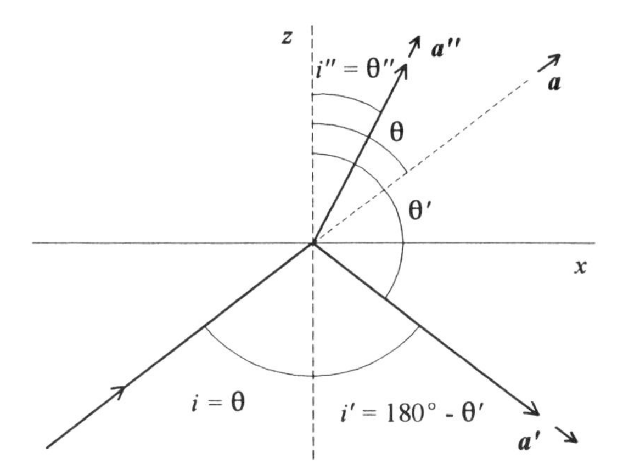

Figure 1.6 General geometry for refraction and reflection.

We can also use the electromagnetic wave theory to indicate why light travels slower in a material medium than in a vacuum. The electric fields associated with a wave traveling through an array of atoms will interact with the charged particles in the atoms (i.e., the electrons). The charged particles will be accelerated by the applied electric field. The oscillating charges re-radiate the incoming energy at the same frequency. The only difference is that there is a phase or time lag between the emission and re-radiation. This is true in general for harmonic oscillators which are driven at frequencies different from their natural resonance frequencies. Imagine trying to swing a small child on a playground swing. If you try to move the chains of the swing back and forth very quickly, the child lags behind your motion. When you push forward, the child moves backward. The re-radiated wavefront lags behind the position it would have had if the atoms were not present. That is, the wave travels slower in a material medium than in a vacuum.

1.3 QUANTUM OPTICS

Since we have established that light is an electromagnetic wave which propagates according to simple geometrical laws, you might think that we know all there is to know about the nature of light. But we haven't discussed how light is produced or detected. This is the subject of quantum optics.

One way to generate light is to heat an object until it glows. Candles, incandescent lights, and the sun are examples of thermal sources. Actually, all objects, at any temperature, emit some form of electromagnetic radiation. Radiation which depends on the temperature of the emitting object but not on the material itself is called *blackbody radiation*. A blackbody is an object which absorbs all of the light energy which falls on it. A good way to experience a blackbody is to look into the open mouth of an empty bottle of copy machine toner. For a blackbody to remain at a fixed temperature, in thermal equilibrium with its surroundings, it must radiate as much energy as it absorbs.

The physical optics explanation of the source of blackbody radiation is that all objects are made up of atoms which vibrate according to their temperature. A static charge creates a static electric field. But if that charge vibrates (accelerates), then the electric field at a given point will periodically get stronger or weaker as the charge moves closer or further away. The accelerating charge is the source of an electromagnetic wave.

Unfortunately, the physical optics theory runs into trouble when compared with measurements of blackbody radiation. In 1900, Max Planck was able to come up with a theory to match experiments by proposing that the radiation could only be emitted in discrete bundles. These bundles or particles of energy are called *quanta* or *photons*.

Further evidence for the particle nature of light came from Einstein's explanation of the photoelectric effect in 1905. In this phenomenon, light is absorbed by a metal and the energy reappears as the kinetic energy of electrons emitted by the metal. Since the electron kinetic energy is coming from the electromagnetic wave energy, you would expect the maximum electron energy to depend on how energetic is the electromagnetic wave or on how much total energy is absorbed. However, in the photoelectric effect, the maximum kinetic energy does not depend on the intensity of the light or the length of time that the metal is exposed to the light. The maximum kinetic energy depends only on the wavelength of the light. As the wavelength increases, the energy of the emitted electrons decreases. For light with wavelengths longer than some cutoff wavelength, no electrons are emitted at all.

These observations can be explained if the interaction between the light and the metal is not analogous to a wave–particle interaction but rather a particle–particle interaction. If a wave interacts with a particle, then the amplitude of the wave and the duration of the interaction controls how much energy can be transferred. But if two particles interact, then the energy transfer depends only on how much energy the initial particle carries. Therefore, we must think of each photon as a particle which carries an amount of energy which is proportional to its frequency or inversely proportional to its wavelength

$$E = h\nu = \frac{hc}{\lambda} . \tag{1.34}$$

The proportionality constant h is called *Planck's constant* and has a value of 6.62×10^{-34} J · m. Now, for an electron to be emitted, the incoming photon must carry enough energy to free the electron from the metal. Photons with long wavelengths just can't carry enough energy.

One very important application of quantum optics is understanding atomic spectra. As you should know, atoms are made of negatively charged electrons orbiting a positive nucleus. The electrons are constantly accelerating toward the nucleus. According to classical physics, accelerating charges should radiate energy. This would mean that the electron should quickly radiate away all of its orbital energy and fall into the nucleus. The fact that this doesn't happen can be explained by quantum mechanics. Continuous emission of electromagnetic energy is not permitted. The electrons can only exist in certain energy states and can move between states only by absorbing or emitting a photon whose energy is given by Eq. 1.34. A hot gas will have electrons in high-energy states due to atomic collisions. When these electrons drop to lower-energy states, photons of specific energy and wavelengths are emitted. This is the origin of atomic and molecular spectra. The wavelengths and sources of some useful spectral lines are given in Table 1.2.

Spectral lines are emitted by thin, hot gases. As the gas density increases, the spectral lines broaden. For very dense gases, liquids, and solids, the spectral lines merge. The resultant distribution of emitted energy approaches the blackbody radiation distribution we discussed at the beginning of this section.

1.4 OPTICAL MATERIALS

Before we begin our detailed discussion of geometrical optics, we need to discuss the optical materials through which the light will be propagating. Good optical materials have several important characteristics. First, of course, is that they must be able to transmit or reflect light at the desired wavelengths. Optical materials must be able to accept a smooth polish on their surface. If the surface is not smooth compared to the wavelength of light, then the surface will scatter the light rather than reflect or refract it according to the laws of geometrical optics. Materials should be mechanically and chemically stable. They should have a homogeneous index of refraction, or in the case of gradient index materials a well behaved index of refraction. Finally, the bulk of the material should be free of bubbles and defects which would scatter light.

Materials which meet these requirements in the visible region include glass, some plastics, and certain crystals. We will restrict our discussion and usage to glass because of the wide variety of optical properties of this class of materials. Appendix A contains a table of glasses with important optical properties and a glass map. An explanation of both the table and the glass map will be presented below.

Table 1.2 Selected spectral lines.

Designation	Color and Element	Wavelength (nm)
r	red helium	706.5188
C	red hydrogen	656.2725
d	yellow helium	587.5618
e	green mercury	546.0740
F	blue hydrogen	486.1327
g	blue mercury	435.8343
h	violet mercury	404.6561

The most important optical property of a material is its index of refraction. It is the only optical property which enters directly into any of the four laws of geometrical optics. Unfortunately, the index of refraction depends not only on the material, but also on the wavelength of the light. The dependence of index on wavelength is called *dispersion*. Dispersion arises because of differences in the amount of phase lag that a particular atom has to various wavelengths of incident light. Normally, materials have a higher index of refraction in blue light than in red light (Figure 1.7). Short wavelengths travel slower in optical materials than do long wavelengths, which means that blue light bends more when refracted.

Glasses may be characterized by their index of refraction at a particular wavelength and by the amount of dispersion in that type of glass. Unless otherwise indicated, yellow helium light will be used as the standard wavelength. The d-line index will be indicated as n_d or just plain n. The d line is chosen because it is very near the wavelength at which the human eye is most sensitive.

The amount of dispersion can be characterized by taking the difference in index between two different wavelengths

$$\delta n = n_F - n_C . \tag{1.35}$$

The red hydrogen line C and blue hydrogen line F are used to define dispersion because they nearly span the visible region of the electromagnetic spectrum.

Another common way to indicate the amount of dispersion is with the dimensionless Abbe value

$$V = \frac{n_d - 1}{n_F - n_C} . \tag{1.36}$$

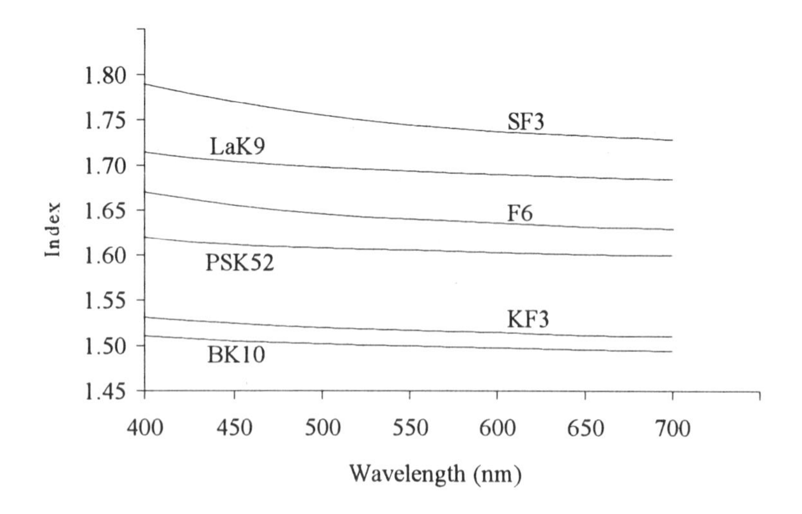

Figure 1.7 Index of refraction as a function of wavelength for randomly selected glasses.

High Abbe values indicate very little dispersion and low Abbe values indicate significant amounts of dispersion. The reason for expressing the dispersion as the Abbe value will be given in the discussion of chromatic aberrations in Section 4.6.

We now have a minor dilemma. We need to know n_d, n_F, and n_C to find V, but knowing n_d and V is not, by itself, enough to determine n_F and n_C. The solution to this dilemma is to note that all of the glass dispersion curves have a wavelength dependence which can be approximated fairly accurately by

$$n(\lambda) = A + \frac{B}{\lambda^2} + \frac{C}{\lambda^4}\,. \tag{1.37}$$

This is Cauchy's formula. The C coefficient turns out to be very small for most glasses in the visible wavelength region. Therefore, we can obtain sufficiently accurate results by setting $C = 0$ and keeping just the first two terms. Substituting Eq. 1.37 into Eq. 1.36 gives

$$B = \frac{(n_d - 1)\lambda_C^2 \lambda_F^2}{V(\lambda_C^2 - \lambda_F^2)}\,. \tag{1.38}$$

The value of the first coefficient can now be determined

$$A = n_d - \frac{B}{\lambda_d^2}\,. \tag{1.39}$$

The first coefficient A is dimensionless. The B coefficient has dimensions of length squared. Finally, the index at any wavelength can be determined by using the coefficients determined above and the appropriate wavelength in equation .

How accurate is the procedure just described? We can assume that the n_d has at best six significant figures based on the table in Appendix A. Indeed, the index of a particular piece of glass may be measured to an accuracy of $\pm\ 3 \times 10^{-5}$. The numbers in the tables are averages over many batches of glass. An individual sample of glass may have an index which varies by as much as ± 0.001 from the average given. Similarly, Abbe values may vary by as much as ± 0.8 %. We will ignore these manufacturing tolerances and take the indices and Abbe values in Appendix A at face value. Therefore, the procedure for calculating indices at any wavelength described above is only as good as the least significant input which is the Abbe value. You should not trust your results past four significant figures.

Example 1.2

What is n_C for BK7 glass?

From Appendix A we have $n_d = 1.51680$ and $V = 64.17$. Then

$$B = (5.2366313 \times 10^5 \text{ nm}^2) \frac{(1.51680) - 1}{(64.17)} ,$$

which gives $B = 4.2173773 \times 10^3 \text{ nm}^2$. Next,

$$A = (1.51680) - \frac{(4.2173773 \times 10^3 \text{ nm}^2)}{(587.5618 \text{ nm})^2} ,$$

or $A = 1.5045838$. Finally,

$$n_C = (1.5045838) + \frac{(4.2173773 \times 10^3 \text{ nm}^2)}{(656.2725 \text{ nm})^2} .$$

The result, $n_C = 1.514$ (keeping only four significant figures), compares favorably with the value given in the manufacturer's catalog $n_C = 1.51432$. ■

The glass map shown in Appendix A graphically displays glasses according to their nominal index n_d and Abbe value V. Note that the Abbe values decrease to the right. Each glass in Appendix A is named by the manufacturer. But we can identify glasses with a generic glass number. The glass number is a six-digit number which identifies a glass by its n_d and V:

$$glassnumber = xxxyyy , \tag{1.40}$$

where

$$xxx = (n_d - 1) \times 1000 \tag{1.41}$$

and

$$yyy = V \times 10 . \tag{1.42}$$

As discussed previously, three digits are really all that are needed when specifying a glass because of manufacturing variability.

Example 1.3

Find the glass number for BK7 glass.

From the table in Appendix A we find that $n_d = 1.5168$ and $V = 64.17$. Therefore,

$$xxx = ((1.5168) - 1) \times 1000 = 517$$

and

$$yyy = (64.17) \times 10 = 642 .$$

Therefore, the *glass number* is 517642. Note that some round off has occurred. ■

Example 1.4

Find the name of the glass 643480.

First, note that $n_d = 1.643$ and $V = 48.0$. With this information, look for the glass nearest these coordinates on the glass map. The answer is BaF9. ■

Why are there so many different glasses? The index and Abbe value offer two degrees of freedom for the optical system designer. As we shall see, the designer can choose glasses to improve the image quality of his system. In many ways the "best" glass would have very large n_d and large V. As you can see from the glass map, this is the region where there aren't any glasses. A lot of the art of lens design involves overcoming material limitations. Having a wide choice of "inferior" glasses is a big help.

1.5 NONIMAGING APPLICATIONS

In this section, four applications of the laws of geometrical optics are presented to illustrate the wide range of phenomena which can be explained by these simple laws of nature. None of these applications involve imaging. Imaging applications are the subject of the remainder of the text.

1.5.1 Optical Fibers

The law of refraction works for any angle of incidence when the light is propagating from a medium of low index to a medium of high index. But when the light starts in the high-index medium, angles of incidence larger than some critical angle produce a value for the sine of the refracted angle which is larger than one. Figure 1.8 shows that as the incidence angle increases, the refracted angle approaches $90°$. The refracted angle cannot become larger than $90°$, but the incident angle can certainly be greater than the critical angle. Since the light cannot penetrate into the lower-index medium, it must be reflected back into the high-index medium. This is called *total internal reflection*. The critical angle is given by

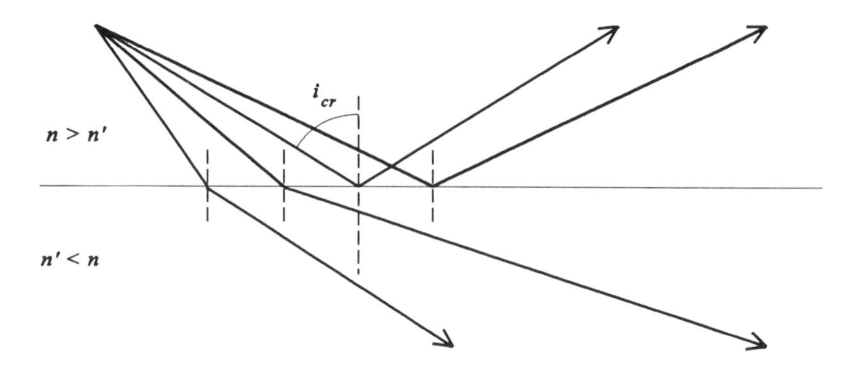

Figure 1.8 The critical angle of refraction.

$$i_{cr} = \arcsin\left(\frac{n'}{n}\right) . \qquad (1.43)$$

Total internal reflection is the mechanism by which light is transported along an optical fiber. Light enters the fiber at a small angle as shown in Figure 1.9. This ray of light will have a large angle of incidence with the sides of the fiber and will be reflected back into the fiber. If the fiber is bent too sharply, then the angle of incidence can become less than the critical angle and the light can escape the fiber.

1.5.2 Plane Parallel Plate

A ray of light which enters a tilted plane parallel plate leaves the plate parallel to its initial direction but displaced a distance d as shown in Figure 1.10. The distance d

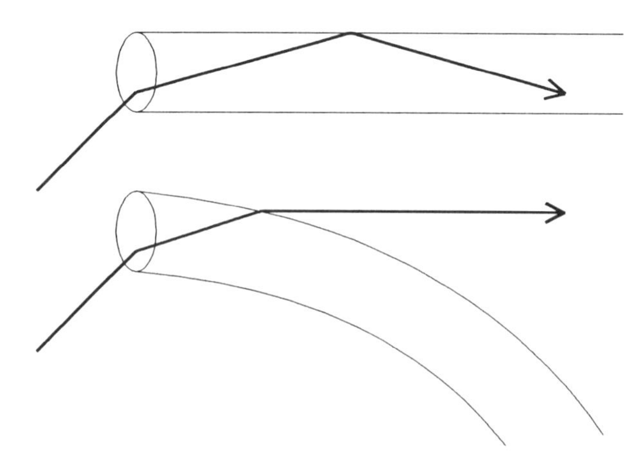

Figure 1.9 Light rays in optical fibers.

can be found using geometrical optics. First, to prove that the ray comes out parallel, note that the angle of refraction at the first surface is equal to the angle of incidence at the second surface because the surfaces are parallel. Applying Snell's law to the second surface shows that the angle of refraction at the second surface is the same as the angle of incidence at the first surface. Therefore, because the surfaces are parallel, the rays are parallel.

The distance d depends on the length of the ray inside the plate l, as well as on the difference in the angle of incidence and angle of refraction

$$d = l \sin (i_1 - i_1') \; . \tag{1.44}$$

An identity relation from trigonometry can be used to expand this equation into a more useful form

$$d = l (\sin i_1 \cos i_1' - \cos i_1 \sin i_1') \; . \tag{1.45}$$

The length of the ray depends on the thickness of the plate and the refracted angle:

$$l = \frac{t_1}{\cos i_1'} \; . \tag{1.46}$$

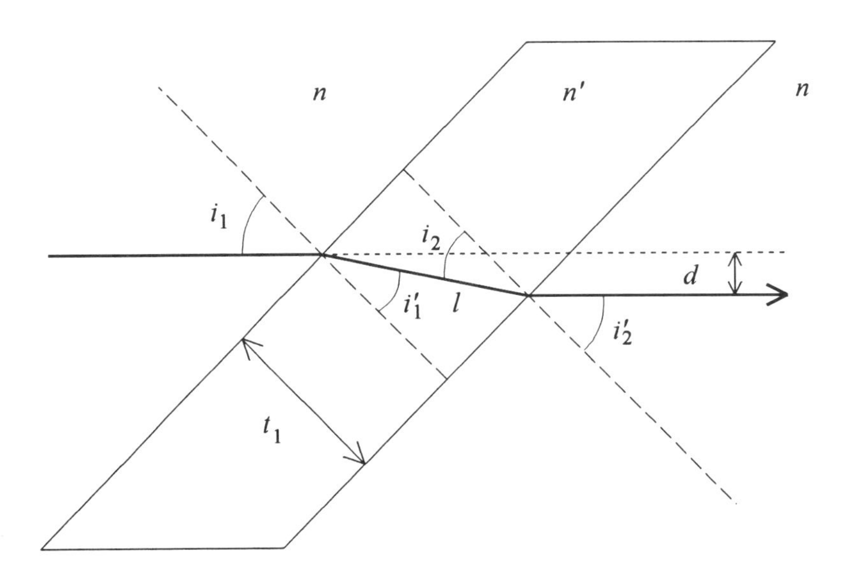

Figure 1.10 Displacement of a ray when refracted by a plane parallel plate.

Combining these two equations and using the law of refraction once more gives

$$d = t_1 \sin i_1 \left(1 - \frac{n \cos i_1}{n' \cos i_1'} \right) . \tag{1.47}$$

The displacement is given as a function of the incidence angle and the refracted angle. A very useful trick is to note that

$$n' \cos i' = \sqrt{n'^2 - n^2 \sin^2 i} . \tag{1.48}$$

Therefore,

$$d = t_1 \sin i_1 \left(1 - \frac{n \cos i_1}{\sqrt{n'^2 - n^2 \sin^2 i_1}} \right) . \tag{1.49}$$

For small incident angles, Eq. 1.49 shows that the displacement varies nearly linearly with the incident angle. This suggests that a rotating plate could be used to scan a beam of light at a linear rate across a surface as long as the angle of rotation remains small.

1.5.3 Prisms

If the sides of the plate are not parallel but meet with some angle A, then a prism is formed. A prism can be used to accurately determine the index of refraction of the material forming the prism. The ray emerging from the prism is not parallel with the incident ray as with a parallel plate, rather its direction is deviated by an angle D (Figure 1.11). Again, geometrical optics can be used to determine the deviation angle as a function of the incidence angle. It turns out that this function has a minimum which can be directly related to the index of refraction of the prism.

Figure 1.11 shows that D is an exterior angle to the triangle EHF. The interior angles at E and F are given by the differences between the incidence angle and refracted angle

$$\angle HEF = i_1 - i_1' \tag{1.50}$$

and

$$\angle HFE = i_2' - i_2 . \tag{1.51}$$

Since the exterior angle of a triangle equals the sum of the two opposite interior angles, we have

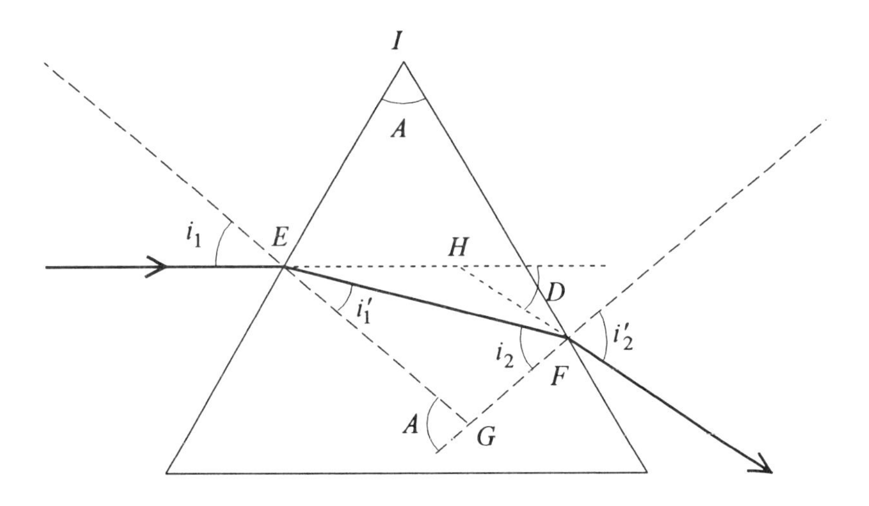

Figure 1.11 Geometry of a ray passing through a prism.

$$D = i_1 - i_1' + i_2' - i_2 \; . \tag{1.52}$$

The two sides of the prism from the apex to the intersection points E and F along with the two normals form a quadrangle $EGFI$. Two of the interior angles are right angles, and the third is the apex angle A. This means that the exterior angle at G is the same as the apex angle. Now, using the triangle EFG it is easy to see that

$$A = i_1' + i_2 \; . \tag{1.53}$$

Combining these two equations gives

$$D = i_1 + i_2' - A \; . \tag{1.54}$$

If the deviation angle is measured as a function of the incidence angle i_1, a minimum value is found. Since there is only one minimum, any ray entering the prism at a different incidence angle would deviate by more than this minimum value. Put another way, the minimum deviation angle determines a particular incidence angle. Now the law of reversibility may be invoked. The reversed ray must traverse exactly the same path as the original ray. Symmetry implies that the reversed ray is also deviated by the minimum amount. Therefore, the reversed ray must enter at the same incidence angle as the original ray. The reversed ray incidence angle is, of course, the same as the original ray's exit angle. That is, $i_2' = i_1$. Thus, at minimum deviation

$$D_{\min} = 2 i_1 - A \tag{1.55}$$

and

$$A = 2 i_1' . \tag{1.56}$$

Rearranging these two equations gives

$$i_1 = \frac{D_{\min} + A}{2} \tag{1.57}$$

and

$$i_1' = \frac{A}{2} . \tag{1.58}$$

Of course, these two angles are related by the law of refraction, so that

$$\frac{n'}{n} = \frac{\sin i_1}{\sin i_1'} = \frac{\sin\left(\dfrac{D_{\min} + A}{2}\right)}{\sin\left(\dfrac{A}{2}\right)} . \tag{1.59}$$

By carefully measuring the apex angle and the minimum deviation angle, the index of refraction of the material relative to its surroundings can be determined from this equation. Normally the prism is surrounded by air, so that the equation gives just the index of the prism itself.

1.5.4 The Rainbow

A rainbow is created by light which is reflected and refracted by a spherical drop of water. Figure 1.12 shows the path of a typical ray through a cross section of the raindrop. The ray refracts at the first surface producing the angle i' with the normal inside the raindrop. Next, the ray is partially reflected from the back of the raindrop. Both normals, the refraction normal and the reflection normal, are radii of the sphere. They form two sides of an isosceles triangle with the third side being the ray. Therefore, the incidence angle for the reflection at the back surface is equal in magnitude to the refracted angle i'. Since this angle must be smaller than the critical angle, the reflection is not total internal reflection. Most of the light energy leaves the drop through the back surface. Of course, the reflected angle also has this same magnitude. Using the isosceles triangle argument once more shows that the angle of incidence at the third surface is also equal in magnitude to i'. Finally, invoking

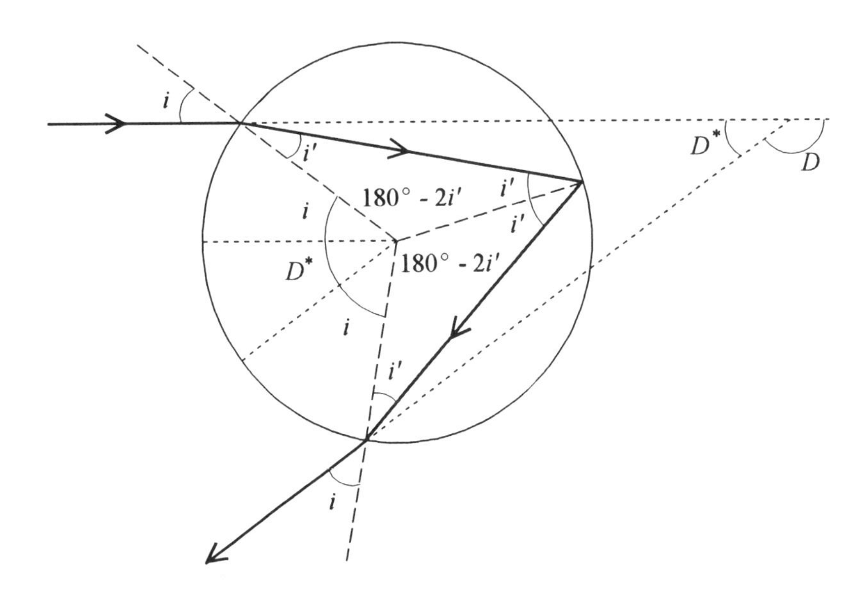

Figure 1.12 Geometry of a ray passing through a spherical drop of water.

Snell's law at this third surface shows that the angle at which the ray leaves the raindrop is the same angle at which it entered.

To analyze the light path, we begin by adding a radial lines parallel to the incoming and outgoing rays. Now, all of the angles around the center of the rainbow can be added together to give a full circle. Thus,

$$360° = D^* + 2i + 2(180° - 2i') . \tag{1.60}$$

Note that the magnitude of two of the angles included here were determined using the fact that the sum of all of the interior angles of a triangle is 180°. This equation can be rearranged to give

$$D^* = 4i' - 2i . \tag{1.61}$$

Since $D = 180° - D^*$, we obtain

$$D = 180° + 2i - 4i' . \tag{1.62}$$

Using Snell's law, the angle of deviation, can be written solely in terms of the incident angle

$$D = 180° + 2i - 4 \arcsin\left(\frac{n}{n'} \sin i \right) . \tag{1.63}$$

Obviously, more than one ray will enter a water drop. Each ray will have its own unique incidence angle ranging from 0° to 90°. Figure 1.13 shows how the deviation angle varies with incidence angle. As you can see from this graph, the deviation angle varies slowly around the minimum value. This means that a fairly large number of incoming rays will leave the raindrop in the same direction. Thus, the minimum deviation angle will be the angle at which the rainbow will reflect the most light. Since the angle of minimum deviation depends on the index of refraction of water, different wavelengths of light will have different minimum deviation angles because of dispersion.

The minimum deviation angle as a function of index of refraction can be determined with a little calculus. First, differentiate Eq. 1.60 to get

$$\frac{dD}{di} = 2 - 4\frac{di'}{di} . \tag{1.64}$$

The derivative of i' with respect to i can be determined by differentiating Snell's law:

$$\frac{di'}{di} = \frac{n \cos i}{n' \cos i'} . \tag{1.65}$$

Since we want the minimum deviation angle, we can set the left hand side of Eq. 1.64 equal to zero. Solving for i (remembering that $n = 1$ for a raindrop in air) gives

$$i = \arcsin\left(\sqrt{\frac{4 - n'^2}{3}}\right) . \tag{1.66}$$

Similarly, the refracted angle can be written

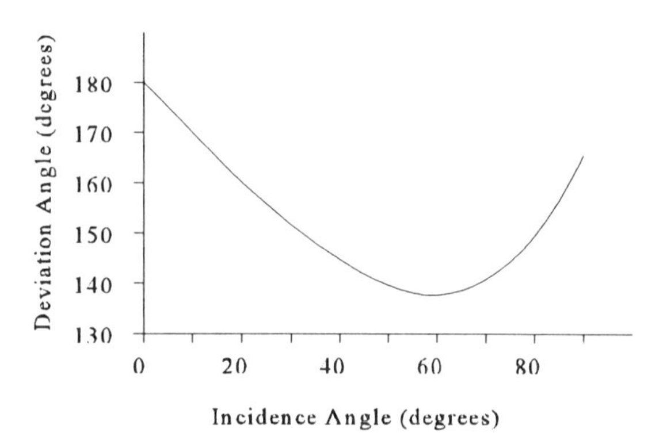

Figure 1.13 Plot of the deviation angle as a function of the incidence angle for $n' = 1.333$.

$$i' = \arcsin\left(\sqrt{\frac{4 - n'^2}{3n'^2}}\right). \tag{1.67}$$

Finally, combining equations gives

$$D_{\text{min}} = 180^\circ + 2\arcsin\left(\sqrt{\frac{4 - n'^2}{3}}\right) - 4\arcsin\left(\sqrt{\frac{4 - n'^2}{3n'^2}}\right). \tag{1.68}$$

The results of applying this equation for various wavelengths are shown in Table 1.3.

This theory can be tested by measuring the location of the rainbow in the sky. Rainbows occur at an angle of approximately 41°. Figure 1.14 shows the geometry.

Secondary rainbows result from light which reflects off the back of the rainbow twice before emerging. The rays we eventually see actually enter the raindrop through the bottom hemisphere. The secondary rainbow forms a larger arc in the sky with a relatively dark band between the two rainbows. All of these features can be explained with geometrical optics (see Greenler).

SUMMARY

Geometrical optics is the study of the propagation of light through systems with dimensions which are large compared with the wavelength of light. The propagation of light under these conditions can be demonstrated with geometrical constructions. Physical optics is the study of the wave properties of light and quantum optics is the study of the particle nature of light.

Geometrical optics consists of four laws: rectilinear propagation, reflection, refraction, and reversibility. Rectilinear propagation means that light travels along

Table 1.3 Minimum deviation angles and rainbow location angles for water at 20° C (Data courtesy of Operation Catapult students Josh Farthing, Kyle Hoskins, and Eric Pollack).

Wavelength (nm)	Minimum Deviation (degrees)	Sky Location (degrees)
404.66	139.89	40.11
435.83	139.48	40.52
491.60	138.86	41.14
546.07	138.66	41.34
576.96	138.56	41.44
671.64	138.02	41.98

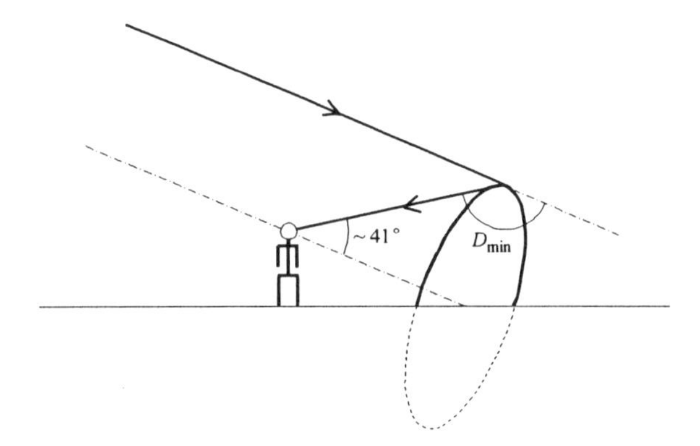

Figure 1.14 Locating rainbows in the sky.

straight lines in a homogeneous, isotropic medium. When light strikes an interface between two media, it can reflect and refract. The law of reflection states that the angle of incidence is equal in magnitude to the angle of reflection and that the reflected ray is in the same plane with the incident ray and the normal to the surface. The law of refraction can be stated as

$$n \sin i = n' \sin i' . \tag{1.69}$$

Again, the incident ray, the normal, and the refracted ray all lie in the same plane. The index of refraction is a property of the medium that is related to the speed of light in the medium

$$n = \frac{c}{v} . \tag{1.70}$$

If light tries to refract from a high-index medium into a low-index medium, it may not be able to because of total internal reflection. The critical angle for total internal reflection is given by

$$i_{cr} = \arcsin\left(\frac{n'}{n}\right) . \tag{1.71}$$

The speed of light in a medium varies with wavelength (dispersion). A common measure of the dispersion of a material is the Abbe value:

$$V = \frac{n_d - 1}{n_F - n_C} . \tag{1.72}$$

Glasses are classified according to their index in d light and Abbe value. The glass number is a six-digit number where the first three digits are

$$xxx = (n_d - 1) \times 1000 , \tag{1.73}$$

and the last three digits are

$$yyy = V \times 10 . \tag{1.74}$$

All of the remaining equations presented in this chapter are special-purpose equations meant for particular situations.

REFERENCES

K. Krane, *Modern Physics* (John Wiley & Sons, New York, 1983).

F. Jenkins and H. White, *Fundamentals of Optics*, 4th ed. (McGraw-Hill, New York, 1976).

E. Hecht, *Optics*, 2nd ed. (Addison-Wesley, Reading, Mass., 1987).

R. Greenler, *Rainbows, Halos, and Glories* (Cambridge, New York, 1980).

PROBLEMS

1.1 Briefly describe the three branches of optics.

1.2 Describe the four laws of geometrical optics.

1.3 What is the speed of light in a vacuum? in water? in diamond?

1.4 Back when I was in high school, I read a science fiction story called "Light of Other Days," by Bob Shaw (now available in *Great Science Fiction of the 20th Century*, compiled by Robert Silverberg and Martin H. Greenberg, Avenal Books, New York, 1987). The story describes a material called *slow glass.* Panes of slow glass were set up in scenic areas and stored the images by slowing the light down so much that it might take a year for light to pass through a 1-cm-thick piece of glass. Calculate the index of refraction for slow glass.

1.5 Look up values (and units) for μ_0 and ϵ_0 in an introductory physics text and calculate the speed of electromagnetic waves in a vacuum. Check to see that the units work out.

1.6 Plot the angle of refraction i' as a function of the incidence angle i for a single refracting surface with $n = 1$ and the following values for n': (a) 1.333, (b) 1.51680, and (c) 1.78831. (*Hint*: Use a spreadsheet or symbolic math program.)

1.7 Plot the angle of refraction i' as a function of the incidence angle i for a single refracting surface with $n' = 1$ and the following values for n: (a) 1.333, (b) 1.51680, and (c) 1.78831.

1.8 What is the manufacturer's name for 626357 glass? What is its n_d and V?

1.9 What is the manufacturer's name for 581408 glass? What is its n_d and V?

1.10 Calculate the Abbe number for a glass with $n_C = 1.50763$, $n_d = 1.51009$, and $n_F = 1.51567$. What is the name of this glass? What is its glass number?

1.11 Calculate the Abbe number for a glass with $n_C = 1.74730$, $n_d = 1.75520$, and $n_F = 1.77468$. What is the name of this glass? What is its glass number?

1.12 Calculate n_F for a glass with $n_d = 1.62588$, and $V = 35.70$.

1.13 Calculate n_C for glass 624470.

1.14 Make a plot of index as a function of wavelength from 400 nm to 700 nm for KF9 glass.

1.15 Plot the critical angle as a function of index from 1.4 to 2.0. Assume $n = 1.0$ outside.

1.16 Plot ray displacement for a ray entering a plane parallel slab versus incidence angle for a K7 glass plate 2.00 cm thick.

1.17 Plot ray displacement for a plane parallel slab versus incidence angle for a LaK10 glass plate 5.00 cm thick.

1.18 Plot n' vs the minimum deviation angle for a prism in air. Assume that the apex angle is (a) 30°, (b) 60°, (c) 90°.

1.19 Find the minimum deviation angle for a glass sphere made of K3. Assume d light.

1.20 Find the minimum deviation angle for a glass sphere made of LaK10.

1.21 A frequently used method for folding the light path within an optical system is to use a 45–45–90 prism as a mirror. The incoming light has zero incidence angle with a short face. The ray reflects off the hypotenuse by means of total internal reflection and leaves through the other short face at 0°. What is the minimum index of refraction which can be used to make such a system?

1.22 A prism made of 744447 glass is shown in Figure 1.15. A beam of polychromatic light is incident on the left face at an angle of 45°. Trace the path of the ray in *d*, *C*, and *F* light through the prism. Describe the effect that the prism has on the beam. Assume that the bottom face is long enough for the ray in *d* light to strike it at its midpoint.

1.23 Show that when the indices of refraction are different on the two sides of a parallel glass plate refraction takes place as if the plate were not there.

1.24 An isosceles prism is made with FK3 glass. The apex angle is 30°. The side opposite the apex of the prism is placed on a table. At what incidence angle should a ray make with the first face of the prism so that it emerges parallel to the table?

1.25 Derive an equation similar to Eq. 1.66 for the secondary rainbow.

1.26 What index of refraction would give a minimum deviation angle of 180° for a sphere?

1.27 Show that

$$v = \frac{1}{\sqrt{\mu \epsilon}}$$

by substituting the solution to the wave equation (Eq. 1.19) back into the wave equation (Eq. 1.17).

1.28 For a plane parallel slab of glass, show that the displacement varies linearly with incidence angle for small angles. Find a formula for the proportionality.

1.29 For a plane parallel slab of glass, show that the displacement varies linearly with incidence angle for large angles (approximately 90°). Find a formula for the proportionality.

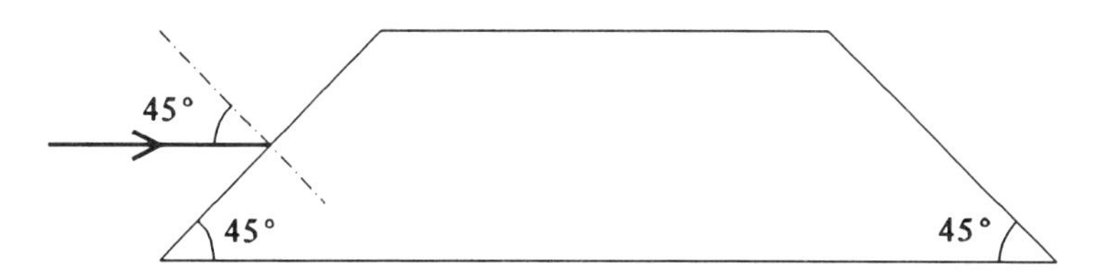

Figure 1.15 Prism for Problem 1.22.

1.30 Another measure of dispersion is called the partial dispersion ratio

$$P_{gF} = \frac{n_g - n_F}{n_F - n_C} . \tag{1.75}$$

Use the partial dispersion ratio, the Abbe value, and n_d to derive formulae for the A, B, and C coefficients in Cauchy's formula. *Note*: Values for partial dispersion are not given in Appendix A, but they are available in glass catalogs.

1.31 Occasionally it is possible to observe a form of ice bow known popularly as a *sun dog* or scientifically as a *parhelia*. Sun dogs are formed by thin plates of ice in the shape of hexagons. The light path through the hexagon which forms a sun dog enters one side, travels through the hexagon parallel with the side next to the entrance side, and leaves through the following side. What is the angle between the Sun and the sun dog? The index of refraction of ice is $n_d = 1.3104$.

1.32 The primary rainbow is formed when light reflects once inside a spherical drop of water. However, only a small fraction of the light reflects; most of the light refracts and leaves the drop. Therefore, there should be a bright zero-order rainbow visible relatively near the sun. Such a rainbow cannot exist; show why.

CHAPTER 2

INTRODUCTION TO IMAGING SYSTEMS

For several years, Fuji film has run television advertisements which portray that it is impossible to distinguish between an object and an image of the object. My personal favorite is the ad which shows four sun bathers and a small child throwing water on them. Only two of the sun bathers react, because the other two are images recorded on film. In this chapter, we introduce the requirements for imaging, and we hope that we will clearly demonstrate why you cannot distinguish an object from a clear image of that object. Simple imaging systems will also be discussed.

I have found that probably the most difficult aspects of geometrical optics for the novice to master are the notation and sign convention which are used. Both of these topics are introduced in this chapter. A summary of the notation and sign convention used throughout this book will be found in Appendix B. The student is urged to become intimately familiar with this material. Note that the sign convention I use in this book is probably different than what you used in introductory physics. I think that the sign convention presented here is easier to use and easier to remember. Most optical system designers use this sign convention.

2.1 GENERAL REQUIREMENTS FOR IMAGING

An object is anything from which electromagnetic radiation can be emitted or scattered. In this book, only electromagnetic radiation which can be sensed by the human eye will be discussed, but the same principles apply to all wavelengths of electromagnetic radiation. The object can be either a primary source (which is self-luminous), or a secondary source (which is illuminated by another source). In either case, electromagnetic waves will spread out from every point on the object. An extended object with a finite lateral extent can be viewed as a collection of point sources from which spherical waves are emitted. The overall wavefront shape and characteristics are determined by the constructive interference of all the spherical waves.

Ultimately, then, only spherical waves diverging from a point source need to be considered in detail. Shifting to the ray picture of light propagation, an object can be defined as a point from which rays of light diverge. The eye "sees" an object by collecting a limited set of these diverging rays.

An image will be formed when rays of light which originated at some object point converge at some other point. In terms of wavefronts, imaging requires a spherical wavefront converging on a point. Once the rays converge at an image point, they will continue along the same direction of propagation. Thus, they subsequently diverge just like rays leaving an object point. Therefore, an image is, in a very real sense, a replica of the object. In fact, the eye has no way of telling whether the diverging rays it collects come directly from an object or from an image.

There are two types of images. A real image is an image where the rays actually converge at a point and then diverge. A virtual image is an image where the rays never actually converge, but instead they diverge as if they came from a particular point (Figure 2.1). Again, the eye has no way of telling which is which. The only significant difference between the two types of images is that a real image can be viewed on a screen located where the rays intersect. Figure 2.2 shows a good way to visualize the difference in the two types of images. If diverging light falls on a screen, the screen is illuminated but no image is formed. If a screen is located at the intersection of converging rays, then only a small point on the screen is illuminated. The bright spot on the screen looks like the point object from which the rays originated. A collection of such points will form an extended image on the screen.

An ideal or perfect image would be an image where every ray leaving an object point later converged on a real image point or appeared to diverge from a virtual image point. Perfect imaging is only an approximation to reality; it does not actually exist and cannot be achieved. Actual, achievable images are those where most of the rays coming from an object point pass through the imaging system and come very

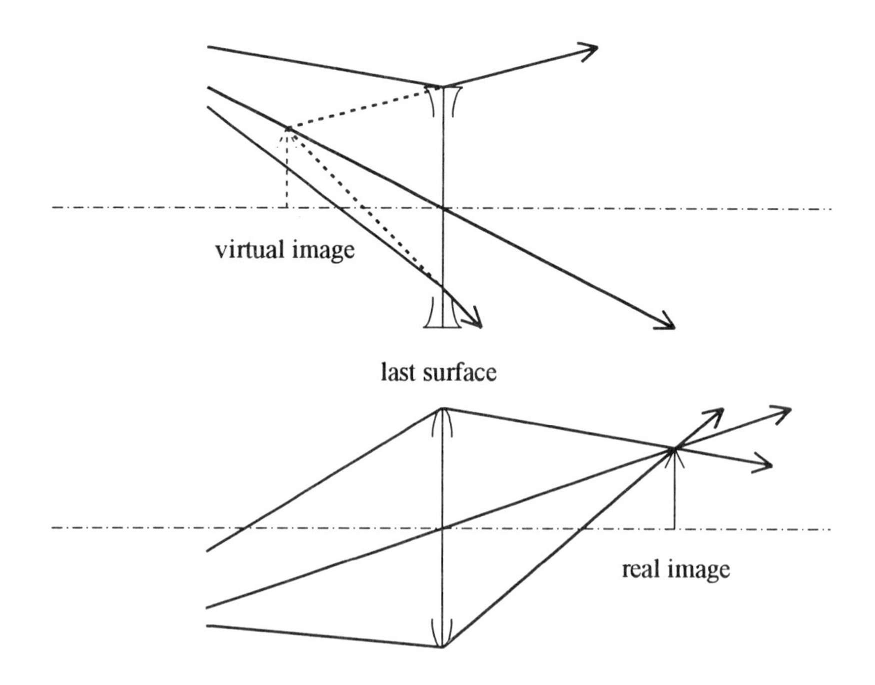

Figure 2.1 Two types of images may be produced by optical systems.

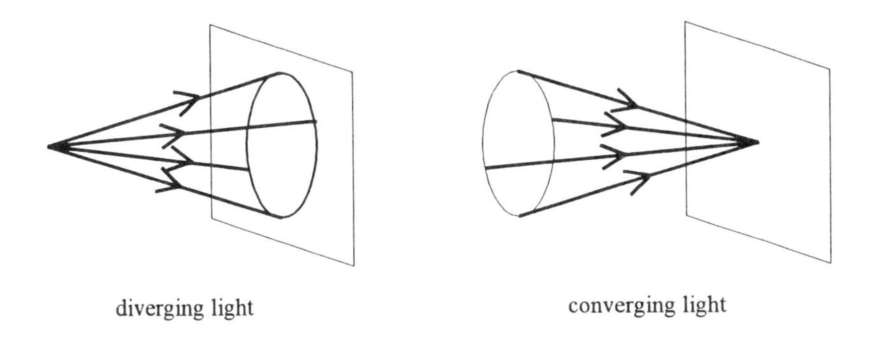

Figure 2.2 Diverging light illuminates a screen while converging light forms an image.

close to intersecting at an image point. The ultimate limit on imaging is set by the wave nature of electromagnetic radiation. In the formation of a real image, for example, waves converge on a point. Unfortunately, these waves can never be exactly spherical after passing through an optical system. The wavefront near the edge of the optical system will be bent by diffraction. Thus the absolute best image that can be formed will still be limited by diffraction. On the other hand, the quality of the optical system often sets the limits on how close the rays come to intersecting at a point. Deviations of ray position from the perfect image point due to the quality of the optical system are called *aberrations*. The study of aberrations will be undertaken later in this text.

A simple device which illustrates this discussion is the camera obscura or pinhole camera shown in Figure 2.3. How can a pinhole form an image? By the definition of imaging which was just given, rays from a single object point must meet at the same or nearly the same image point. The pinhole physically restricts the rays from an object point to meet at nearly the same image point. As the size of the pinhole increases, the imaging effect is reduced; and rather than observing an image, we merely see an illuminated screen. On the other hand, when the size of the pinhole is on the order of a wavelength of light, diffraction takes over and again the size of the image spot increases.

We can make this discussion quantitative by defining the resolution of an imaging system. Closely spaced object points will form images that overlap because the images are not true point images, but rather small circles of light. The minimum separation of the two point objects which results in two distinct image circles is

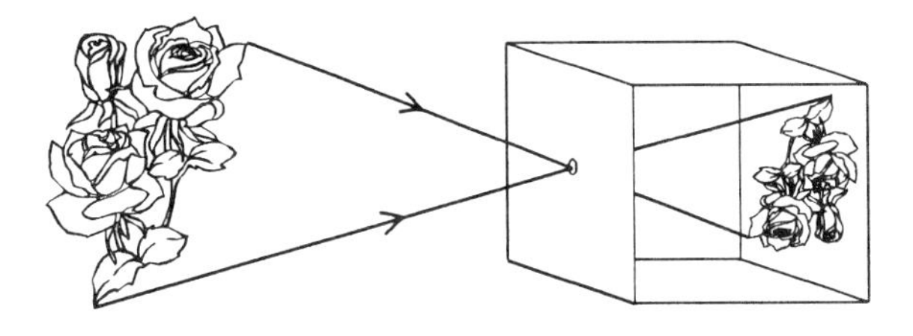

Figure 2.3 Pinhole camera.

called the *resolution* of the system. Resolution is usually measured as the angle θ subtended by the two objects at the optical system. In the geometrical limit, the size of the spot produced by a point object is given by

$$d' = \left(1 - \frac{s'}{s}\right) d , \tag{2.1}$$

where d is the diameter of the pinhole, d' is the diameter of the spot, and s and s' are the object and image distances measured from the pinhole. As a rule of thumb, we can say that the two object points are resolved when the centers of the image circles are separated by at least 5/6 of their diameter. The geometrical resolution of the pinhole camera can be determined from the geometry shown in Figure 2.4 to be

$$\theta_g = \frac{5}{6}\left(1 - \frac{s'}{s}\right)\frac{d}{s'} . \tag{2.2}$$

In the physical optics limit, the resolution is given by Rayleigh's criterion:

$$\theta_p = 1.22\,\frac{\lambda}{d} . \tag{2.3}$$

An explanation of Rayleigh's criterion can be found in almost any text on physical optics. Here, all we need to know is that the resolution angle increases with decreasing pinhole size.

This discussion is mainly of academic interest now. The eye cannot tell an object from a real image or a virtual image if the diffraction and aberration effects are smaller than the resolution limit of the eye. Everyone has experienced optical systems which form images that cannot be distinguished from the object—for example, mirror images in a "house of mirrors" or the commercials for a particular brand of photographic film. That these images are not perfect is not important. For now, in order to understand the basic principles of imaging systems, the

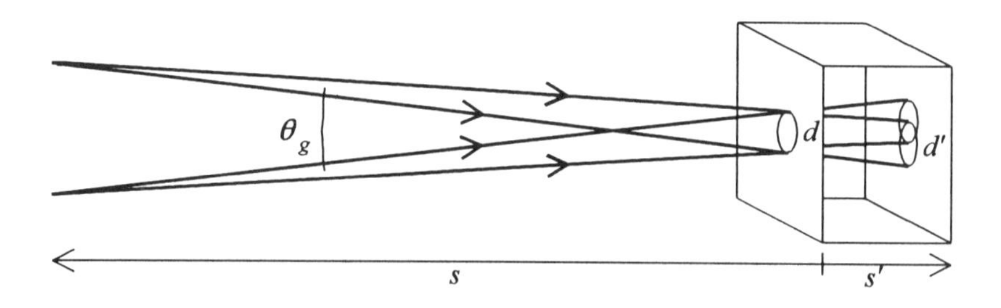

Figure 2.4 Geometrical resolution of a pinhole camera.

approximation is made that the imaging is perfect. How can perfect images be formed? What sort of optical systems can form perfect images? These are the topics of the remainder of this chapter.

2.2 PLANE MIRRORS

Perhaps the simplest image forming system is a plane mirror. Ordinary plane mirrors consist of a flat sheet of glass which is covered on one side with an aluminum or silver coating. Plane mirrors are extremely common imaging systems which give excellent image quality. Diffraction effects caused by the mirror's edges are normally unimportant because of the large size of common mirrors. Aberrations are important only for "fun house" mirrors which are intentionally made to give a distorted image by not being flat. A minor problem with common mirrors is ghost images created by the first glass surface. Also, some light is lost due to scattering at the mirror surfaces, but this doesn't affect the image quality, only its brightness.

The nature, size, and location of plane mirror images can be determined by applying the law of reflection to a number of rays leaving an object point. First, draw an arbitrary ray from the chosen object point until the ray intersects the surface of the mirror. At this point, draw a normal to the surface of the mirror. The angle between the normal and the incident ray is the incidence angle i. The reflected ray can be drawn by copying the incidence angle to the other side of the normal to form a reflected angle. The reflected angle is indicated with the variable i'. The prime indicates a quantity measured after the surface. According to the law of reflection, the magnitude of the angle of incidence is equal to the magnitude of the angle of reflection. But remember that our sign convention on angles states that angles are positive when measured counterclockwise. Since both angles are measured from the normal in different directions, our sign convention demands that the law of reflection be written as

$$i = -i' . \tag{2.4}$$

The procedure outlined above for finding the reflected ray is strictly a geometrical construction process which can be accomplished entirely with straight edge and compass.

Drawing several rays in the fashion described above shows that the reflected rays do not intersect anywhere to form a real image. If, however, the reflected rays are extended behind the mirror as in Figure 2.5, it becomes clear that the rays appear to diverge from a virtual image point. If the construction is done carefully, any number of rays can be drawn from the object in this fashion, and they will all appear to diverge from the same image point. This confirms that the plane mirror creates a perfect image (ignoring diffraction effects).

One special ray which could be drawn is the ray that would form a normal to the plane mirror surface. For this ray, the incidence angle is zero and so is the reflection angle. Therefore, the reflected ray travels directly back to the object. This means that

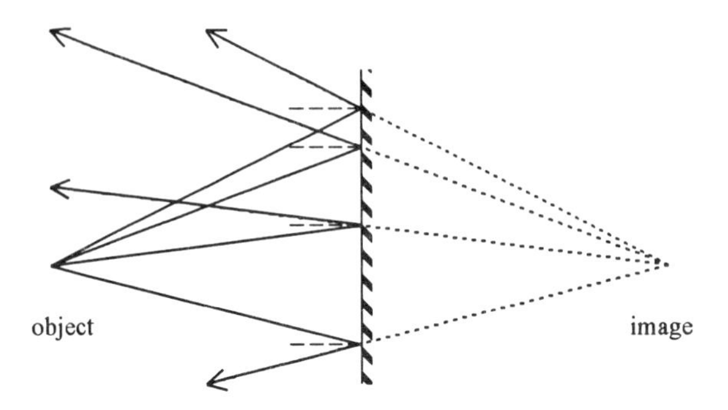

Figure 2.5 Image formation by reflection.

the object and image both lie on a line which is perpendicular to the plane mirror. A careful measurement of the distance s between the object and the mirror along this ray reveals that the image is the same distance s' behind the mirror as the object is in front of the mirror. This can be confirmed by looking at the triangle OAB formed by an arbitrary incident ray, the special ray normal to the mirror, and the mirror itself and the triangle IAB formed by the extensions of the two reflected rays and the mirror (Figure 2.6). These triangles are congruent triangles because they share a common side and have two equal angles. Thus, the lengths of the other sides are also equal.

If we establish a coordinate system with origin at the mirror where the normal ray strikes the surface, then the relationship between object distance and image distance can be expressed as

$$-s = s' . \tag{2.5}$$

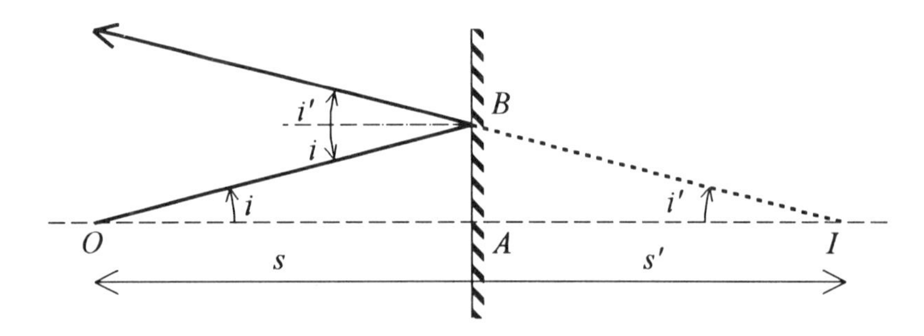

Figure 2.6 Demonstration that the image distance is equal to the object distance for a plane mirror.

The negative sign is needed to satisfy our sign convention because both the object and image distances are measured from the surface. We normally set up systems so that light (at least initially) travels from left to right. Any distance, such as s, which is measured to the left of the origin must be negative. Distances are positive when measured to the right. In general, the object distance s is the distance from the surface of interest to the object. The image distance s' is measured from the surface of interest to the image location.

A shortcut for finding the image position in a plane mirror is to draw a normal to the surface from the object. Extend this normal behind the mirror. Measure the object distance, and locate the image point at an equal distance behind the mirror. This method works even when the normal doesn't actually strike the mirror. If the object point is off to one side of the mirror as shown in Figure 2.7, then the normal must be drawn to an extension of the mirror surface. Figure 2.7 also illustrates that what you see in a mirror is determined by the size of the mirror. The eye must be located in the region between the rays which strike the edges of the mirror for the image to be seen. The mirror acts like a window through which only a limited part of the space behind the mirror can be viewed. Moving the eye changes what images can be seen. The span of objects which can be seen is called the *field of view* (FOV).

Fun house mirrors with curved surfaces were mentioned earlier. Another amusing use of mirrors is to set up two or more mirrors. Images are formed by each mirror, but additionally, images of the images can also be formed. Not only can the eye not distinguish between an object and an image, but neither can optical systems.

To see how multiple mirror systems work, it is helpful to define object space and image space. Object space is the collection of all locations in front of the mirror surface where objects would normally be located. Image space is the collection of all locations behind the mirror where the images are formed. The terms "in front" and "behind" indicate, respectively, light rays before and after they have reflected off the mirror. A pair of object and image points are said to be conjugates of each other. Thus every point in object space has one and only one conjugate point in image space. Another way of saying the same thing is that any point in the object space of a mirror will have a conjugate point in the image space of that mirror. For multiple mirror systems, the image formed by one mirror may or may not lie in the object space of another mirror. If it does, then the second mirror will form an image of the previous image point (Figure 2.8).

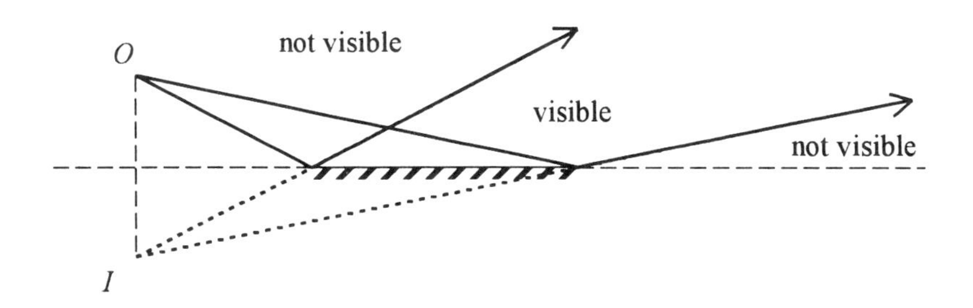

Figure 2.7 Field of view of a plane mirror.

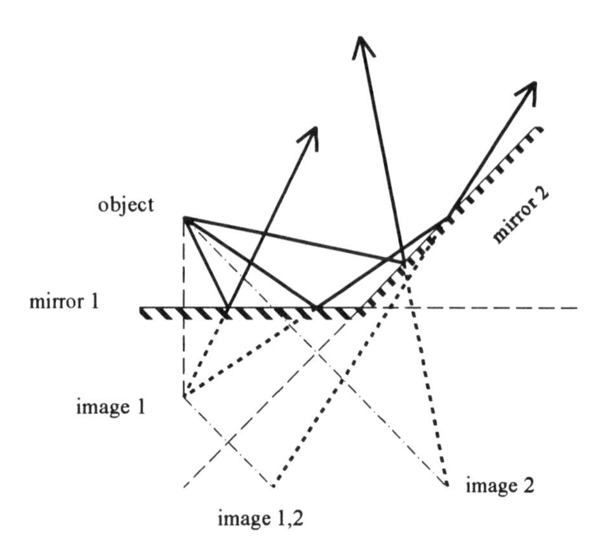

Figure 2.8 Multiple mirror images.

To determine the nature of the image of an extended object, two points on the object can be selected for imaging as shown in Figure 2.9. The additional image point is found as before by measuring an image distance along the new surface normal. It should be fairly obvious that since all the lengths and angles are the same on the object side as on the image side, the image size y' must be the same as the object size y. If we define magnification as the ratio of the image size to the object size,

$$m = \frac{h'}{h}, \tag{2.6}$$

then the magnification of a plane mirror has the value of one.

Another interesting aspect of the images formed by plane mirrors can be shown

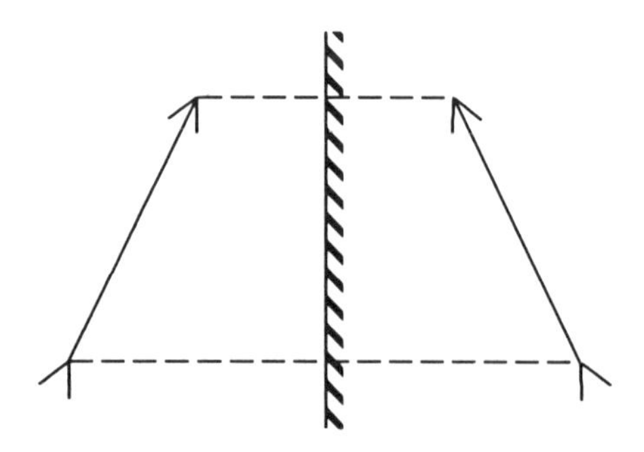

Figure 2.9 Mirror image of an extended object.

by establishing a right-handed coordinate system as the extended object. Since objects which are closer to the mirror form images which are also closer to the mirror, the image looks like a left-handed coordinate system (Figure 2.10). Another way of seeing this is to look at yourself in the mirror. Your image faces you. If you could imagine stepping around behind the mirror and standing in the same position as your image, then your right hand would be where the image of your left hand is located. If you raise your right hand, the image raises its left hand. All mirror images act in this fashion. In fact, the meaning of mirror image in everyday language is that the right and left hands are reversed.

2.3 PLANE REFRACTING SURFACES

Perhaps the next simplest imaging system is a single plane refracting surface consisting of just the interface between two media with different indices of refraction. Typically, one media will be air, but this isn't important. All that is required is that the two indices be different. Examples would include an air–glass interface, the surface of a calm pool of water, or the sides of an aquarium. The object, by convention, will be located on the left side of the interface, but this side could be the side with either the smaller or larger index of refraction. The nature of the image formed by this interface can be determined as with the mirror by drawing several rays and finding where they intersect.

Figure 2.11 shows how to draw a refracted ray. First draw an arbitrary ray from the object to the surface. Next, construct the normal to the surface at the point where the ray strikes the surface. As before, the incidence angle i is measured from the normal to the ray, but constructing the refraction angle is more complicated because the law of refraction is

$$n \sin i = n' \sin i' . \tag{2.7}$$

A distance which corresponds to n sin i can be constructed by drawing an arc of

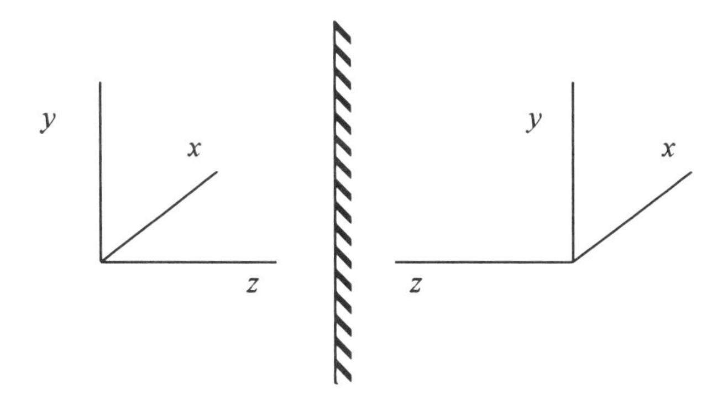

Figure 2.10 A mirror converts a right handed image into a left-handed image.

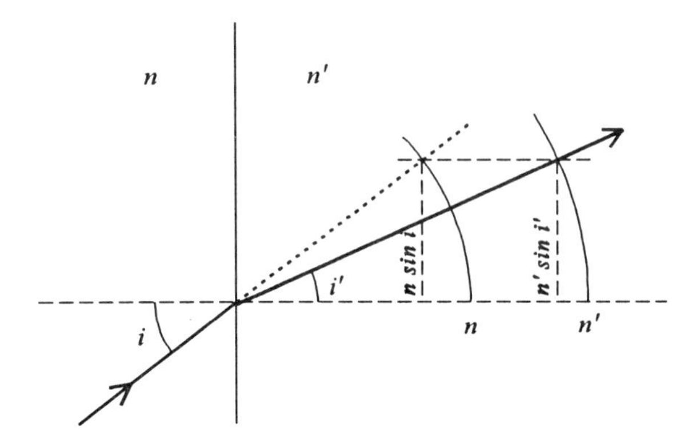

Figure 2.11 Graphical construction technique for a refracted ray.

arbitrary radius centered at the intersection of the ray and the surface. Extend the incident ray until it crosses this circle. The distance from this intersection point to the surface normal will represent $n \sin i$. Next, draw another circle centered at the point where the incident ray hits the refracting surface. This circle must have a radius which is equal to the radius of the first circle times the ratio of the indices n'/n. In other words, the ratio of the radii of the two circles is equal to the ratio of the indices. Copy the distance $n \sin i$ so that it intersects this second circle. This distance also represents the distance $n' \sin i'$, so that Snell's law has been satisfied. The refracted ray extends from the intersection of the incident ray with the surface to the point on the second circle which was just determined.

Additional refracted rays can be constructed using the procedure outlined above. When this is done (Figure 2.12), you should observe that the rays do not all intersect at a single point. It may be surprising to find out that a single refracting surface does not create a perfect image unlike a plane mirror.

It is obvious that objects can be clearly seen through a plane refracting surface,

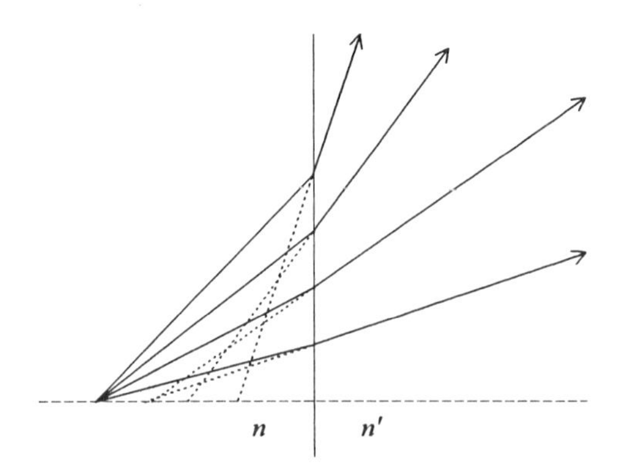

Figure 2.12 Image formation by refraction.

otherwise you couldn't see through a window. When you look at objects under the water of a swimming pool, you may have thought that you were looking at the objects themselves when, in fact, you were observing their images. How can we see clear, sharp images without perfect imaging? The solution to this dilemma is to remember that the eye only intercepts a limited number of rays coming from the object. Since these rays all make about the same incidence angle with the surface, they all appear to diverge from very nearly the same point. This limited group of rays form a nearly perfect image. If the eye moves to a different location, it receives a different set of rays. The image will appear to move as a result. Thus, the image position will depend on the location of the observer for this system.

As a special case, let's look at the location of an image when viewed along the normal to the surface which passes through the object point. A ray perpendicular to the refracting surface can be drawn, as was done for the plane mirror in the previous section (Figure 2.13). For this ray, the incidence angle is zero and so is the refracted angle. Nearby rays (the only ones which will contribute to forming the final image) all have very small angles of incidence and refraction. When angles are expressed in radians, the sine function can be expanded as a power series of the angle. Therefore, Snell's law can be written as

$$n\left(i - \frac{i^3}{3!} + \frac{i^5}{5!} - \dots\right) = n'\left(i' - \frac{i'^3}{3!} + \frac{i'^5}{5!} - \dots\right) . \tag{2.8}$$

For sufficiently small angles, only the first term needs to be retained for accurate results. Thus, Snell's law simplifies tremendously for the ray perpendicular to the refracting surface

$$n\,i = n'i' . \tag{2.9}$$

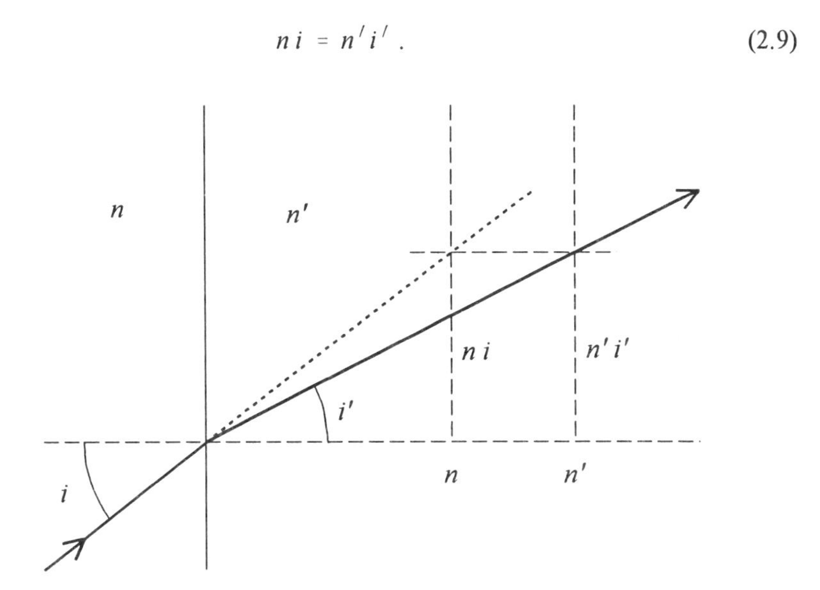

Figure 2.13 Graphical construction technique for refraction at small angles of incidence.

Example 2.1

Just exactly what is meant by "sufficiently small angles"?

The answer depends on the accuracy required to solve a particular problem. If the angle involved is 15° (0.2618 radians), then

$$\sin(0.2618) = 0.2588 \ .$$

The difference between the angle and the sine of the angle is only 0.0030. The percent difference is 1.14%. For angles less than 15°, the error is correspondingly less. ■

Obviously, the mathematics are greatly simplified by using this small-angle approximation. The technique for drawing the exact refracted ray shown in Figure 2.11 is also simplified. Instead of two circles with radii in the ratio of the indices, two straight lines parallel to the refracting surface are drawn. The distances of these lines from the surface are in the ratio of the indices (Figure 2.13). The incident ray is extended until it hits the line which corresponds to the first index of refraction. The distance from this intersection point to the surface normal is copied onto the line corresponding to the second index of refraction. Finally, the refracted ray is drawn between this new location and the point where the incident ray hits the surface. The validity of this procedure can be seen by going back to Figure 2.11. If only very small angles were allowed in Figure 2.11, then the portions of the circles used would begin to look like the straight lines of Figure 2.13.

The small-angle construction technique we just outlined could be used to construct several refracted rays to find the image location. If this is done strictly with rays making small angles with the surface, a couple of problems arise. First, it becomes confusing to draw the rays because they are so close together. Second, the intersection point of the rays is not well defined because of the small intersection angles. To overcome these problems, we can cheat by using the small angle construction technique on rays that do not actually make small-angles with the surface as shown in Figure 2.14. Note that all of these rays do appear to come from a single image point. In effect, all we are doing is stretching the figure in the vertical direction to make the construction easier.

The image distance s' for this special case can be determined using a carefully drawn ray as in Figure 2.14. Alternatively, the small-angle approximation can be used to calculate the image distance. As we've shown, for sufficiently small angles, $\sin i \approx i$. It should be clear that for small angles, $\cos i \approx 1.0$. Therefore, in this approximation we obtain

$$-\frac{y_1}{s} = \tan i = \frac{\sin i}{\cos i} \approx i \qquad (2.10)$$

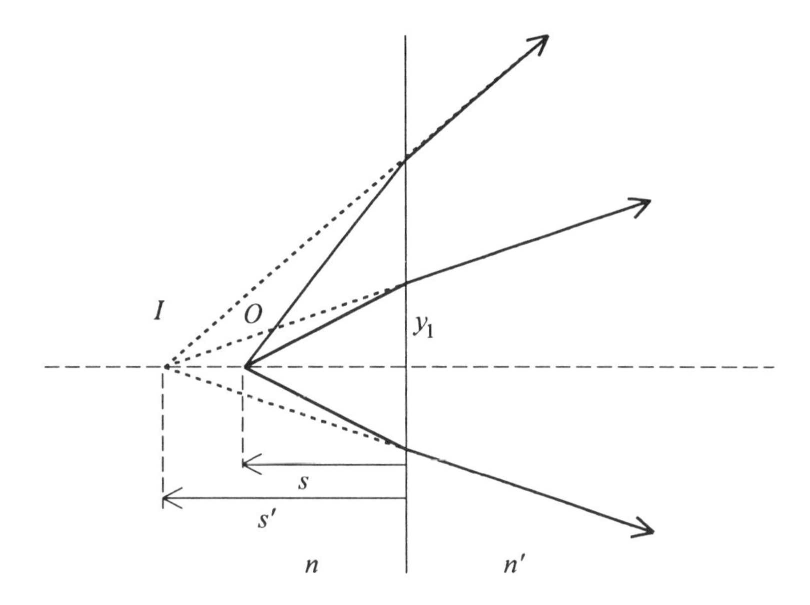

Figure 2.14 Perfect imaging results from the small-angle approximation.

and

$$-\frac{y_1}{s'} = \tan i' = \frac{\sin i'}{\cos i'} \approx i' \ . \tag{2.11}$$

Substituting into Eq. 2.9 , the small-angle version of the law of refraction, gives

$$s' = \frac{n'}{n} s \ . \tag{2.12}$$

Figure 2.15 shows how the magnification can be determined for this special case. A ray from the top of the object is drawn so that

$$h = s \tan i \approx s i \ . \tag{2.13}$$

The refracted ray can be constructed using the small-angle technique. In addition, a second ray from the top of the object can be drawn with zero incidence angle. When this ray hits the interface, the refraction angle is zero. The backward extension of these two rays define the image height y'. From the figure,

$$h' = s' \tan i' \approx s' i' \ . \tag{2.14}$$

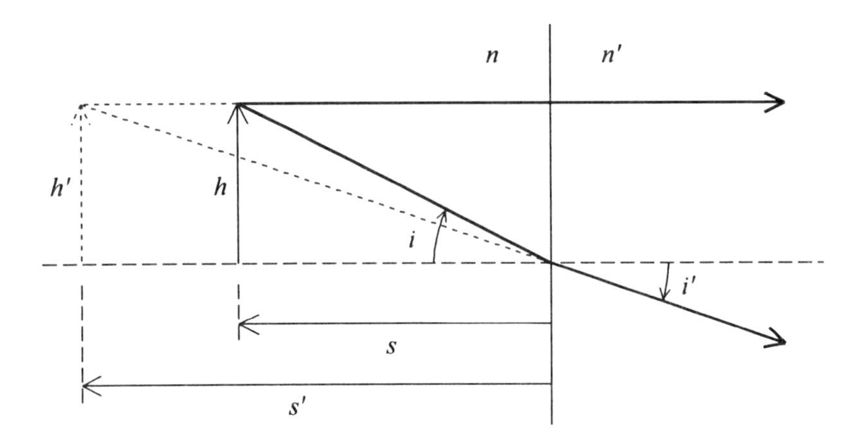

Figure 2.15 Ray drawing used to determine image size for a plane refracting surface.

Combining these two equations with the definition of magnification gives

$$m = \frac{h'}{h} = \frac{s'i'}{si} = \frac{s'n}{sn'} . \tag{2.15}$$

Combining Eq. 2.12 with Eq. 2.15 gives the result that the magnification of a plane refracting surface is one.

We can find the image formed by a thick parallel slab (Figure 2.16) by using the fact that images are optically equivalent to objects. But first, a word about notation. When the ray displacement for a thick slab was calculated in the previous chapter, the index of refraction was the same on either side of the slab. To permit different indices on either side, a subscript notation is introduced here. The index before surface j is n with the subscript $j - 1$. The index after the jth surface is n with the subscript j. Similar notation is used for the object and image distances. From Eq. 2.11, we find that the image created by the first surface is located at

$$s_1' = \frac{n_1}{n_0} s_1 . \tag{2.16}$$

If this image is used as an object for the second surface, then the new object distance is

$$s_2 = s_1' - t_1 , \tag{2.17}$$

where t_1 is the thickness of the slab. This object distance is again used in Eq. 2.11

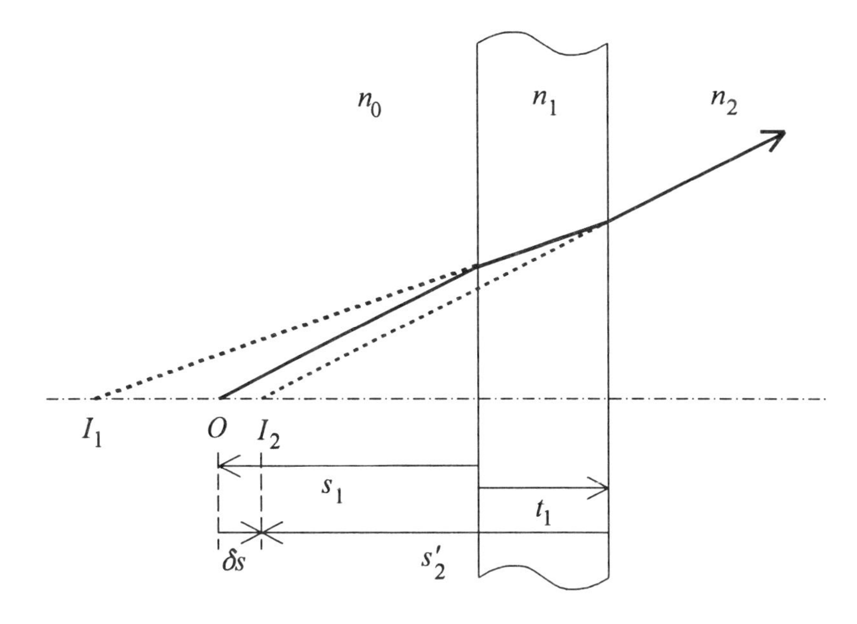

Figure 2.16 Imaging by a thick glass plate.

to give the final image position measured from surface 2:

$$s_2' = \frac{n_2}{n_0} s_1 - \frac{n_2}{n_1} t_1 \ . \tag{2.18}$$

For a slab of glass surrounded by air, the displacement of the image is given by

$$\delta s = (t_1 + s_2') - s_1 = \left(1 - \frac{1}{n_1} \right) t_1 \ . \tag{2.19}$$

Of course, this formula only works for small angles of incidence. If a typical value for the index of refraction of glass $n = 1.5$ is plugged into Eq. 2.19, then the displacement of the image from the object is just one-third the thickness of the glass. The direction of the displacement is toward the glass slab.

2.4 SPHERICAL MIRRORS

From this point on, much of the discussion in this text will be concerned with spherical surfaces, so a word of explanation is in order. Spherical surfaces are used extensively in geometrical optics for several very good reasons. First, spherical surfaces are easy to manufacture. The polishing and grinding of optical components can be accomplished by randomly rubbing the piece of optical glass with either another piece of glass or some sort of tool. An abrasive material placed between the two objects actually accomplishes the grinding. When two surfaces are moved past each other this way, the only shape they can have and remain in contact with each other is a spherical shape (you can think of a planar surface as a special case with infinite radius of curvature). Thus, the natural shape of optical components is spherical. Second, spherical surfaces provide adequate image quality provided that the radius of curvature is relatively large. This will be discussed more fully later. From an analytic point of view, spherical surfaces can be treated with fairly simple mathematics. Additionally, a sphere is a good first-order approximation to the shape of most optical surfaces, even aspheric surfaces. A spherical surface which bows outward is convex and one that bows inward is concave.

An additional restriction we will place on the optical systems we study is that they have axial symmetry. Axial symmetry means that the centers of any spherical surfaces or openings in the system lie along a straight line. This line is called the *optical axis.*

If you draw rays reflecting off a spherical mirror, you should soon realize that the rays do not all cross at one image point (Figure 2.17). Yet, as with the single reflecting surface, we know that spherical mirrors do form sharp images. Just look in the passenger side mirror of a car. Again, we need to restrict the rays somehow. If we deal only with rays that make small angles with the optical axis and never get very far from the optical axis, then the rays take paths through the optical system which are very similar to the path taken by the optical axis. These rays are called *paraxial rays.* Dealing only with paraxial rays guarantees that all rays coming from a particular object point will meet at an image point. In essence, we are temporarily ignoring aberrations.

As mentioned previously, trying to draw rays which are very close together has its drawbacks. We overcome these problems by, once again, stretching the figure in the vertical direction. Stretching the figure causes the spherical reflecting surface to look like a plane surface. In effect, the reflection is taking place in the plane tangent to the sphere through the vertex. The vertex is the point where the optical axis intersects the sphere. Since the mirror is still really a spherical surface, the angles of incidence and reflection must be measured from the normal to the spherical

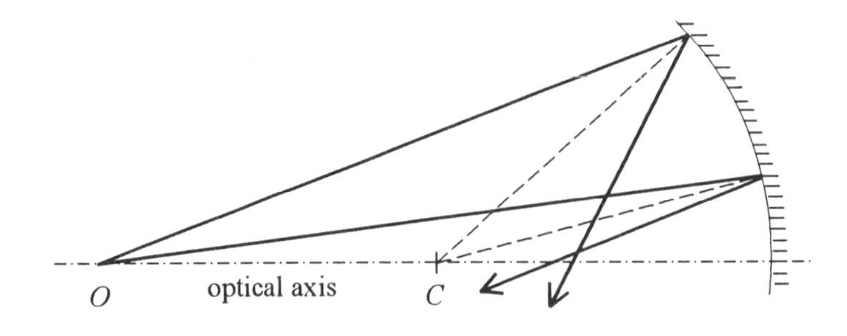

Figure 2.17 Nonparaxial rays reflecting off a spherical mirror.

surface. Figure 2.18 shows that the normal is drawn from the center of curvature of the sphere to the point where the incident ray hits the vertex plane. After the normal is drawn, the angle of incidence can be copied to find the direction of the reflected ray. Repeating this procedure for several rays would show that now all rays from an object point will all meet at a single image point. Try it!

Since all rays pass through the same image point, only one off-axis ray is required to locate the image. The optical axis also defines a ray which leaves the object, strikes the mirror at the vertex with zero incidence angle, and reflects off the mirror with zero reflection angle. Therefore, the image is located where the off-axis ray crosses the optical axis.

An equation can be derived from the geometry of Figure 2.18, which gives the image position for the specified object distance and mirror radius. Recall that the law of reflection in our sign notation (and here the sign is important) is

$$-i = i' . \tag{2.20}$$

The angle of incidence and the angle of reflection can be written in terms of the angle u that the ray makes with the optical axis and the angle β between the normal and the optical axis:

$$-i = \beta - u \tag{2.21}$$

and

$$i' = u' - \beta . \tag{2.22}$$

The angle u is called the *slope angle of the incident ray*. The angle u' is the *slope angle of the reflected ray*. Combining these two equations gives

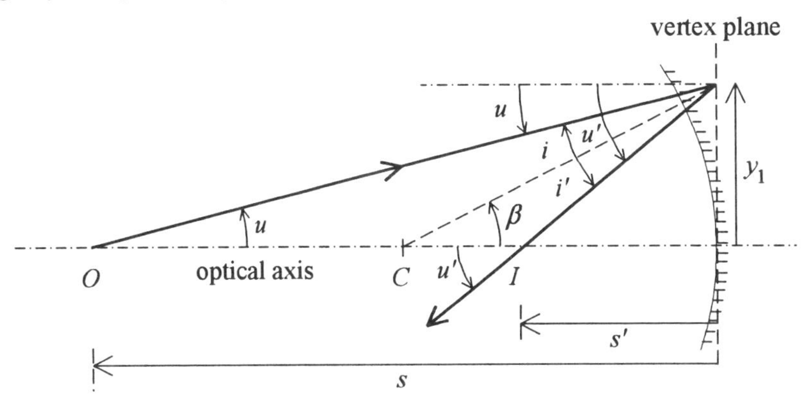

Figure 2.18 Reflection of a paraxial ray by a spherical mirror.

$$2\beta = u + u' . \tag{2.23}$$

Now, the slope angles can be written in terms of the object and image distances. The small-angle approximation is used to give

$$u \approx \tan u = -\frac{y_1}{s} \tag{2.24}$$

and

$$u' \approx \tan u' = -\frac{y_1}{s'} . \tag{2.25}$$

Note that the minus sign is included because y_1 is positive but both s and s' are negative in Figure 2.18. A similar equation applies to the angle β:

$$\beta \approx \tan\beta = -\frac{y_1}{r} . \tag{2.26}$$

Making these substitutions and canceling the common factor of $-y_1$ yields the paraxial equation for a spherical mirror:

$$\frac{1}{s} + \frac{1}{s'} = \frac{2}{r} . \tag{2.27}$$

This equation was derived using the special geometry of Figure 2.18, but it is applicable to all cases of imaging with a spherical mirror.

One such case would be to place the object at an infinite distance from the mirror. In the limit of the object distance going to infinity, the rays become parallel to the optical axis. From Eq. 2.27, the image distance is just half the radius of the mirror. Because the object is at a special point (infinity), this image point is given a special name also. The image point where rays from infinity meet is called the *focal point* F'. Using the principal of reversibility, it is easy to see that an object placed at the focal point F has an image at infinity. This notation indicates that there are two focal points, one before the surface (unprimed) and one after the surface (primed). We will call these the *object space focal point* and the *image space focal point*. For a spherical mirror, these two focal points are physically located at the same point. The distance from the reflecting surface vertex to the focal point is given by

$$f = f' = \frac{r}{2} . \tag{2.28}$$

Example 2.2

A concave mirror with a 12.0-cm radius of curvature is used to image an object which is 24.0 cm from the mirror. Where is the image located?

Since the mirror is concave, the center of curvature is to the left of the vertex of the mirror. Therefore, $r = -12.0$. Plugging into the mirror equation gives

$$\frac{1}{s'} = \frac{2}{(-12.0)} - \frac{1}{(-24.0)} .$$

The image distance is -8.0 cm. ■

The nature of the image of an extended object can be determined easily by using rays through the focal point and center of curvature as special rays. As discussed in the previous paragraph, an incident ray parallel to the optical axis must pass through the focal point after reflection. The mirror has no way of telling that the ray doesn't originate at infinity. As shown in Figure 2.19, the size and position of the image of an extended object can be determined by drawing a ray from the full extension of the object parallel to the optical axis. This ray must pass through the focal point. A second ray, which passes through the focal point before striking the mirror, must head out parallel to the optical axis. Since these two rays came from the same point on the object, where they meet after reflection determines the corresponding image point. This can be checked by drawing another special ray. A ray which passes through the center of curvature of the mirror has zero incidence angle and zero reflection angle. This ray merely returns along its own path. Notice that we can use rays to find the image which do not actually hit the mirror.

An equation for the magnification can be determined by drawing a ray from the

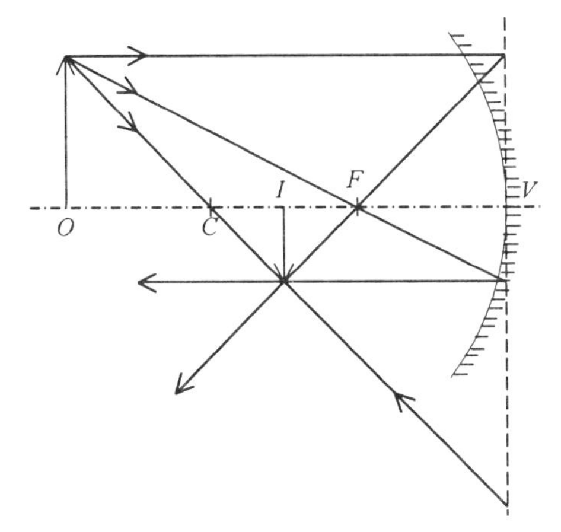

Figure 2.19 Determination of the size and position of an image using the focal point and center of curvature of the mirror.

top of the object to the vertex of the mirror as shown in Figure 2.20. The optical axis is the normal for this ray. Thus, the slope angles u and u' are equal but opposite in sign. The triangle formed by the incident ray, the object and the optical axis is similar to the triangle formed by the reflected ray, the image and the optical axis. Therefore, the lengths of the sides of these triangles are proportional. Using the definition of magnification

$$m = \frac{h'}{h} = -\frac{s'}{s} . \tag{2.29}$$

Example 2.3

If the object in the previous example is 3.0 cm high, how high is the image?

The magnification is

$$m = -\frac{(-8.0)}{(-24.0)} = -0.33 .$$

Therefore, the image size $h' = m\,h = (-0.33) \times (3.0\,\text{cm}) = -1.0\,\text{cm}$. The minus sign indicates that the image is inverted. ■

An extended object which is moved to infinity along the optical axis will appear to be a point source on the optical axis. As discussed earlier, such an object will generate rays which are parallel to the optical axis. But if the object is infinitely large at infinity, then object points can exist off axis. An off-axis point at infinity generates

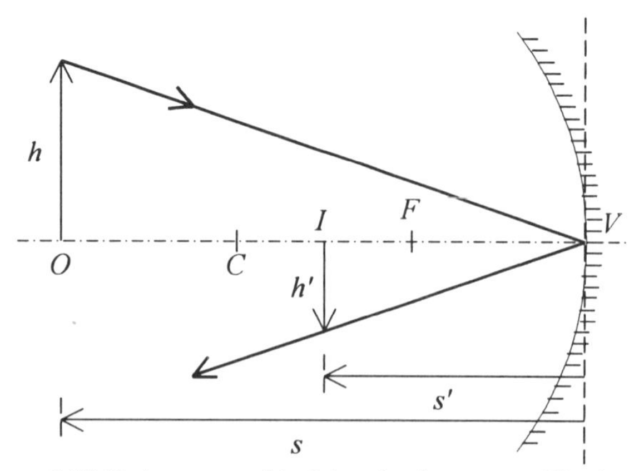

Figure 2.20 Vertex ray used to determine image magnification.

rays which are parallel, but at an angle to the optical axis. As long as this angle is small, the paraxial approximation will hold and all of the parallel rays will meet at a point. By drawing several parallel rays, you can show that the image point lies in a plane through the focal point. The height of the image in the focal plane can be easily determined by drawing a single ray through the center of curvature parallel to all of the other rays coming from infinity (Figure 2.21). The intersection of this ray with the focal plane determines the image position.

The discussion in the previous paragraph provides a shortcut for finding the position of the image of an on-axis object point. As before, an arbitrary ray is drawn from the object point to the mirror (Figure 2.22). Another ray, parallel to the first is drawn through the center of curvature. The intersection of this second ray and the focal plane determine a point through which all parallel rays with this slope angle must pass. Therefore, the reflected ray also crosses the focal plane at this height. The reflected ray can be extended until it crosses the optic axis. The second intersection defines the image position.

2.5 SPHERICAL REFRACTING SURFACE

Imaging by a spherical refracting surface can be analyzed in a fashion similar to that used for the spherical reflecting surface. First, we will see how to draw an exact ray after refracting at a spherical surface.

In Figure 2.23, an arbitrary ray is drawn from the object to the surface. This ray is extended past the surface for later use in determining the direction of the refracted ray. Next, draw two arcs centered on the point of intersection of the ray and the surface. As before, the radii of these arcs are in the ratio of the indices of refraction. Next, draw a line from the center of curvature of the surface to the intersection point. This line defines the normal to the surface at the intersection point and the angle of incidence. The next step is to determine the angle of refraction. Draw a line parallel to the normal which passes through the point where the extension of the incident ray crosses the circle with radius n. Unless there is total internal reflection, the parallel

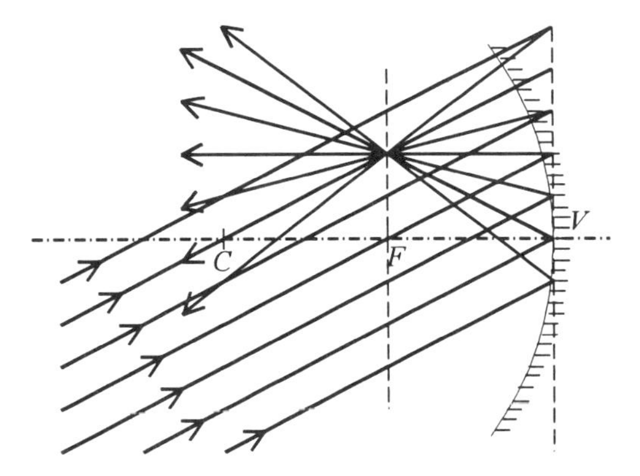

Figure 2.21 Image of an infinitely distant off-axis object point.

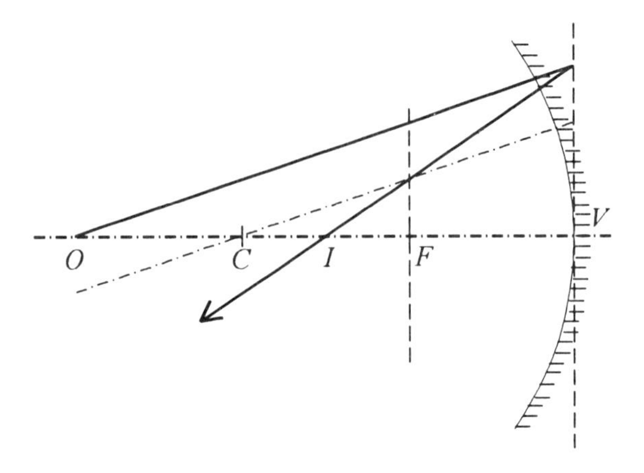

Figure 2.22 Reflection of an arbitrary paraxial ray.

line will cross the n' circle. The refracted ray passes through the intersection of the parallel line and the n' circle.

As with the plane refracting surface, not all rays will come together to form an image. For now, we will, once again, use the paraxial approximation to find the image. The surface of effective refraction will be a plane surface, but the angle of incidence will be determined by the normal to the spherical surface.

Figure 2.24 shows how to draw a ray from the object to the image through a spherical refracting surface. The procedure is very similar to the procedure we just used for an exact ray. First an arbitrary ray is drawn from the object to the effective refracting surface. The ray is extended for later use in constructing the refracted ray. Two additional plane surfaces are constructed perpendicular to the optical axis at distances which are proportional to the indices of refraction. Recall that the incidence angle is determined by the normal to the surface, not by the optical axis. Therefore, a line parallel to the normal is drawn at the intersection of the extended incident ray and the line for the first index. Where this line intersects the line for the second index determines the direction for the refracted ray.

All of the angles associated with a refracted ray are shown in Figure 2.25. An

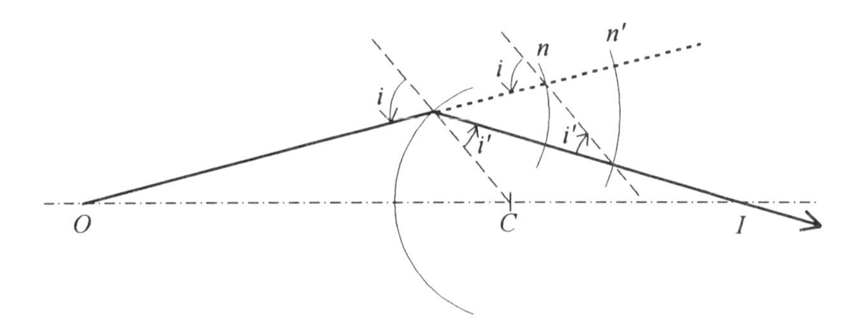

Figure 2.23 Refraction of an exact ray at a spherical surface.

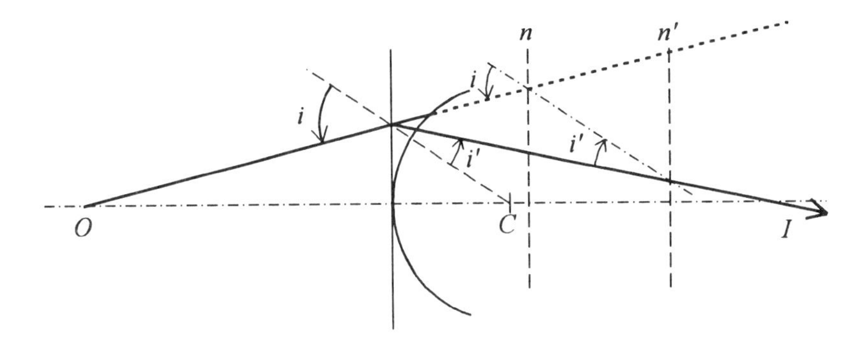

Figure 2.24 Refraction of a paraxial ray at a spherical surface.

equation for the image position can be derived by beginning with Snell's law in the paraxial approximation

$$n\,i = n'i' \,. \tag{2.30}$$

The incident angle can be written in terms of the slope angles of the ray and the normal (watch the signs)

$$i = u - \beta \,. \tag{2.31}$$

Similarly, the refracted angle is

$$i' = u' - \beta \,. \tag{2.32}$$

These equations can be combined to yield

$$n'u' - nu = (n' - n)\beta \,. \tag{2.33}$$

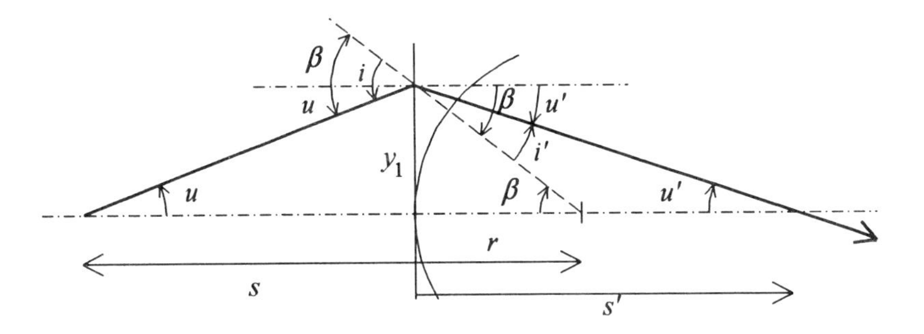

Figure 2.25 Definition of angles and lengths for refractoin at a spherical surface.

Once again, small angle substitutions can be made. Care must be taken to get the signs on the quantities to come out correctly.

$$u \approx \tan u = -\frac{y_1}{s}$$

$$u' \approx \tan u' = -\frac{y_1}{s'} \tag{2.34}$$

$$\beta \approx \tan \beta = -\frac{y_1}{r} .$$

When these substitutions are made, the quantity $-y_1$ appears in every term. Canceling this factor and rearranging gives the paraxial equation for a spherical refracting surface

$$\frac{n'}{s'} - \frac{n}{s} = \frac{(n' - n)}{r} . \tag{2.35}$$

In the limit that the object distance s goes to infinity, all of the incident rays coming from on on-axis become parallel to the optical axis. The image position is given by

$$s' = \frac{n'r}{(n' - n)} = f' . \tag{2.36}$$

Again, this special image position is called a *focal point*. However, unlike the spherical mirror, a spherical refracting surface has two distinct focal points. As before, the focal point that is located in image space is indicated by the prime. The object space focal point can be discovered by letting the image position go to infinity. For this image position, the object distance must be

$$s = -\frac{nr}{(n' - n)} = f . \tag{2.37}$$

The relationship between the two focal points is

$$\frac{f'}{n'} = -\frac{f}{n} . \tag{2.38}$$

Figure 2.26 shows an easier method for finding the image position of an on-axis object using the second focal point. This method is the same as used for the mirror in the previous section. Care must be taken, however, to make sure that the image space focal point is used to draw the refracted ray.

The focal points can also be used to find the size of an extended image. As shown in Figure 2.27, a ray parallel to the optical axis from the extended object must pass through the focal point in image space. At the same time, a ray which passes through the object space focal point must come out parallel to the optical axis. Since these rays came from the same object point, their intersection marks the corresponding image point. This drawing can be checked by drawing a third ray through the center of curvature of the surface. This ray has a zero incidence angle and therefore a zero refracted angle. As expected, it passes through the image point found with the two focal point rays. Any number of arbitrary rays could also be drawn using the methods of either Figure 2.24 or Figure 2.26, all of which will pass through this image point.

The magnification can be determined using the ray through the center of curvature in Figure 2.27. This ray, the optical axis, and the object form a triangle which is similar to a triangle formed by the ray, the optical axis, and the image. The length of the side of the object triangle along the optical axis is given by $-(s-r)$. For the image triangle, the length of the side along the optical axis is $(s'-r)$. Because the triangles are similar, the sides are proportional. But, with our sign convention, both $-(s-r)$ and $(s'-r)$ are positive numbers, while h and h' have opposite signs. We must, therefore, insert another minus sign to get

$$m = \frac{h'}{h} = \frac{(s'-r)}{(s-r)} . \tag{2.39}$$

Note, in the limit that r goes to infinity the refracting surface becomes a plane. In this case, Eq. 2.39 gives a magnification value of one, which agrees with the results of Section 2.3.

Example 2.4

Fred, my pet guppy, seems to prefer to swim at what appears to be a distance of 1.00 inch from the side of his fish bowl. The radius of the bowl is 5.00 inches. Use 1.33 as the index of water and ignore the effect of the glass bowl. How far from the bowl is Fred really swimming? What is the magnification?

To solve this problem, we apply Eq. 2.35, the paraxial equation for a single

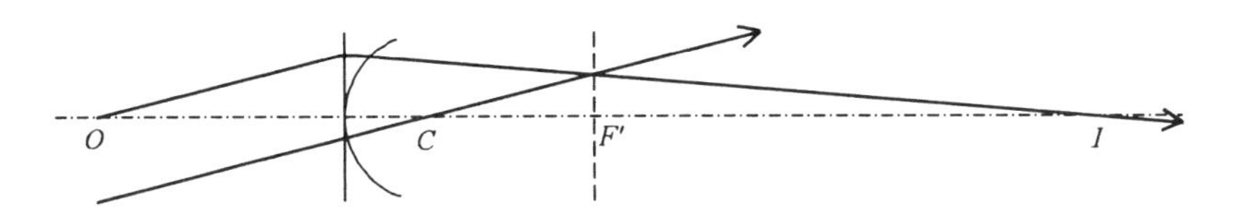

Figure 2.26 Refraction of an arbitrary ray.

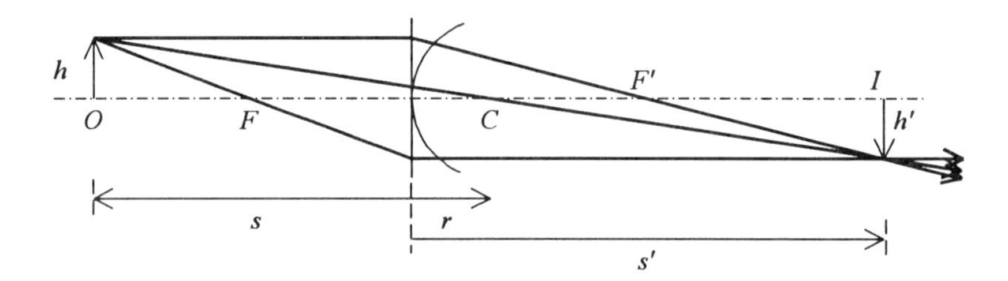

Figure 2.27 Determination of the image size and position using the two focal points and center of curvature of a spherical refracting surface.

refracting surface. The only problem is to properly identify all of the variables. Hopefully, it is obvious that the object is located in water so $n = 1.33$. The image, however, is located in air, not water as you might have guessed. Image space is determined by where the light rays end up. Here, the rays end in air. The image appears to be in the water because it is a virtual image. Therefore, $n' = 1.0$ and $s' = -1.00$. Finally, if we have the light traveling from left to right as we do normally, then the side of the bowl forms a concave surface and $r = -5.00$.

$$\frac{(1.00)}{(-1.00)} - \frac{(1.33)}{s} = \frac{((1.00) - (1.33))}{(-5.00)} .$$

Solving for the object distance gives $s = -1.25$ inches. The magnification is given by

$$m = \frac{(-1.00) - (-5.00)}{(-1.25) - (-5.00)} = 1.07 .$$

■

Typically, optical systems consist of more than one reflecting or refracting surface. The overall nature of the image for such a system can be determined by using the image created by the first surface as the object of the second surface, and so on. As might be expected, equations for combination systems get very complicated so we won't bother deriving them. Instead, you can find the final image by repeatedly using either the mirror equations or the single refracting surface equations. Another method for dealing with systems with multiple surfaces will be discussed in detail in the next chapter.

As an example of repeatedly using single surface equations, consider the thick lens in Figure 2.28. The first surface creates an image at I_1 . The position of this image can be determined by the single refracting surface equation

$$\frac{n_1}{s_1'} - \frac{n_0}{s_1} = \frac{(n_1 - n_0)}{r_1} . \tag{2.40}$$

The presence of the second surface is completely ignored. The image at I_1 is used as the object for the second surface. Just as there are virtual images where the rays never actually converge, there can be virtual objects. Image I_1 is a virtual object for the second surface. The rays which form this image are intercepted by the second surface of the lens and never actually meet there.

The object distance for the second surface is given by

$$s_2 = s_1' - t_1 . \tag{2.41}$$

This distance turns out to be positive which indicates a virtual object. The distance s_2 can be inserted into the single refracting surface equation as with any real object distance to give

$$\frac{n_2}{s_2'} - \frac{n_1}{(s_1' - t_1)} = \frac{(n_2 - n_1)}{r_2} . \tag{2.42}$$

Because of the denominator in the second term, this equation is not easily simplified or combined with the previous single surface equation. Calculations can, however, be done in steps.

2.6 THIN LENSES

A tremendous simplification can be made to the thick lens equations shown above by ignoring the thickness. This approximation is not as ridiculous as it may sound.

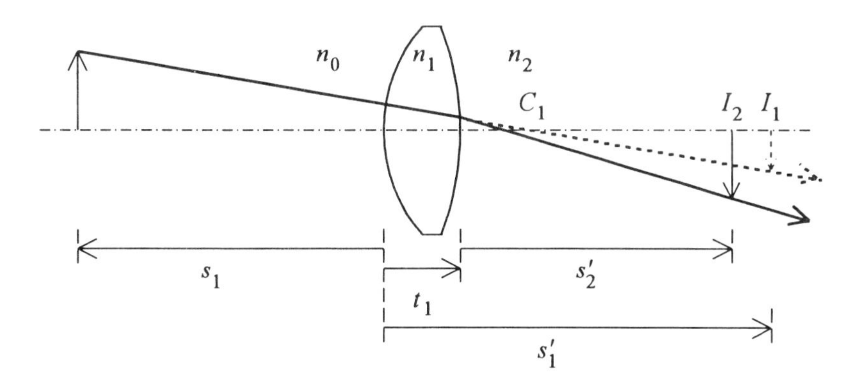

Figure 2.28 Imaging through two refracting surfaces.

Lenses can easily be made with a thickness which is negligible compared with object and image distances. Since the thickness appears only in the term with the image distance from the first surface, then as long as the image distance for the first surface is much greater than the thickness, ignoring the thickness will be a reasonable approximation. The simplification which results is so useful that any loss in accuracy can be justified. The basic workings of virtually every optical system can be understood with this approximation. Thickness can be inserted later when more accuracy is required.

In this approximation, Eq. 2.42 becomes

$$\frac{n_2}{s_2^{/}} - \frac{n_1}{s_1^{/}} = \frac{(n_2 - n_1)}{r_2} . \tag{2.43}$$

Normally, the lens under consideration will be surrounded by air. With this in mind, the indices n_0 and n_2 are both equal to one. Making this substitution and combining Eq. 2.40 with Eq. 2.43 , the result is the thin lens equation

$$\frac{1}{s_2^{/}} - \frac{1}{s_1} = (n_1 - 1)\left(\frac{1}{r_1} - \frac{1}{r_2}\right) . \tag{2.44}$$

This equation indicates that there are two focal points, as with the single refracting surface, one on either side of the lens. Because the indices of refraction are the same on either side of the lens, the focal lengths have the same magnitude. Since the focal lengths differ only in sign, an effective focal length can be defined

$$\frac{1}{f_{eff}} = \frac{1}{f^{/}} = -\frac{1}{f} = (n_1 - 1)\left(\frac{1}{r_1} - \frac{1}{r_2}\right) . \tag{2.45}$$

This equation is called the *lens maker's equation*. If the type of glass is specified, then grinding the glass to certain radii will determine the focal length of the lens. Remember that this equation is merely an approximation which ignores the thickness of the lens. More accurate equations will be derived in the next chapter.

Example 2.5

Calculate the effective focal length for a biconvex lens with surface radii of 15.0 cm and 20.0 cm, respectively. The glass type is F8.

Since the lens is biconvex, the first surface has a positive radius and the second surface has a negative radius. The index of refraction is 1.59551. Plugging these values into Eq. 2.45 gives

$$\frac{1}{f_{eff}} = ((1.59551) - 1)\left(\frac{1}{(15.0)} - \frac{1}{(-20.0)}\right).$$

The effective focal length is 14.4 cm. ■

We now turn to a discussion of techniques for drawing rays through thin lenses. These techniques are similar to the techniques used for drawing rays through single refracting surfaces. In fact, since the lens is thin, both refracting surfaces and the center of the lens are assumed to be at the same location. Thus, when a ray is drawn through a thin lens, it is only bent once. In Figure 2.29, the thin lens is represented as a single line perpendicular to the optical axis with symbols at the top and bottom of the line. The symbols indicate that this lens is thin at the edge and thick in the center and, therefore, has a positive effective focal length. As always, an image point can be found by following two rays from an object point through the lens. One ray is initially parallel to the optical axis. This ray must pass through the image space focal point. A second ray which travels through the first focal point must come out parallel to the optical axis. The intersection of these two rays locates the image point. As this figure shows, the only property of any significance for a thin lens is its focal length. Lenses made of different glasses with different radii of curvature but the same focal length all behave identically in this approximation. Therefore, only the effective focal length is given to specify a thin lens. The thin lens equation is normally expressed as

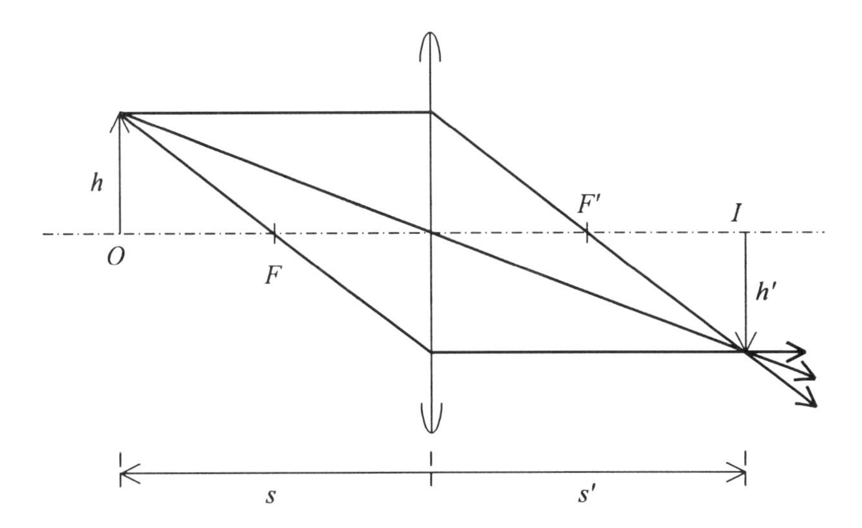

Figure 2.29 Determination of the image size and position using the focal points of a thin lens.

$$\frac{1}{s'} - \frac{1}{s} = \frac{1}{f'} = -\frac{1}{f} . \tag{2.46}$$

This form is called the *Gaussian thin lens equation.*

It is always nice to be able to check the solution to a problem. A third special ray can be used for this. Figure 2.30 shows an enlargement of the central portion of a lens. This section of the lens looks like the thick slab discussed in Chapter 1. We expect a ray through the center of the lens to be displaced but its direction should not change significantly. Using the equation derived in Chapter 1, it is easy to see that the displacement of the ray goes to zero as the thickness of the lens goes to zero. Therefore, in the thin lens approximation, a ray drawn through the center of the lens is not displaced and does not change direction. Adding this ray to Figure 2.29 confirms the image location.

Again, referring to Figure 2.29 the magnification can be determined using the ray through the center of the lens. The object and image heights along with the optical axis and the central ray form two similar triangles. The ratios of the sides of these two triangles gives the magnification

$$m = \frac{h'}{h} = \frac{s'}{s} . \tag{2.47}$$

The location of the image for an on-axis object can not be determined using the three special rays discussed above. Instead, we can draw a totally arbitrary ray from the object point to the lens. Then, a ray parallel to the first ray can be drawn through the center of the lens. The second ray goes straight through the lens as shown in Figure 2.31. The two parallel rays appear to the lens to come from an off-axis object point at infinity. Therefore, the intersection of the central ray with the image space focal plane determines a point through which the refracted ray must also pass.

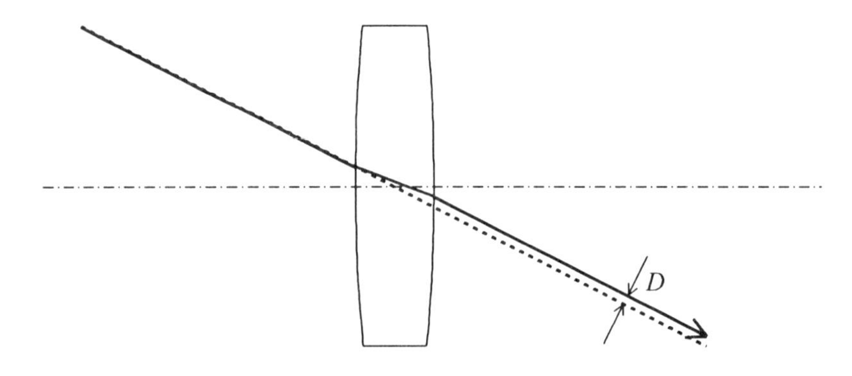

Figure 2.30 Displacement of a central ray through a thick lens.

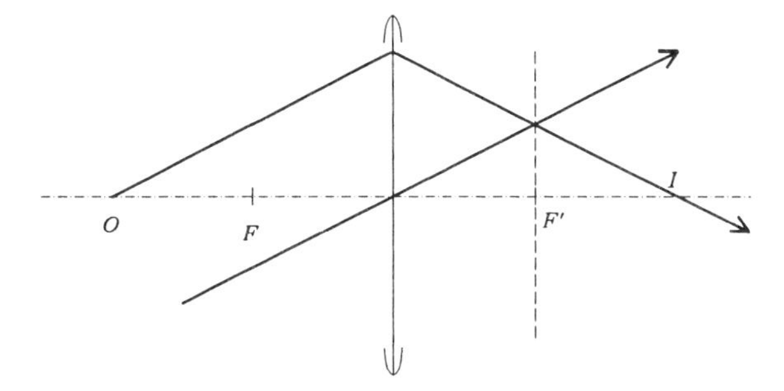

Figure 2.31 An arbitrary ray refracted by a thin lens.

Example 2.6

A 3.0-cm-tall object is located 18.0 cm from a +6.0-cm-focal-length thin lens. Where is the object located, and how big is it?

We can find the image position by applying the thin lens formula

$$\frac{1}{s'} = \frac{1}{(6.0)} + \frac{1}{(-18.0)} .$$

The image distance is +9.0 cm. The magnification is the ratio of the image and object distances

$$m = \frac{(9.0)}{(-18.0)} = -0.5 .$$

Therefore, the image size is − 1.5 cm. The minus sign indicates that the image is inverted. ■

Another equation for determining the image position called the Newtonian thin lens equation can be derived using Figure 2.32. First, the distance from the first focal point to the object is defined as z. The distance from the second focal point to the image is z'. These two distances are signed quantities. In this example, z is negative because the object is to the left of the object space focal point and z' is positive because the image is to the right of the image space focal point. The ray which passes through the first focal point creates two similar triangles as shown in the figure. Thus,

$$-\frac{h'}{h} = \frac{f}{z} . \tag{2.48}$$

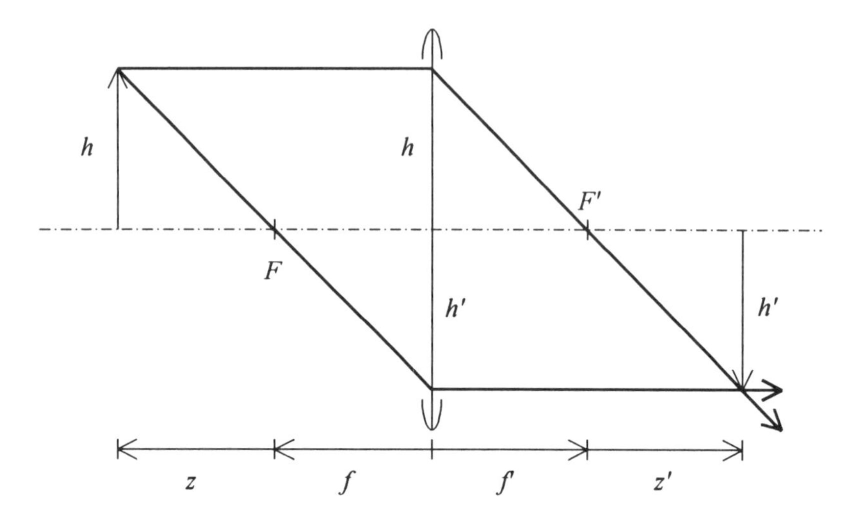

Figure 2.32 Derivation of the Newtonian form of the thin lens equation.

The ray through the second focal point also creates two similar triangles

$$-\frac{h'}{h} = \frac{z'}{f'} . \tag{2.49}$$

These two equations can be combined to give the Newtonian form of the thin lens equation

$$z z' = f f' . \tag{2.50}$$

Really useful and interesting optical systems can be built with combinations of thin lenses. Examples of such systems will be discussed in detail in the next section. Here, it is simply pointed out that combinations of thin lenses are analyzed by finding the position of the image created by the first lens and using this image as either a real or virtual object for the second lens. One way to solve problems with multiple imaging surfaces is with a simple step by step application of the Gaussian thin lens equation. More powerful techniques will be presented in later chapters.

2.7 THE FUNDAMENTAL CONCEPT OF IMAGING

So far, we have talked about three different types of imaging systems—mirrors, single refracting surfaces, and thin lenses—and we have three sets of equations for the three systems. But all three types of systems do basically the same thing. They bend rays coming from object points to form images. We ought to be able to express this process with just one equation which is applicable to all three types of systems. To accomplish this, we introduce the concept of vergence.

In general, we will define vergence as the index of refraction divided by the distance between the optical system and the point from which the rays are diverging. Thus, the object vergence is

$$v = \frac{n}{s} . \tag{2.51}$$

The image vergence is given by

$$v' = \frac{n'}{s'} . \tag{2.52}$$

Vergence can be thought of as the rate at which rays spread out from an object or image point. For small angles (the only kind we have dealt with), the vergence is an angle which is equal to a fixed reference distance (the index of refraction) divided by either the object or image distance. The vergence is negative for diverging rays and positive for converging rays.

We can also define the power of an optical system as the image space index divided by the image space or effective focal length

$$\phi = \frac{n'}{f'} . \tag{2.53}$$

With this definition, the affect of any optical system on object rays becomes clear. The power of an optical system is simply the change in the vergence:

$$\phi = v' - v . \tag{2.54}$$

The magnification of any optical system is simply the ratio of the vergences

$$m = \frac{v}{v'} . \tag{2.55}$$

To get these equations to work with mirrors, we need to make one small addition to our sign convention. We will define the index of refraction after a mirror to be the negative of its normal value. This convention is discussed more fully in the next chapter. Also note that since the vergence is related inversely to the distance, the magnification expressed in the previous equation appears odd, but is correct.

As we will see in Chapter 3, these same ideas are applicable to all optical systems, no matter how complex. The only trick is in determining the optical power and measuring the object and image locations appropriately.

2.8 BASIC APPLICATIONS

The purpose of many optical systems is to form an image on some sort of detector. The detector might be a piece of film, an array of solid state detectors, or even the retina of the human eye. In addition to just forming an image, many optical systems enhance the image that the detector would otherwise see. The enhancement can take the form of a magnified image, a brighter image, or a larger field of view. A simple description of how some optical systems work is given in this section.

2.8.1 The Human Eye

Figure 2.33 shows a horizontal cross section of the right eye as seen from above. The eye is a tough, plastic-like sack (the *sclera*) which is white and opaque. It is filled with a clear jelly-like fluid (the *vitreous humor*) which is under sufficient pressure to maintain the eye's shape and to prevent blood from diffusing into the eye. Along the inside surface of the sack is the light-sensing part of the eye, the *retina*. The entire eye can be rotated to view specific objects by a set of six muscles connected between the sclera and surrounding bone structure.

The transparent front surface of the eye is the *cornea*. Because of the large change in index, most of the imaging power of the eye is at the air–cornea surface. Behind the cornea is a watery fluid called the *aqueous humor*. A muscular body, the *iris*, lies directly in front of the lens. The iris opens and closes to control the amount of light which enters the eye. Under very bright light, the iris closes down to a

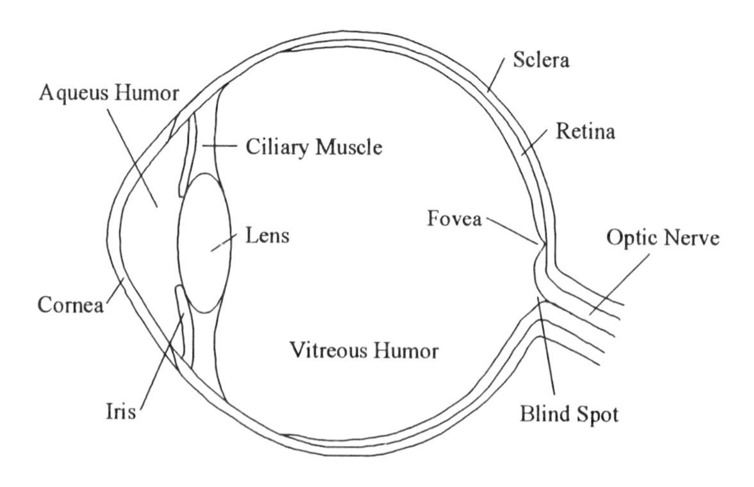

Figure 2.33 Horizontal cross section of the human eye.

diameter of less than 2 mm. With very dim light, the iris opens to about 8 mm in diameter. The iris is also responsible for the color of the eye.

The distance between the cornea and the image plane, the retina, is fixed. Therefore, only objects at one specific distance should be clearly imaged by the eye. Focusing on objects at other distances is accomplished not by moving the cornea in and out as you would with a 35-mm camera, but rather by changing the focal length of the system. This is done by changing the shape of the lens in the eye. The lens is a fibrous, layered body suspended by a set of ligaments attached to surrounding muscles. When the muscles contract, the lens thickens to view nearby objects. When the muscles relax, the lens thins to view distant objects. This effect is called *accommodation*.

The greatest distance at which the eye can focus is called the *far point* and should be at infinity. If the far point is not at infinity, then the eye is nearsighted and needs to have its overall power reduced. The closest distance at which the eye can focus is called the *near point*. The common standard value for the near point is 25 cm. Farsighted eyes cannot focus at this small of a distance and need the overall power of the system increased. As the eye ages, the lens becomes less flexible and loses its ability to accommodate. It is possible that an eye can neither focus at infinity nor at the near point. Bifocal glasses are needed to correct both conditions.

The retina is the light-sensing part of the eye. It consists of several layers. First come blood vessels which nourish all of the cells of the retina. Then come nerve cells which connect the light sensing cells with the brain. The nerves all exit the eye through the retina at the same location. There are no light-sensing cells at this point, which creates a blind spot. There are two types of light-sensing cells: *rods* and *cones*. The names come from the general shape of these cells. The rods are sensitive to very low levels of light, but do not sense color. This is why everything looks gray at night. The cones need more light to function, but sense different wavelengths as different colors. Color vision depends largely on how the signals from the retina are interpreted by the brain which is a subject beyond the scope of this text.

The rods and cones of the retina are nonuniformly spaced. In a very small region called the *fovea*, there are no rods at all and the cones that are present are very densely packed. This is the region of sharpest and most detailed vision. The eye moves constantly so that objects of interest are imaged on the fovea. The cones in the fovea subtend an angle at the lens of approximately 30 arc-seconds. For the eye to see an object, such as a star, light from the object must strike at least one cone. But for the eye to resolve two nearby objects or details of an object, light must fall on two cones separated by at least one cone which does not sense light (Figure 2.34). This means that objects must be separated by at least one arc-minute to be distinguished. This limit represents an ideal case which is seldom realized. A more practical limit is a resolving power of two arc-minutes.

2.8.2 Magnifiers and Eyepieces

As indicated above, the eye has a fixed resolution limit. Therefore, as an object is moved closer to the eye, finer detail may be seen. But the near point, the smallest distance at which the eye can focus, limits the object distance. If the object is brought

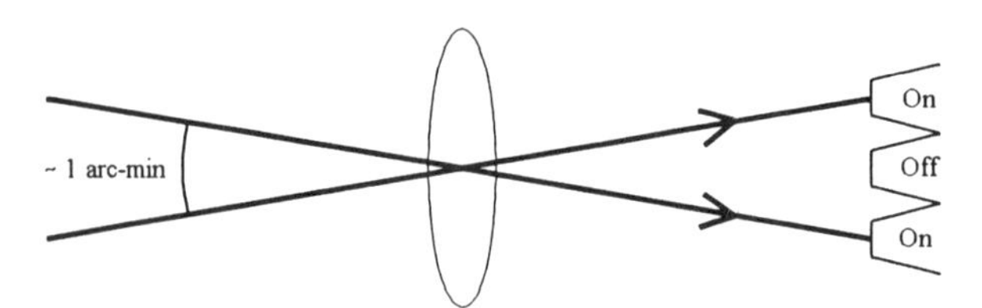

Figure 2.34 Minimum resolution of the eye as determined by cone spacing.

closer to the eye than the near point, detail is lost because of the loss of focus. What is needed is an instrument which can increase the angle that an object subtends at the eye. This instrument should create a magnified image somewhere between the near and far points of the eye. The simplest optical system which can accomplish this task is a single positive lens.

Equation 2.47 shows that to get a magnified image, the image distance must be larger than the object distance. This can be accomplished with either a real or an virtual image. The real image has the disadvantage that the lens must be at a large distance from the eye. Remember that the image must be at least as far away as the near point. Therefore, the lens distance is at least 25 cm plus the image distance s'. The large distance between the lens and the eye greatly reduces the field of view. Also, the image is inverted.

On the other hand, if the image is virtual, then the lens can be close to the eye and the image and object will both be on the other side of the lens, away from the eye. The thin lens formula (Eq. 2.46) indicates that lenses with negative values for f' will always form virtual images. Unfortunately, these images are always smaller than the objects. With a positive lens, if the object distance is slightly smaller than the focal length of the lens, then the image produced will be virtual, erect, and magnified.

Using Eq. 2.46 the object distance can be solved for in terms of the focal length and an image distance of -25 cm. Thus,

$$s = -\frac{25f'}{25 + f'} . \tag{2.56}$$

Let us define the angle θ as the angle subtended by the object when it is located at the near point without the presence of the magnifier. We also define the angle θ' as the angle subtended by the image formed at the near point when the magnifier is used and the object is moved closer to the eye (Figure 2.35). Now we can determine the angular magnification defined as the ratio of the angles θ' and θ

$$M = \frac{\theta'}{\theta} . \tag{2.57}$$

This ratio tells us how much larger the image appears to the eye than the object did

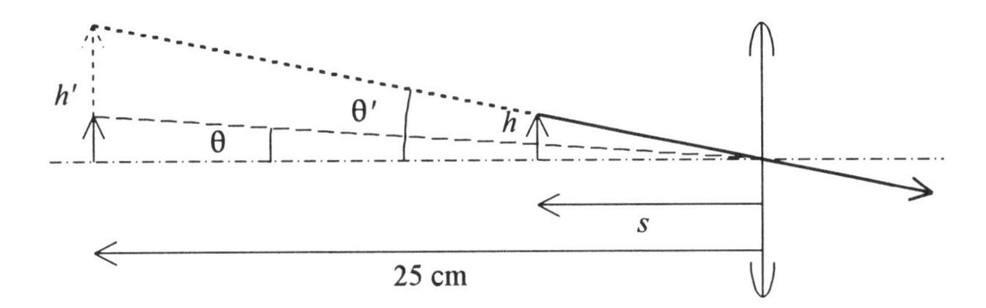

Figure 2.35 Illustration of angular magnification.

without the magnifier. The rays that are used to measure θ and θ′ travel from the extreme edge of the object through the center of the optical system. For a thin lens system, this ray is undeviated as mentioned previously. Therefore,

$$\theta \approx \tan\theta = -\frac{h}{25} \tag{2.58}$$

and

$$\theta' \approx \tan\theta' = -\frac{h'}{25} = \frac{h}{s} . \tag{2.59}$$

Combining Eqs. 2.51, 2.53, and 2.54 gives

$$M = \frac{25 + f'}{f'} . \tag{2.60}$$

If the object to be magnified is moved back so that its location coincides with the front focal point, then the image will be located at infinity. This is desirable because the normal eye is relaxed when viewing at infinity. Also, because the image size increases with increasing distance, the angular magnification doesn't decrease by much when the image is at infinity. Using an analysis similar to that given above, the angular magnification for an image at infinity can be found to be

$$M = \frac{25}{f'} , \tag{2.61}$$

so not much magnification is lost with an image at infinity. Because this value is not much different than the value for an image at the near point and because it is simpler, this is the value which is given when a magnifier is described as having a magnification of 5×.

Equation 2.61 shows that a large magnifying power requires a small focal length. The lens maker's formula (Eq. 2.45) shows that small focal lengths require small radii of curvature. For a biconvex lens made of glass with index of about 1.5, the focal length is about the same as the radii of curvature. But if the radii of curvature of a lens are small, then the transverse diameter of the lens will also be small, which limits the field of view. To keep the aperture at a reasonable size, the lens may be thickened, but then the thin lens approximations and small-angle approximations fail. Combinations of lenses, both thick and thin, may be used to correct the limitations of the approximations. These corrections will be discussed in later chapters. Even with several elements and thick lenses, magnifiers are limited to maximum magnifications of about 20×.

To achieve higher magnifications, magnifiers can be combined with other lenses to create more complex systems. A magnifier used in such a system is called an *eyepiece*, but its function remains the same.

2.8.3 Microscopes

To achieve higher magnifications, what is needed is a large ratio of the image distance to object distance without requiring extremely small focal lengths for any single element. In the previous section, large magnifications with real images were dismissed because the near point distance had to be added to the image distance to get the distance of the lens from *compound microscope* the eye. If this problem is ignored temporarily, then the thin lens equation shows that virtually any magnification is possible with a positive lens. The object distance only needs to be a little larger than the focal length to get very large image distances.

The magnifier discussed in the first section permitted an object to be brought close to the eye while the image is kept beyond the near point. A magnifier can be combined with another positive lens which forms a real, intermediate image near the eye. This combination is called a and is shown in thin lens form in Figure 2.36. The lens nearest the object is called the *objective*. The objective is a short focal length lens which forms an intermediate real magnified and inverted image near the eye. The eyepiece is simply a magnifier which forms a virtual final image beyond the near point. The object for the eyepiece is the intermediate image generated by the objective. In the figure, it may look like the intermediate image is bending and splitting a ray. It is not, of course. The intermediate image is a source of rays for the eyepiece. These rays are just not shown going back to their origin at the object. Moving the eyepiece toward the objective will move the final image from infinity to the near point. The microscope itself is focused by moving both lenses together relative to the object.

The overall magnification of the microscope depends on the separate magnifications of the objective and the eyepiece. Thus,

$$M = m_o M_e \,, \tag{2.62}$$

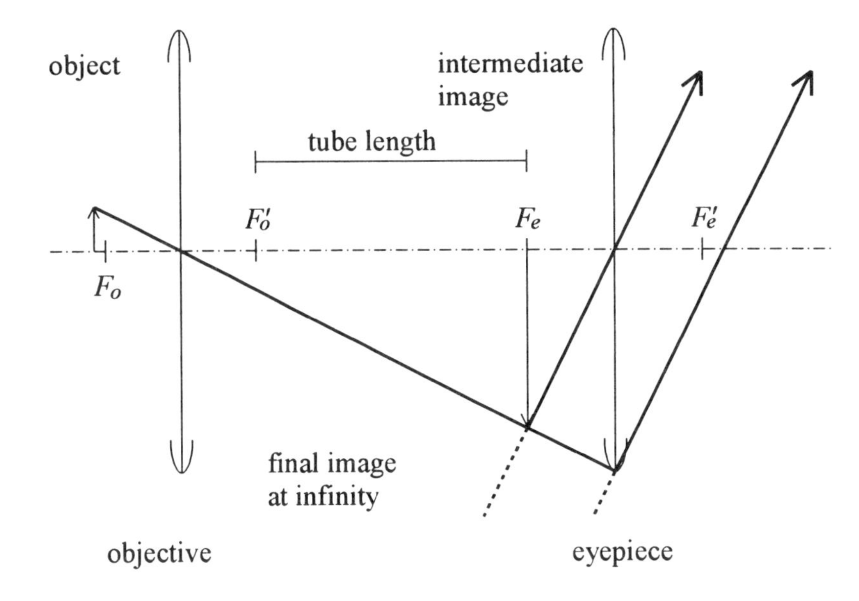

Figure 2.36 General form of the compound microscope.

where m_o is the linear magnification of the objective, and M_e is the angular magnification of the eyepiece. The magnification of the eyepiece can be found using the same formulae that were used for the simple magnifier. The linear magnification of the objective is simply the ratio of the image distance to the object distance, as usual.

It is more useful, however, to use the Newtonian form of magnification given in Eq. 2.49. In this form,

$$m_o = -\frac{z'}{f_o'} \,. \tag{2.63}$$

One reason for using this form is that many microscope manufactures have set a standard value for z' of 16.0 cm. This distance is called the *tube length*, even though the actual tube which holds the eyepiece and objective is longer by at least the combined focal lengths of the two lenses. A standard value for this distance permits both the eyepieces and objectives to be interchanged. The objective is usually marked with its magnifying power, which is given by Eq. 2.61. Thus an objective with focal length 1.6 cm is a 10× objective.

The overall magnification of a standard tube length microscope is simply

$$M = -\frac{25\,z'}{f_e' f_o'} \,. \tag{2.64}$$

As with the simple magnifier, the overall magnification depends inversely on the focal lengths of the lenses. For large magnification, very small focal lengths are needed. Small focal lengths and small object distances lead to the breakdown of the thin lens and small angle approximations. Not surprisingly then, high-powered microscope objectives consist of many lens elements.

2.8.4 Telescopes

The purpose of a telescope is to generate a magnified image of a distant object. The magnification, however, must be angular magnification rather than linear magnification. The real image of Jupiter as viewed on Earth, for example, will never be larger in linear extent than Jupiter itself. Because of the tremendous distance between the Earth and Jupiter, the angle θ that the planet subtends at a telescope here on Earth will be very small. If a small image can be formed much closer to the eye, then perhaps the angle θ' subtended by the image at the eye will be much larger. Figure 2.37 indicates that the image size will be directly proportional to the focal length of the lens. Therefore, a long focal length lens is desired. The problem with this single lens telescope is that the image must still be beyond the near point to be seen clearly by the eye. Since this image is very small and 25 cm is still a fairly large distance, additional angular magnification from an eyepiece is needed.

Figure 2.38 shows a thin lens version of an astronomical telescope with the final image at infinity. As with the microscope, the lens nearest the object is called the *objective* and the lens nearest the eye is the *eyepiece*. Unlike the microscope, however, the objective should have the largest possible focal length to produce the largest possible intermediate image. The separation between the two lenses is the

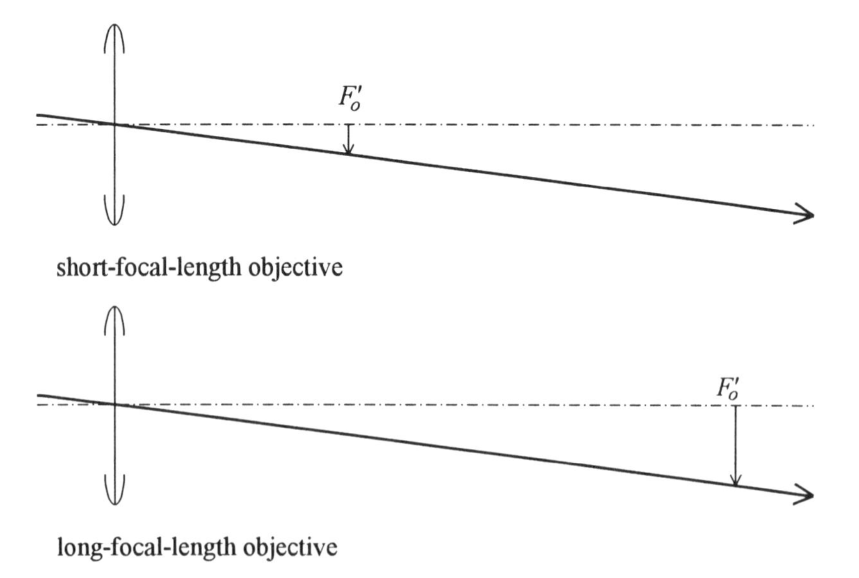

Figure 2.37 Image formation of a distant object for both short- and long-focal-length lenses.

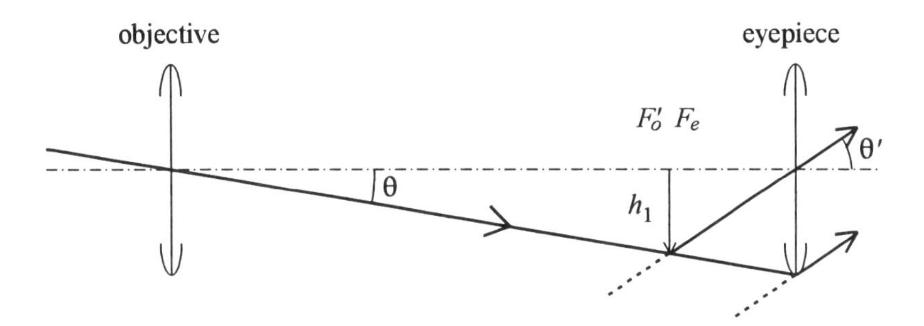

Figure 2.38 General form of an astronomical telescope with final image at infinity.

sum of their focal lengths. The angular magnification is the ratio of θ' to θ. These angles can be determined using the height of the intermediate image h_1

$$\theta \approx \tan\theta = \frac{h_1}{f_o'} , \tag{2.65}$$

and

$$\theta' \approx \tan\theta' = -\frac{h_1}{f_e'} . \tag{2.66}$$

Thus the angular magnification is simply given by the ratios of the focal lengths:

$$M = -\frac{f_o'}{f_e'} . \tag{2.67}$$

Equation 2.67 shows explicitly that a long-focal-length objective is desirable. Up to this point, the lenses we dealt with were short-focal-length lenses. This caused the small-angle approximations to be violated which leads to complicated lens systems. A long-focal-length lens should have none of these problems, so telescope objectives should be simpler lenses. This is generally true. Inexpensive telescopes give reasonable images with only a single thin lens for the objective. Even simpler are reflecting telescopes where the objective is a concave spherical mirror.

SUMMARY

Objects are sources of diverging rays. Images are formed when rays coming from an object point pass through an optical system and then either converge (real image) and subsequently diverge or appear to diverge (virtual image) from a different point.

The only significant distinction between the two types of images is that real images can be projected onto a screen.

A measure of the rate at which rays diverge or converge is called the *vergence*. The object vergence is given by

$$v = \frac{n}{s} , \tag{2.68}$$

and the image vergence is given by

$$v' = \frac{n'}{s'} . \tag{2.69}$$

The optical power of an optical system is simply the change in vergence

$$\phi = v' - v . \tag{2.70}$$

In general, the optical power is related to the image space focal length

$$\phi = \frac{n'}{f'} . \tag{2.71}$$

But different systems have different ways to calculate the image space focal length.

The paraxial imaging equations for a single mirror are

$$\frac{1}{s'} + \frac{1}{s} = \frac{1}{f'} \tag{2.72}$$

and

$$m = -\frac{s'}{s} , \tag{2.73}$$

where

$$f' = f = \frac{r}{2} . \tag{2.74}$$

The object distance is s and the image distance is s'. The image magnification is m. The radius of curvature of the mirror is given by r. The two focal points f (object space) and f' (image space) are actually located at the same place. All distances are measured from the mirror along the optical axis.

The paraxial imaging equations for a single refracting surface are

$$\frac{n'}{s'} - \frac{n}{s} = \frac{n'}{f'} \tag{2.75}$$

and

$$m = \frac{s' - r}{s - r} = \frac{n s'}{n' s} , \tag{2.76}$$

where

$$\frac{n'}{f'} = - \frac{n}{f} = \frac{n' - n}{r} . \tag{2.77}$$

The indices of refraction n and n' are defined in object space and image space. Object space is where the rays originate, and image space is where they end up.

For a single thin lens, the paraxial imaging equations are

$$\frac{1}{s'} - \frac{1}{s} = \frac{1}{f'} \tag{2.78}$$

and

$$m = \frac{s'}{s} , \tag{2.79}$$

where

$$\frac{1}{f'} = - \frac{1}{f} = (n - 1)\left(\frac{1}{r_1} - \frac{1}{r_2} \right) . \tag{2.80}$$

Here, r_1 and r_2 are the radii of curvature of the first and second surfaces of the lens, respectively.

An alternative to the imaging equations presented above is the Newtonian form:

$$z z' = f f' \tag{2.81}$$

and

$$m = -\frac{f}{z} = -\frac{z'}{f'} . \tag{2.82}$$

For these equations, the object distance z is measured from the object space focal point and the image distance z' is measured from the image space focal point.

When a simple magnifier is focused to form an image at infinity, its angular magnification is given by

$$M = \frac{25 \text{ cm}}{f'} . \tag{2.83}$$

The 25-cm length is the nominal near-point distance.

The angular magnification for a compound microscope is

$$M = -\frac{25 \text{ cm } z'}{f_e' f_o'} . \tag{2.84}$$

The Newtonian image distance z' is called the *tube length*. The standard tube length for microscopes is 16 cm.

The angular magnification for a telescope is

$$M = -\frac{f_o'}{f_e'} . \tag{2.85}$$

REFERENCES

P. Tipler, *Physics* (Worth, New York, 1982).

F. Jenkins and H. White, *Fundamentals of Optics*, 4th ed. (McGraw-Hill, New York, 1976).

PROBLEMS

2.1 What is the basic requirement for image formation?

2.2 Describe the difference between a real and a virtual image. Can the eye tell the difference?

2.3 List pairs of equations which give the image position and magnification for a single reflecting surface, a single refracting surface and a thin lens.

2.4 Explain why spherical surfaces are so common in optical system analysis.

2.5 Describe how you determine the whether an image is real or virtual, erect or inverted, enlarged or reduced based on calculations.

2.6 Define image space and object space. Explain what is meant by image space focal point and object space focal point.

2.7 What is the relationship between the image space focal distance and the object space focal distance for (a) a single reflecting surface, (b) a single refracting surface and (c) a thin lens.

2.8 Describe the parts of the human eye and their function.

2.9 Measure your near point by placing a ruler next to your eye and moving your hand toward your eye until it just goes out of focus. I am confident that the majority of people reading this text have a near point which is less than 25 cm. Can you think of an explanation?

2.10 A 10.0-cm-tall object is located 5.00 m from a pinhole camera which is 10.0 cm deep. How large is the image?

2.11 Two 10.0-cm-long plane mirrors are placed at right angles to each other with their edges touching. A point object is placed 6.0 cm above the horizontal mirror and 5.0 cm to the right of the vertical mirror. Find the location of all of the images which are formed. Draw two rays which indicated the limits on the field of view for each image.

2.12 Derive the equation for finding the image produced by a spherical mirror by substituting $n = 1.0$ and $n' = -1.0$ into the equation for finding the image produced by a single refracting surface.

2.13 Assume that a point object is located at a depth of 5.0 cm in a medium with an index of 1.5. The surface between the medium and air is flat. Use paraxial ray drawing to find the image. Check your work with appropriate calculations.

2.14 A point object is placed 5.00 cm from a plane refracting surface. The index in object space is 1.0. In image space, the index is 1.5. Find the image position by drawing exact rays with $40°$ and $45°$ incidence angles and constructing the refracted rays.

2.15 A 2.00-cm-high object is placed 8.00 cm in front of a spherical mirror with radius +8.00 cm. Use paraxial rays through the focal points and center of curvature to find the size and position of the image. Check your work with appropriate calculations.

2.16 A picture is painted on the flat end of a 5.00-cm-long glass rod with index of refraction 1.500. The other end of the rod is curved so that a real image of the picture is formed 5.00 cm from the curved surface. What is the radius of curvature needed to form the image? If the original picture is 1.00 cm across, how large is the image?

2.17 A 1.00-cm-high object is located in glass (n = 1.500), 4.00 cm in front of a spherical refracting surface. The radius of the surface is −2.00 cm. The index of refraction after the surface is 1.000. Calculate the positions of the focal points. Draw rays through the focal points and center of curvature to find the size and position of the image.

2.18 A thin lens is made of SF19 glass with surface radii of 5.00 cm and −10.0 cm on the front and back surfaces, respectively. Find the effective focal length of this lens. An object 2.0 cm high is located 8.00 cm in front of the lens. Draw three special rays to find the image distance and image size.

2.19 An object is located on the optical axis 12.00 cm in front of a negative lens of focal length −4.00 cm. Use the arbitrary ray drawing technique to find the location of the image. Check your work with appropriate calculations.

2.20 An object is located on the optical axis 3.00 cm in front of a lens of focal length +4.00 cm. Use the arbitrary ray drawing technique to find the location of the image. Check your work with appropriate calculations.

2.21 A telescope is made with a 25.0-cm-focal-length objective lens and a 1.00-cm-focal-length eyepiece lens. How far apart are the two lenses if an object at infinity is imaged at infinity? What is the system magnification?

2.22 A terrestial telescope (erect image) is made with an objective lens with a focal length of 100.0 cm and an eyepiece lens with a focal length of −10.00 cm. What is the overall magnification of this telescope and how long is it if the final image is formed at infinity?

2.23 A 50× microscope is to be made with the same two lenses as in Problem 2.21. Assume that the final image is formed at infinity. How far apart must the lenses be spaced? What is the required tube length? Where must the object be located relative to the objective lens?

2.24 Plot pinhole camera resolution as a function of pinhole diameter for λ = 550 nm.

2.25 Plot the difference between i and $\sin i$ as a function of i from 0° to 90°. Plot the difference between $(i - i^3 / 3!)$ and $\sin i$ as a function of i.

2.26 Two 10.0-cm-long plane mirrors are placed at a 75° angle to each other with their edges touching. A point object is placed 6.0 cm above the horizontal mirror and 4.0 cm to the right of the tilted mirror. Find the location of all of the images which are formed. Using the images, draw a ray from the object which strikes the tilted mirror twice.

2.27 Fill in the table below for various spherical mirrors:

	s	s'	r	f'	m
(a)	-20	-20			
(b)		+20			+2
(c)	-5		-20		
(d)		-3	-6		
(e)				-16	+4
(f)	-12				-2
(g)	-12			+6	

2.28 Show that Eq. 2.15 holds for spherical refracting surfaces as well as plane refracting surfaces. Start with the equation for finding the position of the image formed by a single refracting surface.

2.29 Fill in the table below for various spherical refracting surfaces:

	n	n'	s	s'	r	f	f'	m
(a)	1.5	1.0	-20		-10			
(b)	1.0	1.5		+40	+10			
(c)	1.5		-24	-12	+6			
(d)	1.0	1.5	-16			-4		
(e)		1.0		+8		-9	+6	
(f)	1.0	1.5			+12			+2

2.30 A special case of a combination of thin lenses which results in a simple formula is when the thin lenses are in contact. As with the original derivation of the thin lens equation, simplification comes about because the separation between the two lenses is zero. Show that the effective focal length can be written

$$f_{eff} = \frac{f_1' f_2'}{f_1' + f_2'} . \tag{2.86}$$

2.31 Plot the effective focal length of two thin lenses in contact as a function of the focal length of the first lens. Assume that the first lens is the magnifier and the second lens is the eye itself with a focal length of 2.40 cm. Let the focal length of the magnifier vary from 0 cm to 20.0 cm.

2.32 Given a thin lens with 10.0 cm focal length, what is the smallest possible distance between the object and a real image. What is the magnification in this case?

2.33 A microscope consists of a 2.50-cm objective and a 4.00-cm eyepiece. Assuming a standard tube length and a final image at infinity, what is the overall magnification of this microscope? How far from the objective lens should the object be located? If the eyepiece is moved, the final image can be formed at the near point. How much does the eyepiece need to move (and in what direction) to accomplish this image shift?

2.34 A 400-cm object lens is combined with a 4.00-cm eyepiece to make a telescope. What is the magnification of this telescope? The moon is 380,000 km from the earth and has a diameter of 3,476 km. What is the angle subtended by the moon? What is the angle subtended by the image of the moon produced by the telescope described above? What is the linear diameter of the image formed by the objective lens alone?

2.35 Make a plot of the magnification m produced by a thin lens as a function of object position s. Let s vary from zero to three times the effective focal length of the lens f'.

2.36 Derive a Newtonian style paraxial imaging equation for a single mirror.

2.37 Derive a Newtonian style paraxial imaging equation for a single refracting surface.

2.38 A simple magnifier consists of a lens with $f' = +5.00$ cm. Initially, the object is located so that the final image is at infinity. Then the lens is moved so that the final image is at the standard near point. How far, and in what direction did the lens move relative to the object?

2.39 An optical system consists of two thin lenses. The first lens has a focal length of +2.00 cm. The second lens has a focal length of −4.00 cm. The two lenses are separated by 1.00 cm. For a 2.00-cm-tall object located 6.00 cm in front of the first lens, what is the image size and position? Make a scale drawing of this system. For each lens draw three special rays to find the image formed by that lens. Show extensions of rays as dashed lines and draw all rays from the object through to the image.

2.40 An optical system consists of two thin lenses. The first lens has a focal length of −12.0 cm. The second lens has a focal length of +4.00 cm. The two lenses are separated by 5.00 cm. For a 4.00-cm-tall object located 4.00 cm in front of the first lens, what is the image size and position? Make a scale drawing of this system. For each lens draw three special rays to find the image formed by that lens. Show extensions of rays as dashed lines and draw all rays from the object through to the image.

2.41 Design an optical system consisting of a single refracting surface between air and some type of glass. We want the real object to be located 15.0 cm from the surface in air and the real image to be located 45.0 cm from the surface in the glass. The magnitude of the magnification should be 2.00. (a) What is the index of refraction of the glass? (b) What is the radius of curvature of the surface?

2.42 We want to design a thin lens made from glass with an index of refraction of 1.500. We want a real object 6.00 cm from the lens to form an image with $m = +5.00$. (a) Where is the image formed? (b) What is the focal length of the lens? (c) If the radius of curvature of the first surface of the lens is five times the radius of the second surface, what are the radii?

CHAPTER 3

PARAXIAL OPTICS I

I really enjoyed the movie *Aliens*. The heroine of the movie is faced with a terrifying, incomprehensible situation. She tries various tactics to defeat her adversary. But only when she finds the right tool is she victorious.

Now consider trying to find the image size and position for the optical system shown in Figure 3.1 using the techniques discussed so far. It should be clear that we can't use the thin lens equations on this lens system because of the overall thickness of the system and the finite thickness of each element. But it is terrifying to contemplate trying to draw two paraxial rays through each surface to find the intermediate images. This procedure would take a very long time and not be very accurate. Alternatively, we could use the single refracting surface equations from Chapter 2 to find the size and position of the intermediate image produced by each

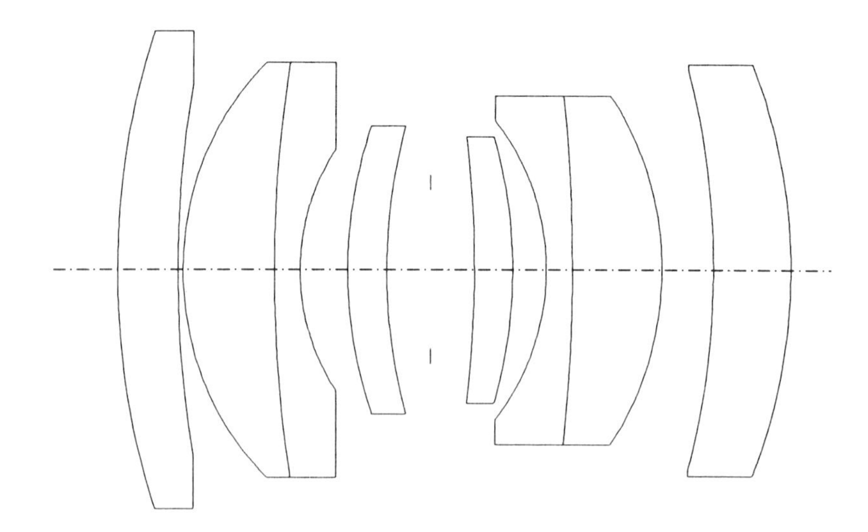

Figure 3.1 An example of a complex optical system. U.S.P. 3,006,249.

surface. This method would be accurate, but would still take several pages of calculations. It is also very easy to make a mistake which could be fatal.

In this chapter, we introduce the *y-nu* ray trace technique which will make it relatively easy to find the image size and position for even the most complicated (and horrible) optical system. The name *y-nu* comes from the names of the intermediate values, the ray height *y* and the reduced slope angle *nu*, that are calculated at each surface. This technique is also useful for finding all of the important first-order properties of real-life optical systems. Another important point is that *y-nu* ray tracing is easily adapted to a computer program. In every respect, *y-nu* ray tracing is just what we need to defeat the worst monsters.

3.1 THE *Y-NU* RAY TRACE

In this section the powerful technique of *y-nu* ray tracing is developed. First, the notation and sign convention are introduced. Then, the *y-nu* ray trace equations are derived. Applying these equations to finding the image size and position is discussed. Finally, setting up and using a ray trace table is demonstrated. A computer program which performs *y-nu* ray tracing given a specific optical system is available from the publisher. This software is discussed in Appendix C.

3.1.1 Sign Convention and Notation

Figure 3.2 shows a generalized optical system and indicates the notation that will be used. Both the sign convention and notation are summarized in Appendix B.

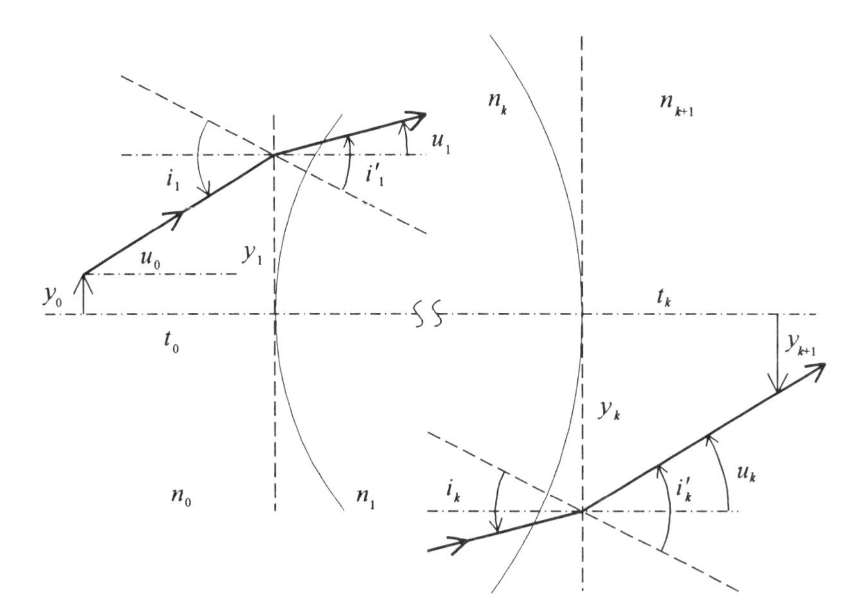

Figure 3.2 Notation for ray tracing through a general optical system.

Subscripts are used to identify which surface is associated with a particular parameter. The object surface is surface 0 (zero). The first surface of the optical system is surface 1, and so on. The last surface of the system is indicated with the subscript k, and the image is the next surface, $k + 1$. As shown in Figures 3.3 and 3.4, a general, arbitrary surface will be denoted with the subscript j. Quantities like the slope angle u and the index of refraction n have the same value everywhere between two surfaces. These quantities will always bear the subscript of the surface which precedes the space. Thus u_0 is the slope angle of a ray after the object surface. The separation between surfaces is also given the subscript of the preceding surface. Quantities which take on a particular value at a surface (like y and i and i') are given the subscript of that surface.

The sign convention for angles and distances is the same as used previously. Thus, an upward sloping ray will have a positive value for u. The distance from the object to the first lens surface t_0 will also be positive since it is measured from the object to the first surface of the system along the positive z-axis. We will show that, in general, t_0 is not the same as the negative object distance $-s$, and t_k, the distance from the last surface to the image, is not the same as the image distance s'. Note that we are still using the paraxial approximation so that effectively the refraction takes place in the plane perpendicular to the optical axis through the vertex of each surface.

3.1.2 Ray Tracing Equations

Tracing a ray of light through an optical system consists of calculating the change in ray slope angle caused by a reflection or refraction, and the change in ray height caused by translation from one surface to the next. The effect of translation is determined first for no better reason than that typically a ray must initially travel from the object to the first surface of the system. As shown in Figure 3.3, the change in height can be calculated from

$$y_j - y_{j-1} = \delta y = t_{j-1} \tan u_{j-1} \approx t_{j-1} u_{j-1} \,. \tag{3.1}$$

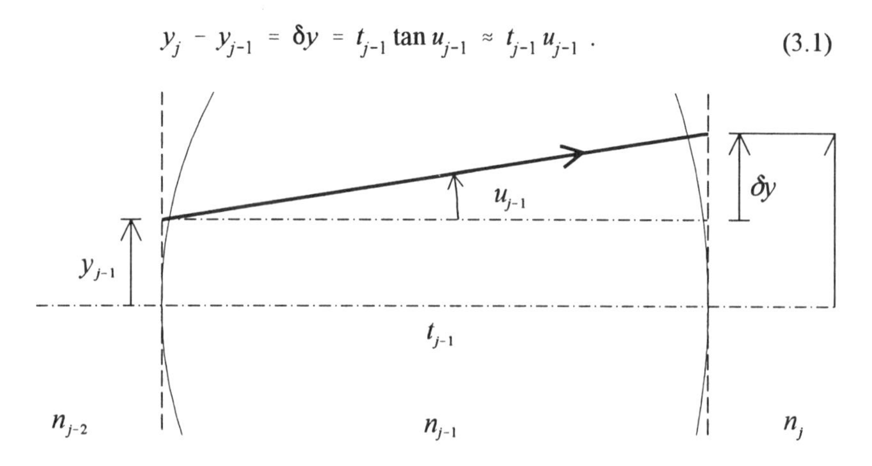

Figure 3.3 Translation of a paraxial ray between two surfaces.

Thus, if the ray height and ray slope angle are known at the current surface, the ray height can easily be calculated at the next surface

$$y_j = y_{j-1} + t_{j-1} u_{j-1} \ . \tag{3.2}$$

Determining the new slope angle after a refraction is only slightly more complicated. Refraction at the jth surface is shown in Figure 3.4. We begin with Snell's law in the paraxial approximation

$$n_{j-1} i_j = n_j i_j' \ . \tag{3.3}$$

Substitutions can be made for the angle of incidence and angle of refraction:

$$i_j = u_{j-1} - \beta_j \tag{3.4}$$

and

$$i_j' = u_j - \beta_j \ . \tag{3.5}$$

The angle between the optical axis and the normal to the surface, β, can be expressed as

$$\beta_j = -\frac{y_j}{r_j} = -c_j y_j \ . \tag{3.6}$$

The curvature of the surface c_j is defined as the inverse of the radius of curvature.

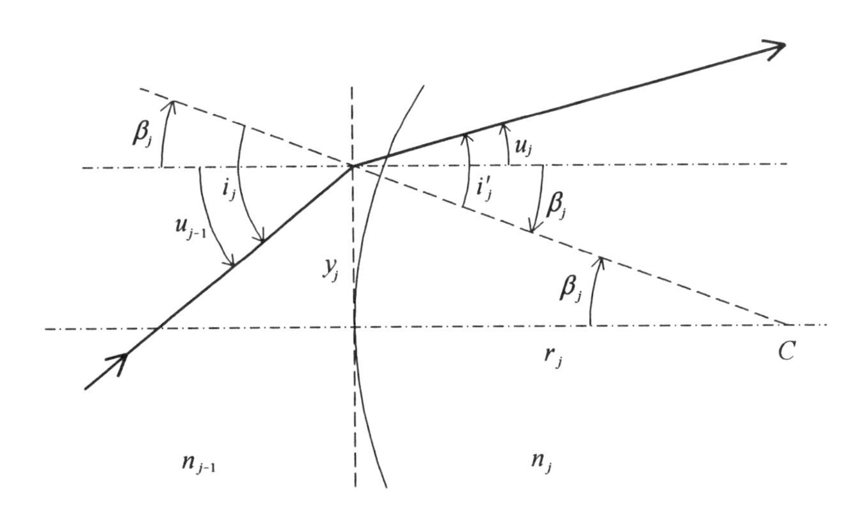

Figure 3.4 Refraction of a paraxial ray at a spherical surface.

Upon substitution we get two new equations for the angle of incidence and the angle of refraction

$$i_j = u_{j-1} + c_j y_j \tag{3.7}$$

and

$$i_j^{\prime} = u_j + c_j y_j \, . \tag{3.8}$$

Substituting the previous two equations into Eq. 3.3 and rearranging terms gives

$$n_j u_j = n_{j-1} u_{j-1} - (n_j - n_{j-1}) c_j y_j \, . \tag{3.9}$$

Repeated applications of Eqs 3.2 and 3.9 permit tracing a paraxial ray through any refractive optical system. These calculations will be performed frequently, so it is worthwhile to simplify the equations as much as possible. The factor which multiplies y_j in Eq. 3.9 is the optical power of the jth surface:

$$\phi_j = (n_j - n_{j-1}) c_j = \frac{n_j}{f_j^{\prime}} \, . \tag{3.10}$$

A further simplification can be accomplished by introducing two new variables called the *reduced slope angle* ($\alpha = n\,u$) and the *reduced thickness* ($\tau = t\,/\,n$). The purpose of these variables is to keep the index of refraction from explicitly appearing in Eq. 3.9. The ray trace equations become

$$y_j = y_{j-1} + \tau_{j-1} \alpha_{j-1} \tag{3.11}$$

and

$$\alpha_j = \alpha_{j-1} - \phi_j y_j \, . \tag{3.12}$$

3.1.3 Finding the Image Size and Position

One of the primary uses of the *y-nu* or y-α ray trace is to find the location and size of an image generated by a complex lens system. Recall that we defined an image as the location of the intersection of two rays. It doesn't really matter if the intersection point is found by drawing the rays as we have done previously, or if the location is found mathematically. First, we will use these ideas to find the image location, then the image size.

Since we are dealing only with axially symmetric systems, the optical axis defines a ray from the base of the object through the system to the base of the image. We only need to mathematically trace one other ray from the base of the object through the system. A ray from the base of the object is called an *axial ray*. Whether the object is at infinity or at some finite distance from the lens system affects how we begin the ray trace. If the object is at a finite distance, then naturally, the axial ray begins with $y_0 = 0$. The slope angle for an axial ray can have any arbitrary value, since all rays from an object point must meet at the same image point. A good value for u_0 is 0.1 radians. On the other hand, if the object is at infinity, then the intersection of two rays is also at infinity. This means that a second ray which crosses the optical axis at infinity is parallel to the optical axis. To initiate an axial ray trace from infinity, we set the slope angle $u_0 = 0$ and select an arbitrary ray height at the first surface, say $y_1 = 1$. Either way, you can use one of these sets of initial values and the ray trace equations to calculate the ray height and slope angle at each subsequent surface of the system. Eventually, you will calculate the ray height and slope angle at the last surface. As shown in Figure 3.5, these values can be used to find the distance to the intersection of the ray with the optical axis

$$t_k = -\frac{y_k}{u_k} . \tag{3.13}$$

This equation works whether the ray emerges from last surface above or below the optical axis and whether the slope angle is positive or negative. For an all refractive system, a negative value for t_k indicates that the image is virtual.

The image size can be determined by tracing an additional ray from the top of the

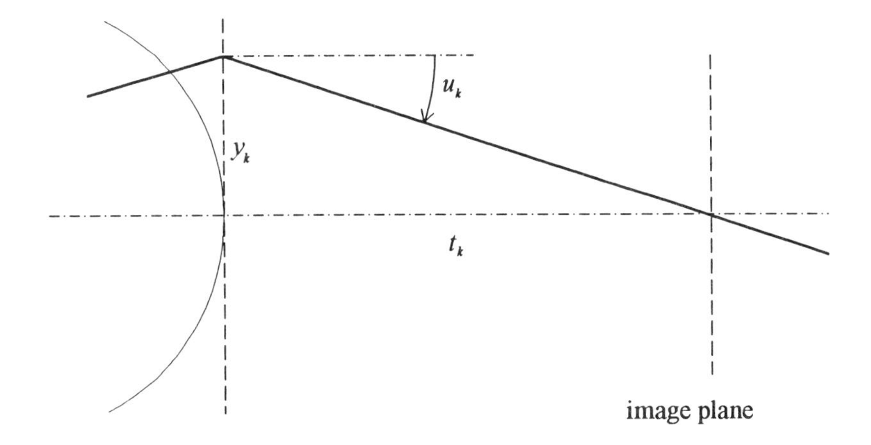

Figure 3.5 Final translation to find the paraxial image plane.

object until it intersects the image plane determined above. The parameters for this second ray are distinguished from the first ray by placing a bar over them. We will call this type of ray an *arbitrary full field ray*. Again, there are two sets of initial values depending on the location of the object. For an object at infinity, the object size is determined by the angle at which the light enters the system. The initial slope angle $\bar{u}_0$ is set accordingly. The height of this ray at the first surface is arbitrary. A convenient value is zero. Later, we will see how to choose a more useful value for $\bar{y}_1$. For a finite object, the initial ray slope is arbitrary (zero is convenient) and the initial ray height is determined by the object size. The image size is found by repeatedly using the ray trace equations to find the ray height and slope angles at each surface. When the last surface is reached, an additional translation to the image plane determines the image size

$$h' = \bar{y}_{k+1} = \bar{y}_k + \tau_k \bar{\alpha}_k \,. \tag{3.14}$$

Another way to find the image size h' which requires less effort will be shown later in this chapter.

3.2 THE RAY TRACE TABLE

The ray trace calculations described above are made even simpler by setting up a ray trace table to do the calculations in a prescribed sequence. The layout of the ray trace table makes it easier to remember the ray trace equations and what calculations to do next. It also standardizes the presentation of ray trace results which makes it easier to find and correct mistakes (not to mention making homework grading easier).

Table 3.1 shows an example ray trace table. The layout and use of this table are described in this section. The actual calculations are discussed in Example 3.1. Typically, the dimensions of the values are not included in the ray trace table. I have included them for illustration in this chapter. As long as all of the distances are given in the same units and the curvatures are given as the inverse of these units, there should be no problem with leaving out the units as is done in all of the remaining chapters. Measuring angles in radians is critical in using Eq. 3.13 for finding the image location. Angles will be given in radians unless otherwise specified.

A ray trace table consists of rows and columns of numbers. The very top row contains numbers that indicate which surface is described in the two columns below the number. The next three rows contain the lens prescription data: curvature, thickness, and index. Since curvature is a property of the surface itself, the numerical value for the curvature of each surface is listed in a column directly below the surface number. On the other hand, thickness and index describe the space between surfaces, so they go in a separate column between the surface numbers. The next two rows contain values for the power and reduced thickness which are calculated from the lens prescription data. Again, since the power is a property of the surface itself while

Table 3.1 Example ray trace table for a thick lens.

	0	1	2	3
c		0.300 cm^{-1}	−0.100 cm^{-1}	0.200 cm^{-1}
t	12.0 cm	2.00 cm	1.00 cm	<u>15.9 cm</u>
n	1.000	1.720	1.617	1.00
$-\phi$		−0.216000	−0.010300	0.123400
τ	12.0 cm	1.162791	0.618429	15.863111
y	0.0 cm	1.200000	1.014884	0.909965
α	0.1 rad	−0.159200	−0.169653	−0.057364
u				−0.057364
i				
y	1.0 cm	1.000000	0.748837	0.610487
α	0.0 rad	−0.216000	−0.223713	−0.148379
u				
i				

the reduced thickness describes the space between surfaces, they are listed in the appropriate columns. Notice that the negative of the power is the value given in the table. This is a direct result of how the table is laid out and makes using the *y-nu* ray trace easier.

Finally, there are several sets of four rows which contain the actual ray trace data. Each group of four rows can be used to carry out one ray trace. As discussed in the previous section, one set would be used for an axial ray to find the image location and another would be used for an arbitrary full field ray to find the image size. The first row in each group gives the height of the ray at each surface. Obviously the ray height should be placed in the surface column. The next two rows give the reduced slope angle and the actual slope angle between surfaces. The last row lists the incident angle at each surface. This row will not be needed until we discuss third-order optics in Chapter 7.

Example 3.1

An object which is 1.00 cm tall is located 12.0 cm in front of a lens. This lens has three surfaces with curvatures of 0.300 cm^{-1}, −0.100 cm^{-1}, and 0.200 cm^{-1}, respectively. The surfaces are separated by distances of 2.00 cm and 1.00 cm. Two types of glass are used in this lens. The first has an index of refraction 1.720 and the second has an index of 1.617. The problem is to find the location of the image and its size. Table 3.1 is a ray trace table which shows how to accomplish this task.

The first thing to do is to enter the given lens prescription data into the appropriate places in the table. Notice that curvatures for the object (surface 0) and image (surface 4) are not given. These are assumed to be planar surfaces, and the curvatures are not needed for calculating a surface power anyway. Initially, the thickness between the last surface and the image surface (underlined) is not given. This is one of the two quantities we are trying to determine. The indices after the object and after the last surface are assumed to be 1.000, indicating a lens in air.

The next step is to calculate the power and the reduced thickness. The reduced thickness between each surface is found by dividing the physical distance by the index of refraction. Notice the values needed for this computation are in the column directly above the reduced thickness and that they are in the proper order, t over n. For example, after surface 1 we have

$$\frac{t_1}{n_1} = \frac{(2.00)}{(1.720)} = 1.162791 \ .$$

I normally keep at least six significant figures in the intermediate calculations to avoid potential round off errors.

The negative of the power is found by taking the index before the surface and subtracting the index after the surface. Multiply this difference by the curvature of the surface to get the negative power. For surface 2 we have

$$-\phi_2 = (n_1 - n_2)c_2 = ((1.720) - (1.617))(-0.100) = -0.0103 \ .$$

Calculating the positive power would require using the right-hand index first, which is the reverse of the way we are used to doing things. Besides, as you will see, using the negative power is easier.

Now we are ready to actually do a ray trace. *The key to successful ray tracing is to choose the correct initial values.* As discussed previously, an axial ray is needed to find the image position. Since the object is not located at infinity, the object ray height is set to 0.0 cm and the reduced slope angle is set to an arbitrary value of 0.1 rad. We now have enough information to find the ray height at the first surface. This calculation consists of taking the old ray height and adding to it the old reduced slope angle times the reduced thickness. The multiplicative factors in this calculation occur in the same column and the results are added from left to right. To find the ray height at the first surface we have

$$y_1 = y_0 + \tau_0 \alpha_0 = (0.0) + (12.0)(0.1) = 1.20 \ .$$

Exactly the same type of calculation is used to find the new reduced slope angle. The old reduced slope angle is added to the product of the new ray height and the negative power

$$\alpha_1 = \alpha_0 + (-\phi_1)y_1 = (0.1) + (-0.216)(1.20) = -0.1592 .$$

Now you can see why we insisted on listing the negative power in the ray trace table. If the actual slope angle is needed, it can be determined by taking the reduced slope angle and dividing by the index in the column directly above. Similar calculations are done for each remaining surface.

The distance from the last surface to the image plane is calculated by taking the last ray height and dividing by the last slope angle and then changing the sign on the result

$$t_3 = -\frac{y_3}{u_3} = -\frac{(0.909965)}{(-0.057364)} = 15.9 \text{ cm} .$$

Since this value is a final result, only three significant figures are kept. This value should be put in the appropriate column of the t row.

Finally, the image size can be determined by tracing a ray from the top of the object at an arbitrary reduced slope angle (0.0 rad in this example). The same technique is used to calculate the ray parameters through the system. This ray trace ends with an additional translation using the image location

$$y_4 = y_3 + \tau_3\alpha_3 = (0.610487) + (15.863111)(-0.148379) = -1.74 \text{ cm} .$$

I've found that in tracing the second ray (the arbitrary full-field ray) I sometimes mistakenly use the values directly above the current rows as the negative power and reduced separation. This, of course, is an error since the rows above the current rows contain the ray height and reduced slope angle of the axial ray. One way around this difficulty is to do the axial ray trace using the second set of ray trace rows. Since this ray is traced first, the first set of ray trace rows are blank while the axial ray is being traced. Then I use the first set of ray trace rows to trace the arbitrary full field ray and avoid any confusion. ■

After doing a few ray traces, using a table like this becomes almost second nature. Once the ray trace is mastered, the calculations can be made even easier by using a spreadsheet program or a special-purpose program like the one available from the publisher of this text.

Occasionally, a ray must be traced backward through the system from image space to object space. This might be required, for example, to find the object space focal point. A backward ray trace can be accomplished with the same equations as a forward ray trace or with the ray trace table. The equations are solved for y_{j-1} and α_{j-1} instead of y_j and α_j. Therefore, only one term changes sign in each relationship:

$$y_{j-1} = y_j - \tau_{j-1} \alpha_{j-1} \tag{3.15}$$

and

$$\alpha_{j-1} = \alpha_j + \phi_j y_j \,. \tag{3.16}$$

In the ray trace table, a backward ray trace is carried out from right to left by subtracting the multiplicative terms instead of adding. A backward ray tracing example is given in Section 3.4.

It should be pointed out that the ray trace equations and the ray trace table work for thin lenses as well as thick lenses. The only modification is that data for the individual surfaces (the first three lines) are not needed. Surface data are replaced by data for each lens as a whole on the second two lines. The subscripts refer to each thin lens, rather than to each surface. The power of a thin lens ϕ is equal to the inverse of the effective focal length. And the separation between lenses (assuming the lenses are in air) replaces the reduced thickness.

Example 3.2

We want to find the image size and location produced by an optical system which consists of three thin lenses. The object is located at infinity and has an angular size of 0.100 radians. The focal lengths of the lenses are +6.66667 cm, −3.33333 cm, and +5.000 cm, respectively. The separation between each lens is 1.50 cm.

The first step in solving this problem is to set up the lens prescription part of the ray trace table (as shown in Table 3.2). Since we don't know (or need) the curvatures or indices of the lenses, we can skip the first three lines of the table. The negative power is calculated by taking the negative inverse of the focal length for each lens. The first lens gives

$$-\frac{1}{f_1'} = -\frac{1}{6.66667 \text{ cm}} = -0.15 \text{ cm}^{-1} \,.$$

The reduced separation is the same as the physical separation because the lenses are assumed to be in air.

Next, a ray from the base of the object is traced to find the image location. Since the object is at infinity, this ray comes in parallel to the optical axis ($\alpha_0 = 0.00$ rad). All rays parallel to the optical axis should cross at the same point in the image plane, so one particular ray ($y_1 = 1.00$ cm) is selected. The ray trace then proceeds and ends in the normal way.

Finally, a ray from the top of the object ($\alpha_0 = -0.1$ rad) is traced to find the image size. The height of this ray at the first surface is arbitrary (chosen to be 0.00 cm here). ■

Table 3.2 Example ray trace table for a system of thin lenses.

	0	1	2	3
c				
t				
n				
$-\phi$		-0.15	0.3	-0.2
τ		1.5	1.5	9.24
y	1.0	1.000000	0.775000	0.898750
α	0.0	-0.150000	0.082500	-0.097250
u				-0.097250
i				
y		0.0	-0.150000	-0.367500
α	-0.1	-0.100000	-0.145000	-0.071500
u				
i				

3.3 REFLECTING SURFACES

Y-nu ray traces are used to analyze systems containing mirrors as well as lenses. In fact, with one clever substitution, exactly the same equations that are used for refraction are used for reflection.

Figure 3.6 shows the effect of a spherical mirror on a paraxial ray. Starting with the law of reflection

$$-i_j = i_j' \tag{3.17}$$

and making the usual substitutions for these angles gives

$$u_j = 2\beta_j - u_{j-1} \ . \tag{3.18}$$

Equation 3.18 can be made identical to Eq. 3.9 by multiplying by the index of refraction. In order to get the signs correct, the index after the reflection is made equal to the negative of the index before the reflection

$$n_j = -n_{j-1} \ . \tag{3.19}$$

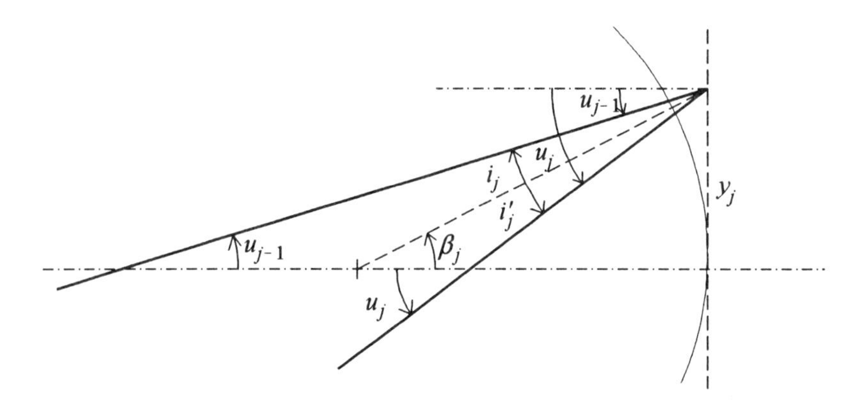

Figure 3.6 Reflection of a paraxial ray by a spherical mirror.

This substitution is really not as crazy as it looks. First of all, it works. No changes need to be made in the *y-nu* ray trace equations to accommodate reflection, and no changes need to be made in the ray trace table method. Second, recall that the index of refraction is related to the velocity of light in a particular medium. On reflection, the direction of the velocity of light changes direction. Typically, such a direction change is indicated by a sign change.

Thus, the equation for the reduced slope angle of a ray after reflection at a mirror is given by

$$\alpha_j = \alpha_{j-1} - \phi_j y_j \, , \tag{3.20}$$

where the power of the reflecting surface is

$$\phi_j = \frac{n_j}{f_j'} = -\frac{2\,n_j}{r_j} = (n_j - n_{j-1})\,c_j \, . \tag{3.21}$$

Not only does the equation for reflection take on the same form as the equation of refraction, the equation for translation is also the same. Note, however, that the next surface is to the left. This means that the separation distance t_{j+1} is negative. As long as the ray is traveling to the left, both the separations and indices must be taken as negative. With this sign convention, the separation between the last surface and the image plane is still given by Eq. 3.13. Now, however, a negative value for t_{j+1} indicates a real image.

It may look like we are changing the sign conventions that have been used previously. Actually, we are preserving these conventions. Distances are still positive to the right and negative to the left of a surface. Curvatures are still positive if the center is to the right of the surface. And slope angles are positive if measured counterclockwise from the optical axis to the ray. The only real difference is the sign change of the index of refraction.

Combinations of refractions and reflections can be easily handled. For example, Figure 3.7 shows a thick lens in front of a spherical mirror. The refracting surfaces are given two identifying subscripts since the ray passes through a second time after striking the mirror. The ray table setup is shown in Table 3.3. A system which has both reflecting and refracting surfaces is said to be *catadioptric*.

Example 3.3

Once again the problem is to find the image size and position, but this time the system consists of a thick lens and mirror. The first lens surface has a curvature of 0.300 cm^{-1} and the second surface has a curvature of −0.100 cm^{-1}. The mirror also has a curvature of −0.100 cm^{-1}. The lens is 2.00 cm thick and has an index of 1.63. The distance between the lens and the mirror is 5.00 cm. The object is located 18.0 cm in front of the lens and is 1.00 cm tall.

The ray trace table set up proceeds as usual by filling in the lens prescription data on the first three lines (Table 3.3). This system has five surfaces because the light must pass through the lens twice. Thus

$$c_1 = c_5 = 0.300 \text{ cm}^{-1} ,$$

$$t_1 = -t_4 = 2.00 \text{ cm} ,$$

$$n_1 = -n_4 = 1.63 .$$

A similar set of relations exists for surfaces 2 and 4.

Next, the negative power and reduced separation are calculated in the normal fashion. Notice that the power and reduced separation of a surface are both independent of the direction that the light travels through the surface. Use this fact as a check on your work.

The ray traces are initiated in the normal fashion. When calculating the image position, however, you must be sure to use the slope angle u and not the reduced

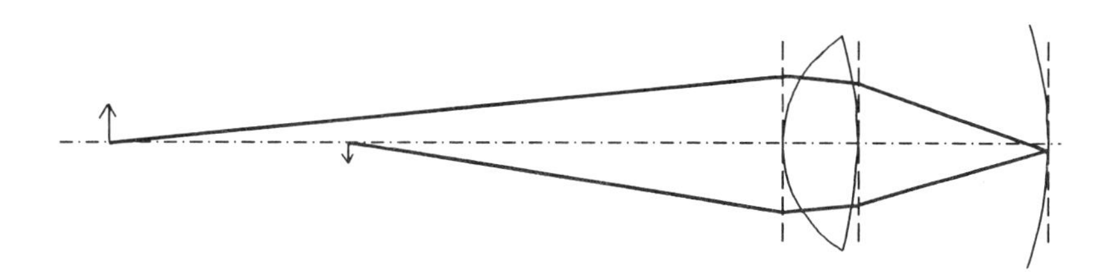

Figure 3.7 Paraxial ray trace through a catadioptric system.

Table 3.3 Paraxial ray trace table for a catadioptric system.

	0	1	2	3
c		0.300	-0.100	-0.100
t	18.0	2.00	5.00	-5.00
n	1.000	1.630	1.000	-1.000
$-\phi$		-0.18900	-0.06300	-0.20000
τ	18.0 cm	1.22699	5.00000	5.00000
y	0.0	1.80000	1.50528	-0.16989
α	0.1	-0.24020	-0.33503	-0.30106
u	0.10000	-0.14736	-0.33503	0.30106
i				
y	1.0	1.00000	0.768098	-0.41885
α	0.0	-0.18900	-0.23739	-0.15362
u	0.00000	-0.11595	-0.23739	0.15362
i				

Table 3.3 (*continued*) Paraxial ray trace table for a catadioptric system (column 3 is repeated for continuity).

	3	4	5	6
c	-0.100	-0.100	0.300	
t	-5.000	-2.00	-11.5069	
n	-1.000	-1.630	-1.000	
$-\phi$	-0.20000	-0.06300	-0.18900	
τ	5.00000	1.22699	11.5069	
y	-0.16989	-1.67516	-1.91506	
α	-0.30106	-0.19552	0.16643	
u	0.030106	0.11995	-0.16643	
i				
y	-0.41885	-1.18695	-1.28369	0.60086
α	-0.15362	-0.07884	0.16378	
u	0.15362	0.048369	-0.16378	
i				

slope angle α since the two angles now differ in sign. Here the image position is found by

$$t_5 = -\frac{y_5}{u_5} = -\frac{(-1.91506 \text{ cm})}{(-0.16643 \text{ rad})} = -11.5 \text{ cm} .$$

The negative sign does not indicate a virtual image. Because of the change of direction of propagation, this image is real. The image size is found in the normal way. ■

3.4 CARDINAL POINTS

As noted earlier, $-t_0$ and t_k are not the same as s and s'. If particular values for $-t_0$ and t_k are substituted into the thin lens equation for s and s', then an effective focal length f' could be found. Unfortunately, if a different set of values for $-t_0$ and t_k are used, then a different value for f' is found. The problem is not fundamental, however. Both the *y-nu* ray trace and the thin lens equation use the same approximations except for assuming that the separations between surfaces are negligible. This merely means that s and s' are measured from the effective plane of refraction or reflection, while $-t_0$ is measured from the first actual surface backward to the object, and t_k is measured from the last actual surface. The thin lens equation should be usable, if planes of effective refraction can be found for a thick or complex lens system. Such planes do indeed exist and are called *principal planes*.

Several parallel rays can be traced through a complex lens to show that the principal planes actually do exist (Figure 3.8). If the incoming rays are extended until they intersect the outgoing rays, the points of intersection all fall on a straight line which represents a plane perpendicular to the optical axis. This plane of intersection is a plane of effective refraction since all of the individual refractions have the same effect as one refraction in this plane.

The location of the second or image space principal plane can be found by drawing rays as shown, or by a simple three step numerical procedure. Since the incoming rays in this case are parallel, they represent rays coming from an object at infinity, and the image point is the second focal point of the system. The distance from the last surface of the system to the image plane is calculated as before, but now it is called the *back focal length*:

$$l_{F'} = t_k = -\frac{y_k}{u_k} . \tag{3.22}$$

Note that the back focal length is equal to the formula given above only for an axial ray trace from an object at infinity. The effective focal length is measured from the

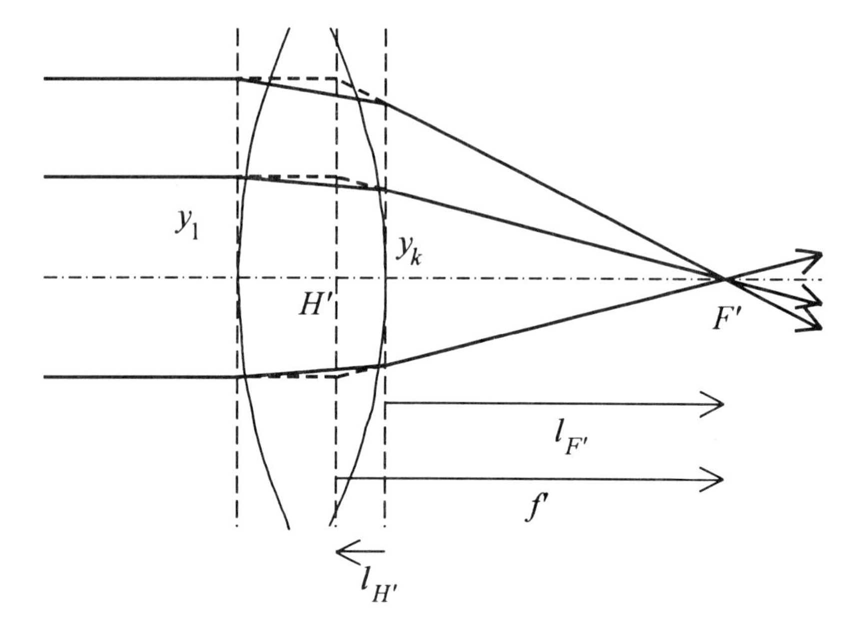

Figure 3.8 Location of the second principal plane and definition of back focal length.

effective plane of refraction, and it can be found using

$$f_{eff} = f' = -\frac{y_1}{u_k} . \tag{3.23}$$

The difference in these two values gives the distance of the principal plane from the last surface of the system

$$l_{H'} = l_{F'} - f' . \tag{3.24}$$

A similar procedure can be used to find the first focal point and first principal plane by tracing a parallel ray backward through the system (Figure 3.9). This ray trace finds the front focal length

$$l_F = -t_0 = -\frac{y_1}{u_0} . \tag{3.25}$$

The effective focal length is found in the same manner as before:

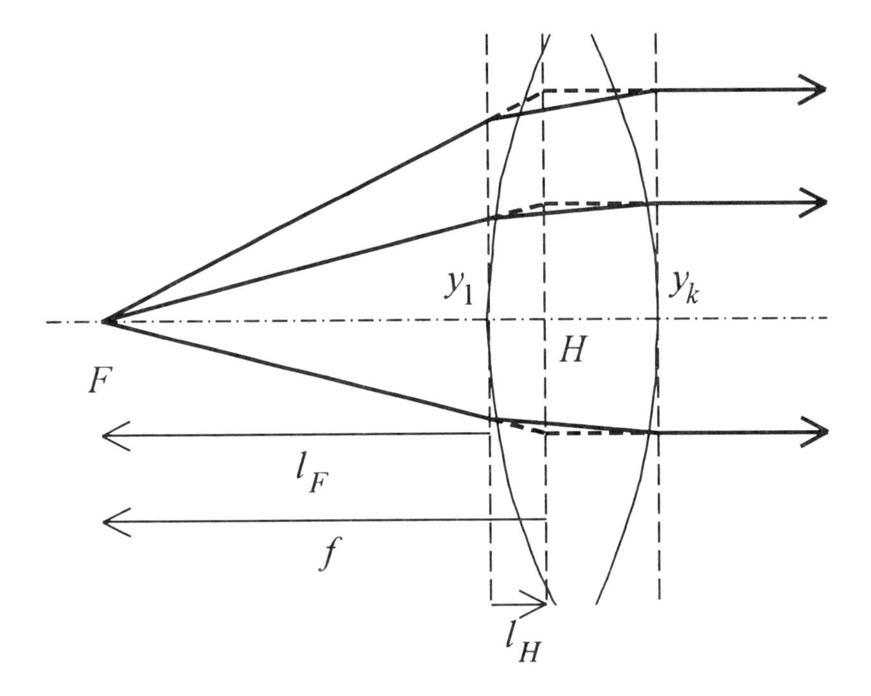

Figure 3.9 Location of the first principal plane and the definition of front focal length.

$$f = -f_{eff} = -\frac{y_k}{u_0}\,; \tag{3.26}$$

this should also have the same value as before. Finally, the difference between the front focal length and the effective focal length gives the position of the first principal plane:

$$l_H = l_F - f\,. \tag{3.27}$$

Example 3.4

Let's return to the system given in the first example of the chapter and find the locations of the principal points and focal points for this system.

Table 3.4 shows the layout of the ray trace table used to solve this problem. The first ray trace shown comes in parallel to the optical axis and is used to find the image space principal point and focal point:

$$l_{F'} = -\frac{y_3}{u_3} = -\frac{(0.61049\ \text{cm})}{(-0.14838\ \text{rad})} = 4.11\ \text{cm}\ ,$$

$$f' = -\frac{y_1}{u_3} = -\frac{(1.00\ \text{cm})}{(-0.14838\ \text{rad})} = 6.74\ \text{cm}\ ,$$

$$l_{H'} = l_{F'} - f' = (4.11\ \text{cm}) - (6.74\ \text{cm}) = -2.63\ \text{cm}\ .$$

The second ray trace starts at surface 3 and proceeds backward through the system. This ray is equivalent to a ray which starts at the front focal plane and leaves the system parallel to the optical axis. Since we don't know the focal plane position (that's what we're trying to find), we must do a backward ray trace starting with what we do know. As described earlier, a backward ray trace is accomplished by working from right to left and subtracting instead of adding terms. For example,

$$\alpha_2 = (0.0\ \text{rad}) - (1.00\ \text{cm}) \times (0.1234\ \text{cm}^{-1}) = -0.1234\ \text{rad}\ ,$$

$$y_2 = (1.00\ \text{cm}) - (-0.1234\ \text{rad}) \times (0.61843\ \text{cm}) = 1.07631\ \text{cm}\ .$$

Table 3.4 Ray traces used to find principal planes.

	0		1		2		3	
c			0.300 cm^{-1}		−0.100 cm^{-1}		0.200 cm^{-1}	
t		12.0 cm		2.00 cm		1.00 cm		
n		1.000		1.720		1.617		1.00
$-\phi$			−0.216000		−0.010300		0.123400	
τ		12.0 cm		1.162791		0.618429		
y	1.0 cm		1.00000		0.74884		0.61049	
α		0.0 rad		−0.21600		−0.22371		−0.14838
u								−0.14838
i								
y			1.20691		1.07631		1.0 cm	
α		0.14838		−0.11231		−0.12340		0.0 rad
u		0.14838						
i								

With this ray trace we can find the object space principal point and focal point

$$l_F = -\frac{y_1}{u_0} = -\frac{(1.20691\ \text{cm})}{(0.14838\ \text{rad})} = -8.13\ \text{cm}\ ,$$

$$f = -\frac{y_3}{u_0} = -\frac{(1.00\ \text{cm})}{(0.14838\ \text{rad})} = -6.74\ \text{cm}\ ,$$

$$l_H = l_F - f = (-8.13\ \text{cm}) - (-6.74\ \text{cm}) = -1.39\ \text{cm}\ .$$

■

The principal planes can be used with the focal points to greatly simplify drawing rays through a complex optical system (Figure 3.10). As discussed above, a ray parallel to the optical axis in object space will effectively refract at the second principal plane and pass through the second focal point. A ray which passes through the first focal point will refract at the first principal plane and end up parallel to the optical axis in image space. If these two rays originate at an object point, their intersection will determine the corresponding image point.

Figure 3.10 can also be used to show that the thin lens equation derived in Chapter 2 works for any optical system if the object distance and image distance are measured from the principal planes. In this figure, triangles *QRS* and *FHS* are similar. The ratio of their sides are therefore equal:

$$\frac{h - h'}{s} = -\frac{h'}{f}\ . \tag{3.28}$$

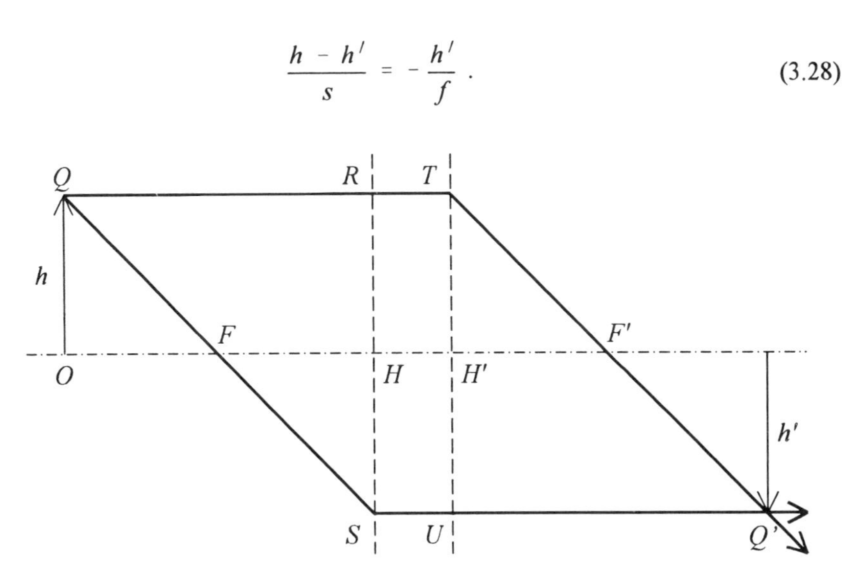

Figure 3.10 Paraxial ray drawing using only the principal planes to find the image.

Also, triangles $Q'TU$ and $F'TH'$ are similar; therefore

$$\frac{h - h'}{s'} = \frac{h}{f'} . \tag{3.29}$$

Now, multiply Eq. 3.28 by n_0 and Eq. 3.29 by n_k and subtract

$$\frac{n_k(h - h')}{s'} - \frac{n_0(h - h')}{s} = \frac{n_k h}{f'} + \frac{n_0 h'}{f} . \tag{3.30}$$

By definition, as s goes to infinity, s' goes to the second focal position f'. Making these substitutions in Eq. 3.30 and rearranging a little gives the relationship between the two focal lengths of the system:

$$\frac{n_k}{f'} = -\frac{n_0}{f} . \tag{3.31}$$

Finally, using this relation in Eq. 3.30 gives the thin lens equation

$$\frac{n_k}{s'} - \frac{n_0}{s} = \frac{n_k}{f'} = \phi . \tag{3.32}$$

To find the image size, take the ratio of Eq. 3.28 and Eq. 3.29. This will give the magnification of the system:

$$m = \frac{h'}{h} = \frac{n_0 s'}{n_k s} . \tag{3.33}$$

As shown previously, two ray traces are needed to find the principal planes. Additional ray traces are not necessary to find the image size and position. Merely use the object distance s and the thin lens equations derived above.

Example 3.5

Let us once again return to the problem given in Example 3.1. Our task is to find the image size and position, but now we will use the cardinal points instead of specific ray tracing. The positions of the cardinal points for this system were determined in the previous example.

The first step is to find the object distance

$$s = -(t_0 + l_H) = -((10.0\ \text{cm}) + (-0.06079\ \text{cm})) = -9.9392\ \text{cm}\ .$$

Next, use the thin lens formula to find the image distance

$$\frac{1}{s'} - \frac{1}{(-9.9392\ \text{cm})} = \frac{1}{(8.4502\ \text{cm})}\ ,$$

$$s' = 56.4\ \text{cm}\ .$$

Now the image distance must be converted to distance from the last surface:

$$t_k = l_{H'} + s' = (-0.6973\ \text{cm}) + (56.4\ \text{cm}) = 55.7\ \text{cm}\ .$$

The magnification is given by

$$m = \frac{s'}{s} = \frac{(56.4\ \text{cm})}{(-9.9392\ \text{cm})} = -5.675$$

and the image size is

$$h' = m \times h = (-5.675)(0.1\ \text{cm}) = -0.57\ \text{cm}\ .$$

Any small differences between these results and the results of Example 3.1 are due to round-off errors in the calculations. ■

It is always nice to be able to check the drawing by using a third ray. For thin lenses, the third ray went through the center of the lens and was undeflected. Obviously, for a thick or complex lens system, no ray (other than the optical axis) will pass through undeviated. But there will always be a ray which leaves the system with the same slope angle that it entered (Figure 3.11). In object space this ray intersects the optical axis at what is known as the first nodal point. In image space the intersection is the second nodal point. Any ray in object space which appears to pass through the first nodal point will appear to pass through the second nodal point with the same slope angle in image space.

The position of the nodal points can be found if the locations of the first and second focal points are known. The distance of the second nodal point from the last surface is

$$l_{N'} = l_{F'} - N'F'\ , \tag{3.34}$$

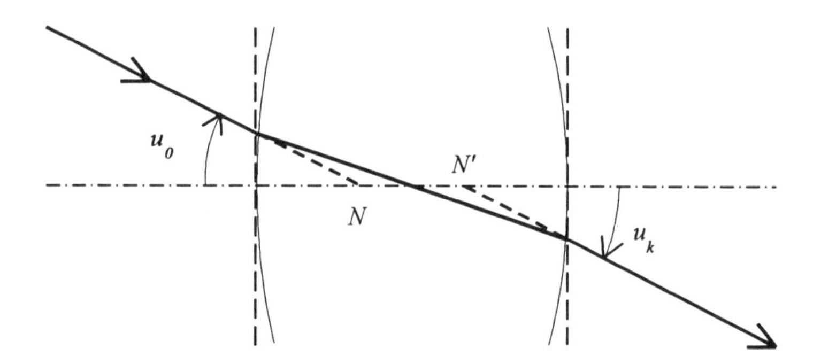

Figure 3.11 Definition and location of nodal points.

where $N'F'$ is the distance from the second nodal point to the focal point. The distance $N'F'$ can be shown to be equal to $-f$. Recall that parallel rays in object space will intersect at the second focal plane. Figure 3.12 shows a ray drawn parallel to the ray from the object through the first focal point. This new ray is aimed at the first nodal point. It emerges from the second nodal point with the same slope angle and intersects in the focal plane at point B. This means that triangles $N'F'B$ and FHA are congruent. Therefore, $N'F' = -f$ (remember sign convention). We can make a similar construction for rays in image space, and working backwards show that $NF = -f'$. The nodal point locations are

$$l_N = l_F + f' \tag{3.35}$$

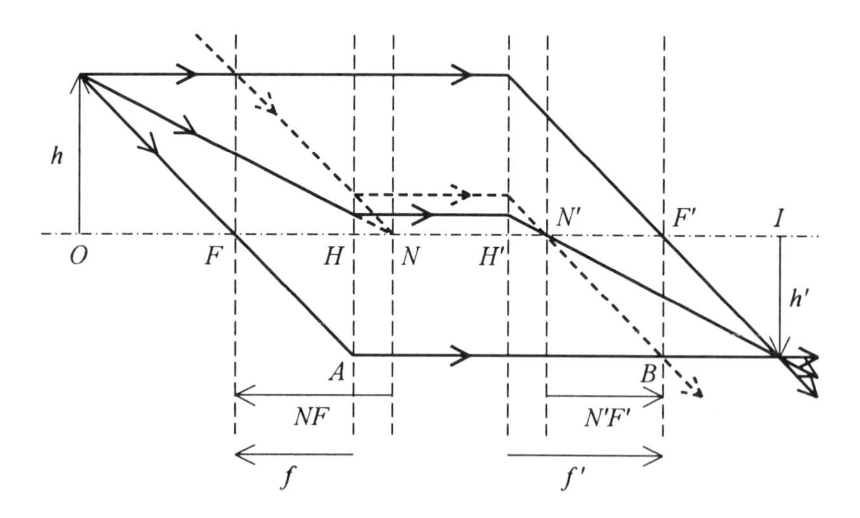

Figure 3.12 Graphical determination of nodal point locations using principal planes and focal planes.

and

$$l_{N'} = l_{F'} + f . \tag{3.36}$$

Comparing Eqs. 3.35 and 3.36 with the equations for the principal plane positions (Eqs. 3.24 and 3.27) shows that if $f = -f'$, then the two sets of planes are coincident. Recall that this is true if $n_0 = n_k$. Otherwise

$$\frac{n_k}{f'} = -\frac{n_0}{f} . \tag{3.37}$$

In other words, if the indices in image space and object space are not the same, the nodal points are shifted away from the principal planes toward the space with the higher index of refraction.

Example 3.6

If we put water in the space after the last surface of the system given in Example 3.1, the power of this surface will be reduced. As shown in Table 3.5, the only effect of this change is that the last slope angle is reduced to −0.14181 rad. Therefore, the back focal length will be slightly longer:

$$l_{F'} = -\frac{y_3}{u_3} = -\frac{(0.61049 \text{ cm})}{(-0.14181 \text{ rad})} = 4.30 \text{ cm} .$$

The value for f' will change:

$$f' = -\frac{y_1}{u_3} = -\frac{(1.00 \text{ cm})}{(-0.14181 \text{ rad})} = 7.05 \text{ cm} .$$

Therefore, the second principal plane position will also change:

$$l_{H'} = l_{F'} - f' = (4.30 \text{ cm}) - (7.05 \text{ cm}) = -2.75 \text{ cm} .$$

The front principal plane location is found by

$$l_F = -\frac{y_1}{u_0} = -\frac{(1.08878 \text{ cm})}{(0.18904 \text{ rad})} = -5.76 \text{ cm} ,$$

Table 3.5 Example ray trace table for a thick lens.

	0	1	2	3
c		0.300 cm^{-1}	-0.100 cm^{-1}	0.200 cm^{-1}
t	12.0 cm	2.00 cm	1.00 cm	
n	1.000	1.720	1.617	1.333
$-\phi$		-0.216000	-0.010300	0.05680
τ	12.0 cm	1.162791	0.618429	
y		1.00000	0.74884	0.61049
α	0.0	-0.21600	-0.22371	-0.18904
u				-0.14181
i				
y		1.08878	1.03513	1.0
α	0.18904	-0.04614	-0.05680	0.0
u	0.18904			
i				

$$f = -\frac{y_3}{u_0} = -\frac{(1.00 \text{ cm})}{(0.18904 \text{ cm})} = -5.29 \text{ cm} ,$$

$$l_H = l_F - f = (-5.76 \text{ cm}) - (-5.29 \text{ cm}) = -0.47 \text{ cm} .$$

Another way to find f' is to consider the overall power of the system which must be the same no matter which way light travels through the system:

$$\phi = \frac{n_k}{f'} = -\frac{n_0}{f} .$$

From this relation we get

$$f' = -f\frac{n_k}{n_0} = -(-5.29 \text{ cm})\frac{(1.333)}{(1.000)} = 7.05 \text{ cm} .$$

Next, we want to find the positions of the nodal points

$$l_{N'} = l_{F'} + f = (4.30\ \text{cm}) + (-5.29\ \text{cm}) = -0.99\ \text{cm}$$

and

$$l_N = l_F + f' = (-5.76\ \text{cm}) + (7.05\ \text{cm}) = 1.29\ \text{cm}\ .$$

■

Besides being interpreted as planes of effective refraction, the principal planes have another meaning. Since the object distance s and image distance s' are measured from these planes, if s approaches zero, this means that the object approaches the first principal plane. As s goes to zero, s' also goes to zero. Therefore, the second principal plane is conjugate to the first. In addition, the magnification goes to unity as s goes to zero. Another way of defining the principal planes is as conjugate planes of unit magnification.

Similarly, the nodal points can be defined as conjugate planes of unit angular magnification.

The set of special points associated with an optical system which includes the focal points, principal points, and nodal points are collectively called *cardinal points*. Note that in this group, only the focal points are not conjugates of each other. The focal points are conjugates with points at infinity. If needed, other cardinal points could be defined such as points of negative unit magnification, or double angular magnification, etc. These points are not needed for ray drawing or for simplifying the system, so they are rarely used.

3.5 SPECIFIC APPLICATIONS

In this section, two special cases will be looked at in detail. Equations for the effective focal length of a combination of two thin lenses in air separated by some distance will be derived. The locations of the principal planes will also be determined. Similar calculations will be made for a single thick lens with different media on either side of the lens.

3.5.1 Two Thin Lenses

Table 3.6 shows ray traces for finding the first and second focal points of a combination of thin lenses in symbolic form. From the first ray trace, the back focal length is found to be

$$l_{F'} = \frac{1 - t_1 \phi_1}{\phi_1 + \phi_2 - t_1 \phi_1 \phi_2}\ . \tag{3.38}$$

Table 3.6 Symbolic paraxial ray trace for two thin lenses.

	0	1	3
$-\phi$		$-\phi_1$	$-\phi_2$
τ		t_1	
y		1.0	$1 - t_1\phi_1$
α	0.0	$-\phi_1$	$-\phi_1 - \phi_2(1 - t_1\phi_1)$
y			1.0
α	$\phi_2 + \phi_1(1 - t_1\phi_2)$	ϕ_2	0.0

The effective focal length or the distance from the second principal plane to the second focal point is given by

$$f_{eff} = f' = \frac{1}{\phi_1 + \phi_2 - t_1\phi_1\phi_2} . \tag{3.39}$$

Recall that the power of a thin lens is equal to the inverse of its effective focal length. Then Eq. 3.39 becomes

$$f_{eff} = \frac{f_1' f_2'}{f_1' + f_2' - t_1} . \tag{3.40}$$

Using Eq. 3.24, the location of the second principal plane can be shown to be

$$l_{H'} = -\frac{t_1 f_2'}{f_1' + f_2' - t_1} . \tag{3.41}$$

A backward ray trace is used to find the first focal plane and first principal plane. Again, from Table 3.6, the front focal length can be determined

$$l_F = -\frac{1 - t_1\phi_2}{\phi_1 + \phi_2 - t_1\phi_1\phi_2} . \tag{3.42}$$

Of course, the effective focal length is the same ($f_{eff} = -f$). And the position of the first principal plane is given by

$$l_H = \frac{t_1 f_1'}{f_1' + f_2' - t_1} . \tag{3.43}$$

The principal planes are separated by

$$HH' = t_1 + l_{H'} - l_H \, . \tag{3.44}$$

The effect of changing the separation between the two lenses can be seen clearly in Figure 3.13. For this graph the effective focal length of both lenses is 1.0 cm. When the two lenses are in contact, the effective focal length is 0.5 cm as expected. As the separation increases, the focal length increases. When the separation is 2.0 cm, the focal length becomes infinite. This type of system is said to be afocal. As the separation is further increased, the effective focal length becomes negative, and the system acts like a diverging rather than a converging lens.

Another point to note is that for separations less than 2.0 cm, the principal planes are crossed. The first principal plane H actually comes after the second principal plane H'. As shown in Figure 3.14, the principal planes still work, but in an odd way. The actual rays still travel from left to right through the system, but the idealized rays or ray extensions that have their effective refraction at the principal planes must travel from right to left. In effect, such systems have a negative thickness.

3.5.2 Thick Lens

Table 3.7 shows a symbolic ray trace for a thick lens. Notice that the indices on either side of the lens have been explicitly retained as n_0 and n_2 to show their effect on the position of the nodal points. The effective focal length is determined as always:

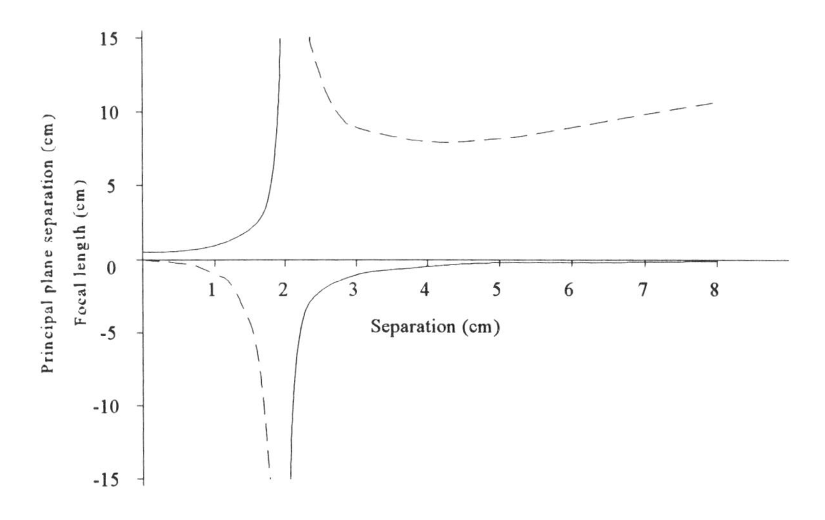

Figure 3.13 Focal length and principal plane separation for two thin lenses.

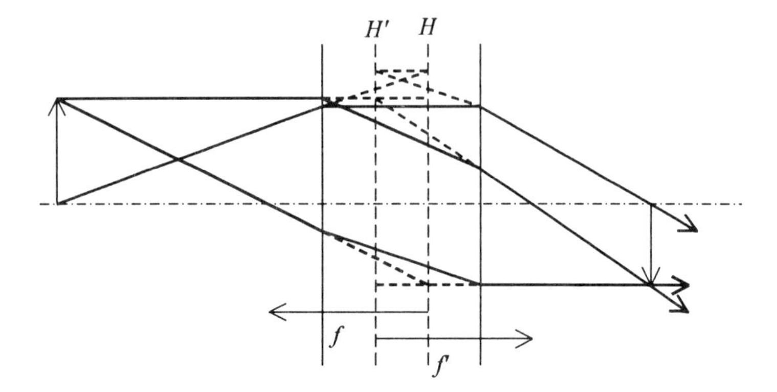

Figure 3.14 Paraxial ray drawing with crossed principal planes.

$$f_{eff} = f' = -\frac{y_1}{u_2} = -\frac{n_2}{-\phi_1 - \phi_2\left(1 - \frac{t_1}{n_1}\phi_1\right)} . \tag{3.45}$$

We can use this result to find the lens maker's formula for a thick lens. If we assume the lens is in air, then $n_0 = n_2 = 1$. Furthermore, let $n_1 = n$. Then

$$\frac{1}{f'} = (n - 1)\left(\frac{1}{r_1} - \frac{1}{r_2} + \frac{(n-1)t_1}{n r_1 r_2}\right) . \tag{3.46}$$

This equation obviously reduces to the thin lens version of the lens maker's equation as the thickness goes to zero.

Back to the general case. We can determine the back focal length using the symbolic ray trace and the standard formula:

$$l_{F'} = -\frac{y_2}{u_2} = -\frac{\left(1 - \frac{t_1}{n_1}\phi_1\right) n_2}{-\phi_1 - \phi_2\left(1 - \frac{t_1}{n_1}\phi_1\right)} . \tag{3.47}$$

This can be simplified by using the formula for the focal length:

Table 3.7 Symbolic paraxial ray trace for a thick lens.

	0	1		2	
c		c_1		c_2	
t	t_0		t_1		t_2
n	n_0		n_1		n_2
$-\phi$		$-\phi_1$		$-\phi_2$	
τ			$\frac{t_1}{n_1}$		
y		1.0		$1 - \frac{t_1}{n_1}\phi_1$	
α	0.0		$-\phi_1$		$-\phi_1 - \phi_2\left(1 - \frac{t_1}{n_1}\phi_1\right)$
y		$1 - \frac{t_1}{n_1}\phi_2$		1.0	
α	$\phi_2 + \phi_1\left(1 - \frac{t_1}{n_1}\phi_2\right)$		ϕ_2		0.0

$$l_{F'} = \left(1 - \frac{t_1}{n_1}\phi_1\right) f' = \left(1 - \frac{(n - 1)\, t_1}{n\, r_1}\right) f' . \tag{3.48}$$

Next, we determine the location of the second principal plane as

$$l_{H'} = l_{F'} - f' = - \frac{t_1}{n_1}\phi_1 f' = - \frac{(n - 1)\, t_1}{n\, r_1} f' . \tag{3.49}$$

The location of the second nodal point is

$$l_{N'} = l_{F'} + f = \left(1 - \frac{t_1}{n_1}\phi_1\right) f' - \frac{n_0}{n_2} f' . \tag{3.50}$$

And the separation between the second principal plane and the second nodal point is given by

$$H'N' = f' + f = \left(1 - \frac{n_0}{n_2}\right) f' . \tag{3.51}$$

The shift in the nodal point position away from the principal plane position is toward the side of the system which has the highest external index of refraction for a positive lens.

Similar equations can be derived for the first cardinal points:

$$l_F = \left(1 - \frac{t_1}{n_1}\phi_2\right) f = \left(\frac{(n - 1)t_1}{n r_2} - 1\right) f' , \tag{3.52}$$

$$l_H = -\frac{t_1}{n_1}\phi_2 f = -\frac{(n - 1)t_1}{n r_2} f' , \tag{3.53}$$

$$l_N = \left(1 - \frac{t_1}{n_1}\phi_2\right) f - \frac{n_2}{n_0} f , \tag{3.54}$$

$$HN = H'N' = \left(1 - \frac{n_0}{n_2}\right) f' . \tag{3.55}$$

It is very important to note that the positions of the principal planes depend on the power of the opposite surface and the sign of the overall effective focal length. That is, the second principal plane position is calculated using the power of the first surface and vice versa. The approximate positions of the principal planes can be found by sketching the path of rays as shown in Figure 3.15. To go along with these estimates of principal plane location, an estimate can be made as to the separation between the principal planes:

$$HH' = t_1 - l_H + l_{H'} = t_1\left(1 - \frac{\phi_1 + \phi_2}{n} f'\right) . \tag{3.56}$$

This equation can be simplified by noting that if t_1 is not too large, the inverse of the effective focal length is equal to the sum of the powers of the surfaces. Also, since most optical materials have an index of refraction of roughly 1.5, this value can be used for the index to get

$$HH' = t_1\left(1 - \frac{1}{n_1}\right) \approx \frac{t_1}{3} . \tag{3.57}$$

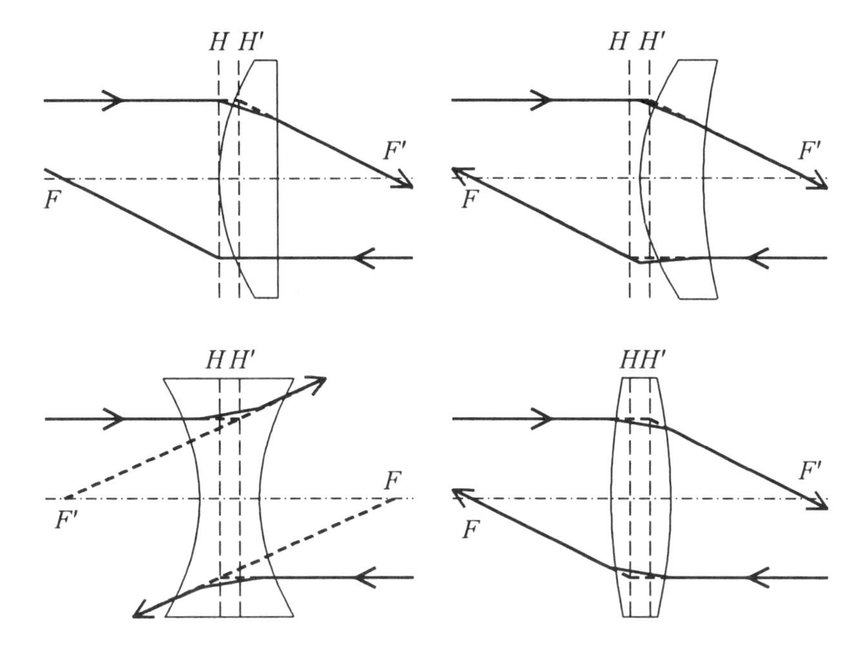

Figure 3.15 Examples of principal plane positions for thick lenses.

As Eq. 3.57 shows, there is no way for the second principal plane H' to precede the first principal plane H in a thick lens. Examples of principal plane locations for a variety of thick lenses are shown in Figure 3.15. Note that the locations of the principal planes depends on the relative powers of the two surfaces but the separation does not.

3.6 OPTICAL INVARIANTS

The fact that the *y-nu* ray trace equations are linear in y and u has some very significant consequences. For example, consider the axial ray used to find the image location. If the initial slope angle is doubled, then all of the subsequent ray heights and slope angles will also be doubled. We will use this fact extensively in the next chapter. On the other hand, the ray from the top of the object to the top of the image is not a simple multiple of the axial ray. Such a pair of rays are said to be linearly independent. We can use the linear independence to our advantage.

Consider tracing any two linearly independent rays through an optical system. Let the first ray be denoted with the standard notation for ray height and slope angle. The second ray will be denoted with a bar over every quantity. Thus, for the second ray, Eqs. 3.11 and 3.12 become

$$\bar{y}_j = \bar{y}_{j-1} + \tau_{j-1}\,\bar{\alpha}_{j-1}\ , \tag{3.58}$$

$$\bar{\alpha}_j = \bar{\alpha}_{j-1} - \phi_j \bar{y}_j . \tag{3.59}$$

If the two refraction / reflection equations are rearranged to solve for the power of the jth surface, then

$$\bar{\alpha}_j y_j - \bar{\alpha}_{j-1} y_j = \alpha_j \bar{y}_j - \alpha_{j-1} \bar{y}_j . \tag{3.60}$$

The two translation equations can be substituted for the y's that are multiplied by the α's before the surface so that each term will have the same subscripts throughout. This yields

$$\alpha_{j-1} \bar{y}_{j-1} - \bar{\alpha}_{j-1} y_{j-1} = \alpha_j \bar{y}_j - \bar{\alpha}_j y_j . \tag{3.61}$$

Since the quantity defined by Eq. 3.61 is the same on either side of the surface, and j represents a general surface, this quantity does not change from surface to surface as the rays pass through the system. The value found above must be invariant throughout the system. In fact, this is the equation for a quantity called the *optical invariant* Λ:

$$\Lambda = \alpha_j \bar{y}_j - \bar{\alpha}_j y_j . \tag{3.62}$$

Please note that despite the name the value of the optical invariant depends on the rays selected to calculate it, and not on any intrinsic property of the optical system. If two different rays are traced, very likely the value for Λ will be different. The optical invariant does not vary from one surface to the next for two particular rays. The value that Λ assumes says something about the relationship between the two rays. If the two rays are linearly independent, then Λ has a constant nonzero value. If the second ray was merely a multiple of the first ray, then Λ would be equal to zero. In the next chapter, two very special rays will be used for all subsequent calculations of the optical invariant.

Example 3.7

Let's calculate the optical invariant for several different cases. We've already done two ray traces through the system given in Example 3.1 (Table 3.1), so let's use them to calculate Λ at two different surfaces:

$$\Lambda = \alpha_0 \bar{y}_0 - \bar{\alpha}_0 y_0 = (0.1)(1.0) - (0)(0) = 0.1 ,$$

$$\Lambda = \alpha_3 \bar{y}_3 - \bar{\alpha}_3 y_3$$

$$= (-0.057364)(0.610487) - (-0.148379)(0.909965) = 0.1 .$$

In Example 3.4 (Table 3.4), we traced two different rays through this same system. These rays will give a different value for the optical invariant:

$$\Lambda = \alpha_1 \bar{y}_1 - \bar{\alpha}_1 y_1 =$$

$$(-0.21600)(1.20691) - (-0.11231)(1.0) = -0.14838 .$$

■

For now, a special application of Eq. 3.62 will be made for the two rays used to find the image size and position. In object space, $y_0 = 0$ and $\bar{y}_0 = h$. While in image space, $y_{k+1} = 0$ and $\bar{y}_{k+1} = h'$. The optical invariant calculated in both spaces is

$$\Lambda = h n_0 u_0 = h' n_k u_k . \tag{3.63}$$

From this relation, the magnification can be determined:

$$m \equiv \frac{h'}{h} = \frac{n_0 u_0}{n_k u_k} = \frac{\alpha_0}{\alpha_k} . \tag{3.64}$$

We now know how to find the image size and position by tracing just one ray through the system. The ray is an axial ray. The height of the ray and the slope angle at the last surface determine the image position, and the ratio of the α's determines the magnification. Thus, the optical invariant is just one more way to make the calculations easier.

Example 3.8

Let us return to Example 3.1 for the last time. Using the idea of the optical invariant means that we don't really need to trace an arbitrary full field ray to find the image size. If, instead, we merely find the ratio of the reduced slope angle for the axial ray in object and image space, then ratio of the values of the arbitrary full field ray heights is determined by Eq. 6.64:

$$m = \frac{\alpha_0}{\alpha_k} = \frac{(0.1 \text{ rad})}{(-0.057364 \text{ rad})} = -1.74 .$$

The image size is now easily determined:

$$h' = m \times h = (-1.74)(1.0 \text{ cm}) = -1.74 \text{ cm} .$$

■

SUMMARY

Paraxial or *y-nu* ray tracing can be used to determine the first-order properties of an optical system. The method consists of two basic steps. The first step is translation between two surfaces:

$$y_j = y_{j-1} + \tau_{j-1}\,\alpha_{j-1} \,. \tag{3.65}$$

The second step is refraction or reflection at the surface:

$$\alpha_j = \alpha_{j-1} - \phi_j y_j \,. \tag{3.66}$$

In these equations, the height of the paraxial ray at the jth surface is y_j. The reduced separation is

$$\tau_{j-1} = \frac{t_{j-1}}{n_{j-1}} \,, \tag{3.67}$$

where t_{j-1} is the actual physical separation between the $j-1$ surface and the jth surface. The index of refraction specified is for the space after the corresponding surface. The reduced slope angle is defined as

$$\alpha_j = n_j u_j \,, \tag{3.68}$$

where u_j is the slope angle of the ray in the space after the jth surface. Finally, the optical power of the surface depends on the type of surface. For a single refracting surface we have

$$\phi_j = (n_j - n_{j-1})\,c_j \,. \tag{3.69}$$

For a spherical mirror in air we have

$$\phi_j = -\frac{2}{r} \,. \tag{3.70}$$

For a thin lens in air we have

$$\phi_j = \frac{1}{f'} \,. \tag{3.71}$$

In general,

$$\phi_j = \frac{n_j}{f_j'} . \tag{3.72}$$

To use the *y-nu* ray trace equations to determine properties of an optical system, the most critical step is to choose the proper initial values. The initial values determine which ray you are tracing.

To find the location of an image, you must trace an axial ray. Axial rays begin at the point where the object plane crosses the optical axis. Thus, for a finite object, $y_0 = 0$ cm. The reduced slope angle is arbitrary, but a convenient value is $\alpha_0 = 0.1$ rad. If the object is at infinity, then the ray must be parallel to the optical axis. Therefore, $\alpha_0 = 0$ rad. Now, the ray height is arbitrary. A convenient value is $y_0 = 1.0$ cm. In either case, the image location is determined by

$$t_k = -\frac{y_k}{u_k} . \tag{3.73}$$

There are two methods for finding the image size. The most obvious method is to trace a second ray from the top of the object to the image plane. For a finite object, $y_0 = h$ and the reduced slope angle can be set arbitrarily to $\alpha_0 = 0$. If, however, the object is at infinity, then the object size must be infinite. Here, the initial reduced slope angle is defined by the half angle of the field of view, $\alpha_0 = u_{HFOV}$. The specific off-axis ray is determined by arbitrarily setting the height of the ray at the first surface, for example, $y_1 = 0$. The image size is found by doing a translation from the last surface of the system to the image plane. Then, $h' = y_{k+1}$. An alternate method for finding the image size is to find the magnification using the concept of the optical invariant:

$$m = \frac{\alpha_0}{\alpha_k} . \tag{3.74}$$

The cardinal points of an optical system can also be found by means of paraxial ray tracing. To find the image space cardinal points, an axial ray from an object at infinity is traced. Then

$$f_{eff} = f' = -\frac{y_1}{u_k} , \tag{3.75}$$

$$l_{F'} = -\frac{y_k}{u_k}, \tag{3.76}$$

$$l_{H'} = l_{F'} - f', \tag{3.77}$$

$$l_{N'} = l_{F'} - \frac{n_0}{n_k} f'. \tag{3.78}$$

For the image space cardinal points, an axial ray is traced, starting from an image at infinity, backwards through the system. Using this ray we have

$$f = -\frac{y_k}{u_0}, \tag{3.79}$$

$$l_F = -\frac{y_1}{u_0}, \tag{3.80}$$

$$l_H = l_F - f, \tag{3.81}$$

$$l_N = l_F - \frac{n_k}{n_0} f. \tag{3.82}$$

Paraxial ray tracing can also be used to find formulae for specific types of systems.

Finally, the optical invariant is expressed as

$$\Lambda = \alpha_j \bar{y}_j - \bar{\alpha}_j y_j. \tag{3.83}$$

REFERENCES

F. Jenkins and H. White, *Fundamentals of Optics*, 4th ed. (McGraw-Hill, New York, 1976).

R. Kingslake, *Lens Design Fundamentals* (Academic Press, Orlando, FL, 1978).

D. C. O'shea, *Elements of Modern Optical Design* (Wiley, New York, 1985).

PROBLEMS

3.1 What is the main purpose of the *y-nu* ray trace table?

3.2 Write down the ray trace equations for translation and refraction.

3.3 Define mathematically the following: reduced separation, reduced slope angle, surface power.

3.4 Describe the ray that is used to determine image location.

3.5 Describe two different methods for finding the image size once the image position is known.

3.6 Figure 3.5 shows one ending to an arbitrary axial ray trace. There are three other possible endings depending on the signs of the ray height and slope angle at the last surface. Sketch figures to illustrate these endings.

3.7 After completing an appropriate *y-nu* ray trace, how do you tell when an image is real or virtual? What determines when the image is erect or inverted?

3.8 Describe what changes need to be made when using the *y-nu* ray trace method with mirrors.

3.9 Describe the rays that are needed to find the locations of the cardinal points.

3.10 What is the minimum number of rays which must be traced through a system to determine the positions of all of the cardinal points and the image size and location?

3.11 Define principal points and nodal points.

3.12 Write down the lens maker's formula for thick lenses.

3.13 Explain what is meant by optical invariant.

3.14 A thick lens is made from F1 glass. The radius of the first surface is 5.000 cm and the radius of the second surface is −5.000 cm. The lens is 1.500 cm thick. Use the *y-nu* ray trace to find the size and location of the image produced by this lens if the object is 20.00 cm in front of the lens and 3.00 cm tall. Describe the nature of the image.

3.15 Find the size and location of the image produced by the lens described in Problem 3.14 if the same object is moved to just 2.00 cm from the lens. Describe the nature of the image.

3.16 A thick lens is made from F1 glass. The radius of the first surface is 5.000 cm. The second surface is planar. The lens is 0.750 cm thick. If the second surface is silvered to form a mirror, find the size and location of the image produced by this lens for an object 20.00 cm from the lens and 3.00 cm tall. Describe the nature of the image.

3.17 Find the size and location of the image produced by the lens described in Problem 3.16 if the same object located 2.00 cm from the lens. Describe the nature of the image.

3.18 Consider an equiconvex thin lens made from SF10 glass. If the radius of curvature of the front surface is 87.39 cm, what is the focal length using the thin lens formula? If a 4.00-cm-high object is located 90.00 cm in front of this lens, what is the image size and position? Describe the nature of the image.

3.19 Again consider an equiconvex lens made from SF10 glass with radius of curvature of 87.39 cm but now the lens is 0.500 cm thick. If a 4.00-cm-high object is located 90.00 cm in front of this lens, what is the image size and position? Describe the nature of the image.

3.20 Consider the lens described in Problem 3.19 one more time but now with a thickness of 5.000 cm. If a 4.00-cm-high object is located 90.00 cm in front of this lens, what is the image size and position? Describe the nature of the image.

3.21 Find the locations of the principal planes for the system described in Problem 3.19. Find the front and back focal lengths and the effective focal length of the system.

3.22 The lens from Problem 3.19. is used as part of the side of an aquarium. Find the nodal points for this system. What is the image size and position for a 1.00-cm-high guppy swimming 10.0 cm from the lens? Take the index of refraction of water to be 1.333.

3.23 Use only one ray trace and the optical invariant to find the size and position of the image for the optical system given in Problem 3.19.

3.24 Trace two rays through the thin lens system given below to find the image size and position. Object height = 2.0.

Surface	f (cm)	t (cm)
0		10.00
1	4.00	2.00
2	-6.00	1.50
3	10.00	

3.25 Trace two rays through the system given below to find the image size and position. Object height = 2.0.

Surface	r (cm)	t (cm)	n
0		30.00	1.000
1	6.050	0.270	1.720
2	3.182	0.270	1.000
3	3.401	0.750	1.620
4	-31.536		1.000

(Only for the brave: Compare your *y-nu* ray trace results with repeated applications of the paraxial equations for a single refracting surface.)

3.26 The lens from Problem 3.25 is used as part of an underwater camera system. How far from the film must the lens be for an object 10.0 meters from the lens in air? What is the image position when the object is 10.0 meters away underwater? Use 1.333 as the index of refraction of water.

3.27 Calculate the value of the optical invariant Λ at each surface including the object and image for the system given in Problem 3.25. Use the rays traced previously for Problem 3.25.

3.28 Trace two rays through the system given below to find the image size and position. Object size = 1.00.

Surface	r (cm)	t (cm)	n
0		16.00	1.000
1	7.660	0.6716	1.506
2	-3.144	0.2048	1.591
3	-7.343	0.000	1.000
4	2.047	0.0845	1.506
5	-4.659	0.2884	1.591
6	4.508		1.000

3.29 Find the locations of the principal planes for the system described in Problem 3.28. Find the front and back focal lengths and the effective focal length of the system.

3.30 Find the locations of the principal planes for the system described in Problem 3.28. Find the front and back focal lengths and the effective focal length of the system.

3.31 In a laboratory experiment (Figure 3.16), the back focal length of a lens system is determined to be +4.0 cm. The front focal length is -12.0 cm. An object is placed +16.0 cm from the lens and its image is observed at +20.0 cm. See Figure 3.16 below. (a) What is the effective focal length of this lens

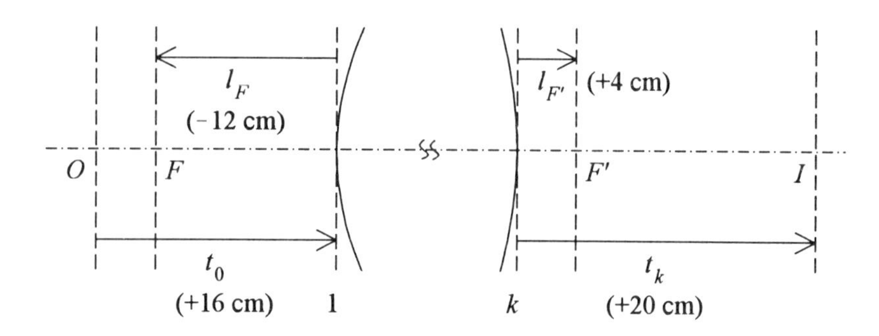

Figure 3.16 Problem 3.31.

system? (b) Where are the principal planes located relative to the first and last surfaces respectively? (c) What is the system magnification?

3.32 A test lens system (Figure 3.17) is mounted 20 cm from a collimated source. An image is formed 4.0 cm from the last surface of the system. Then the lens is reversed so that the first surface is now the last surface. Also, the collimated source is changed to a point source located 12.0 cm from the new first surface of the test system. Under these conditions, the image is located 24.0 cm from the new last surface and is magnified by a factor of −1.5. (a) What is the

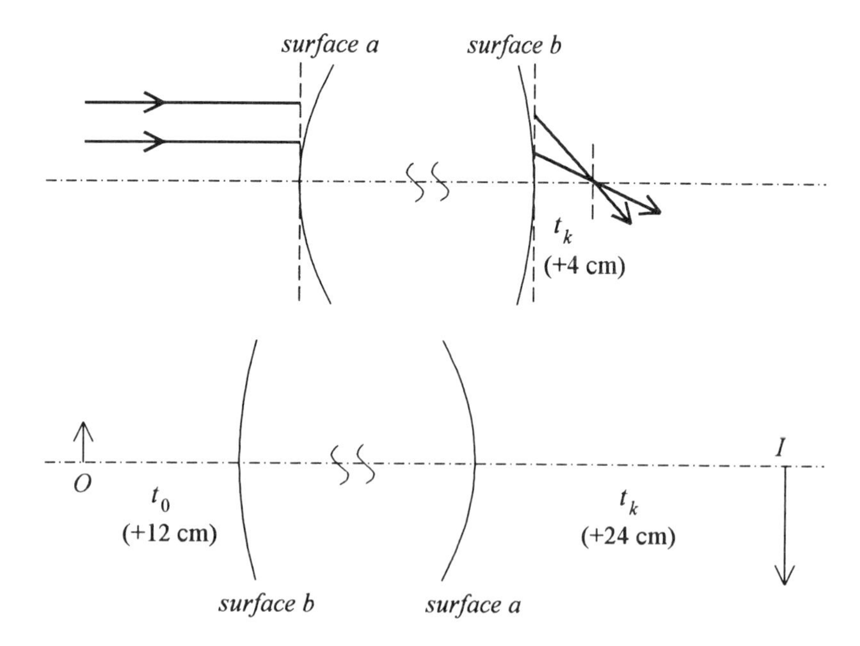

Figure 3.17 Problem 3.32.

effective focal length of this lens system? (b) Where are the principal planes located relative to the new first and last surfaces?

3.33 For the thick lens system given below, trace three rays which are initially parallel to the optical axis. The ray heights at the first surface should be +2.00 cm, +1.00 cm, and −0.50 cm. Show that all three rays give the same back focal length, effective focal length and second principal plane position.

Surface	c (cm^{-1})	t (cm)	n
1	0.01653	2.70	1.720
2	0.03143	2.70	1.000
3	0.02940	7.50	1.620
4	−0.003171		1.000

3.34 For the thick lens system given below, find the principal planes and effective focal length with two *y-nu* ray traces. Then determine the image size and position using the thin lens equations. Check your work by doing two more *y-nu* ray traces to determine the image size and position. Object size = 1.5. Make a scale drawing of this system.

Surface	c (cm^{-1})	t (cm)	n
0		5.000	1.000
1	1.5949	0.111	1.744
2	0.0000	0.074	1.649
3	0.9881	0.278	1.000
4	−0.9881	0.074	1.649
5	0.0000	0.111	1.744
6	−1.5949		

3.35 For the thick lens system given below, find the principal planes and effective focal length with two *y-nu* ray traces. Compare the overall length of this optical system (from the first surface to the second focal plane) with the effective focal length. A system whose effective focal length is greater than its overall length is a telephoto lens. Is this lens a telephoto lens or an inverted telephoto?

Surface	c (cm^{-1})	t (cm)	n
1	2.8752	0.0756	1.623
2	−2.3804	0.0252	1.699
3	0.2012	0.0806	1.000
4	4.2662	0.0867	1.547
5	−1.5871	0.0302	1.613
6	6.9589		1.000

3.36 For the thick lens system given below, find the principal planes and effective focal length with two *y-nu* ray traces. Compare the overall length of this optical system (from the first surface to the second focal plane) with the effective focal length. A system whose effective focal length is greater than its overall length is a telephoto lens. Is this lens a telephoto lens or an inverted telephoto?

Surface	c (cm^{-1})	t (cm)	n
1	1.2210	0.071	1.620
2	3.1646	0.106	1.000
3	2.0747	0.114	1.697
4	-0.5794	0.143	1.000
5	-0.2437	0.049	1.617
6	2.1692	0.097	1.000
7	0.2760	0.046	1.617
8	2.0450	0.194	1.611
9	-2.1786		1.000

3.37 It is not easy to generalize a spread sheet to handle an arbitrary number of surfaces, but you can create spread sheets for specific numbers. Set up a spread sheet to find image size and position automatically for any optical system with two surfaces. Set up a separate spread sheet to handle systems with three surfaces.

3.38 Show that an alternate way of writing the optical invariant is

$$\Lambda = \alpha_{j-1}\bar{y}_j - \bar{\alpha}_{j-1}y_j \,. \tag{3.84}$$

3.39 Show that an alternate way of writing the optical invariant is

$$\Lambda = n_{j-1}\,u_{j-1}\,\bar{i}_j - n_{j-1}\,\bar{u}_{j-1}\,i_j \,. \tag{3.85}$$

3.40 Using the optical invariant show that

$$(u_j - u_{j-1})\,\bar{i}_j = (\bar{u}_j - \bar{u}_{j-1})\,i_j \,. \tag{3.86}$$

CHAPTER 4

PARAXIAL OPTICS II

In the previous chapter, we used the *y-nu* ray trace as a tool to determine image size and position as well as the cardinal points of a lens system. We will continue the study of the first-order properties of optical systems in this chapter.

Because of the finite lateral extent of any real optical system, not all of the rays which originate at an object point can actually pass through the optical system to form the image. In the previous chapters, this important consideration was ignored. Which rays actually make it through the system is important to determine for two reasons. First, the fraction of the light emitted by the object which gets through to form the image determines the brightness of the image. Second, in real systems not all rays can be treated as paraxial rays. Nonparaxial rays may not meet at the paraxial image point. Therefore, the quality of the image formed depends strongly on which rays are used. If significant numbers of nonparaxial rays are present, the image quality may be greatly reduced.

Image quality can also be reduced by chromatic aberrations. In general, aberrations are caused by differences between where paraxial rays are supposed to go and where actual rays really go. Remember that paraxial rays merely approximate real rays at small slope angles near the optical axis. Chromatic aberration depends on the changes in the index of refraction with wavelength (dispersion). Since paraxial properties of an optical system, such as focal length, depend on the index of refraction, chromatic aberration arises even for strictly paraxial optics. For this reason, chromatic aberration is discussed at the end of this chapter. The other forms of aberration will be discussed after exact ray tracing has been introduced.

4.1 LENS THICKNESS

Before we can begin discussing the optical effects of the lateral extent of a lens system, we need to determine the magnitude of the lateral extent. The lateral extent of a lens is the distance from the optical axis to the edge of the lens or the edge of the mount which holds the lens, whichever is smaller. Each surface can have its own lateral extent depending on the shape of the lens. This distance is called the *clear radius* of the surface R. A subscript will be used to denote to which surface the clear radius is related. The clear radius is the maximum distance from the optical axis that a ray can pass through the surface. Figure 4.1 shows that R_1 is defined by the lens holder, R_2 is defined by the edge of the second lens, and R_3 is again defined by the lens holder. The lens needs to be a little larger its holder, so that the holder has an edge to grasp. Making the lens any larger, however, merely wastes glass. In practice, therefore, there is very little difference in the location of the edge of the mount and the edge of the lens. In this text, I will not distinguish between the edge of the lens and the edge of the mounting surface. I will call both the lens rim.

The clear radius of a lens is related to the thickness of the lens. The absolute maximum clear radius is equal to the radius of curvature of the surface r. In this case, the lens is a sphere or hemisphere. Usually, lenses are considerably thinner and the clear radii are correspondingly less.

The thickness of a positive lens consists of three possible parts (Figure 4.2). The two surface curvatures create sags z_1 and z_2 on either side. Also, there needs to be an edge thickness t_e. The edge thickness prevents the lens from having a knife edge. Not only are knife edges difficult and dangerous to handle, but they chip easily. For a standard size lens (on the order of 2.5 cm clear radius), the edge thickness should be at least 1 mm. For very large lenses, lens thickness must be increased for additional lens strength. Increasing the edge thicknesses without good reason wastes glass. The sags can be determined by using the Pythagorean theorem

$$r^2 = (r - z)^2 + R^2 . \tag{4.1}$$

By rearranging this equation, the sag can be determined as

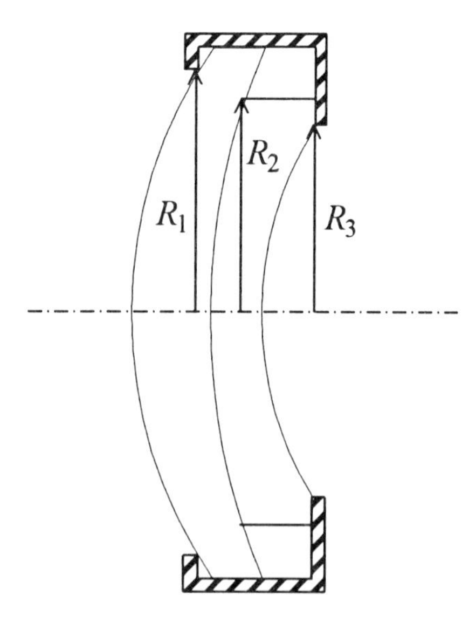

Figure 4.1 Definition of clear radius.

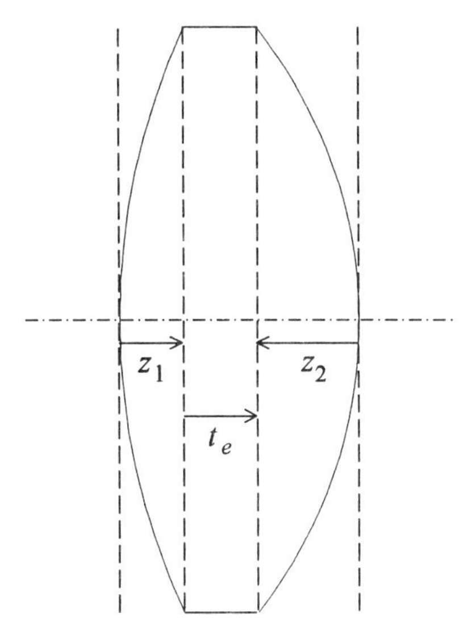

Figure 4.2 Determination of lens thickness.

$$z = r \pm \sqrt{(r^2 - R^2)} \ . \tag{4.2}$$

Note that the sag is a signed quantity. It is positive when measured from the surface toward the right just like the radius of curvature. The sign on the square root is determined by which sign gives the smaller absolute value of the sag. When the radius is positive, use the negative sign on the square root. When the radius is negative, use the positive sign. The sign choice can be taken care of by rewriting Eq. 4.2 as

$$z = r - r\sqrt{1 - \frac{R^2}{r^2}} \ . \tag{4.3}$$

As shown in Figure 4.2, the overall lens thickness is given by

$$t_1 = z_1 + t_e - z_2 \ . \tag{4.4}$$

Since negative lenses are thicker at their edges than at the their centers, edge thickness is not a concern. As a rule of thumb, the axial thickness of a negative lens should be between 12% and 20% of the clear radius of the lens. The larger percentage is used for larger lenses.

Example 4.1

Determine the minimum thickness of a biconvex lens with clear radii 2.00 cm. The surface radii are 10.00 cm and −8.00 cm, respectively.

$$z_1 = (10.00) - (10.00)\sqrt{1 - \frac{(2.00)^2}{(10.00)^2}} = 0.20 \text{ cm} ,$$

$$z_2 = (-8.00) - (-8.00)\sqrt{1 - \frac{(2.00)^2}{(-8.00)^2}} = -0.25 \text{ cm} .$$

Combining these results gives

$$t_1 = (0.20) + (0.10) - (-0.25) = 0.55 \text{ cm} .$$

■

In addition to lens rims, rays can be stopped by specially designed plates called *stops*. A stop is merely a thin plate with a hole in it. In the *y-nu* ray trace, a stop is modeled by inserting a surface with zero power ($c = 0$, and $n' = n$). The stop does nothing to the propagation of the ray. But the size of the hole as given by the transverse radius R or diameter D of the stop determines which rays pass and which do not. If the absolute value of the ray height at the stop is greater than the radius R, then the ray is stopped.

Why would anyone put a stop in an optical system since it does not contribute at all to the power of the system? Stops are used to control which rays make it through the system. As mentioned earlier, aberrations can be controlled to some extent by stop size and placement within an optical system.

4.2 THE APERTURE STOP

If the source of rays passing through a system is on the optical axis, then because we are dealing with axially symmetric systems, one particular lens rim or stop will be responsible for stopping rays from passing through to the image. This stop or lens rim is known as the *aperture stop*. You can see the aperture stop by looking into a complicated lens system from some point on the optical axis of the system. You may see several circles which are the images of lens rims or stops in the system. The smallest of these circles is the aperture stop. You cannot see lens rims or the edges of stops which lie beyond the aperture stop. It is important to note that the aperture stop may not be the lens rim or stop which limits rays from an off-axis object point.

In addition, the aperture stop is defined for a particular object point. As shown in Figure 4.3, a different object point may have a different aperture stop.

Part of the importance of the aperture stop is that it determines how much light from the object actually gets through to form the image. The larger the slope angle of the ray which just makes it through, the larger the fraction of the emitted light that gets through. A small aperture stop can also be used to restrict the slope angles so that the paraxial approximation is more nearly satisfied. This could result in improved image quality.

The aperture stop of a system can be found in several different ways. One method will be described in the next section, and two other methods will be described here.

The aperture stop could be found by tracing a set of rays from the object point with gradually increasing slope angles. Eventually, the slope angle will become large enough that the ray will be stopped by some surface in the system. That surface is the aperture stop. The ray with slightly smaller initial slope angle which just barely makes it through the system is called a *marginal ray*. This method requires a lot of work and is not recommended, although it most clearly shows what is meant by the aperture stop. Another problem is that unless the increment in slope angle is quite small, a surface may be marked as the aperture stop when actually a different surface gives a slightly smaller maximum slope angle.

A better way to determine which surface is the aperture stop of a system is to trace an arbitrary ray from the on-axis object point through the system. This is the same ray that was used in the previous chapter to find the image position. Because all of the ray tracing equations are linear, this ray can be scaled. For example, if the initial slope angle is multiplied by 2, then all subsequent slope angles and ray heights will be twice as large. Let A be the scale factor required to generate a marginal ray from the initial ray. The scale factor can be determined by examining the ratios of the stop or lens rim size to the ray height at each surface. The smallest ratio is produced by the surface which is the aperture stop. Thus,

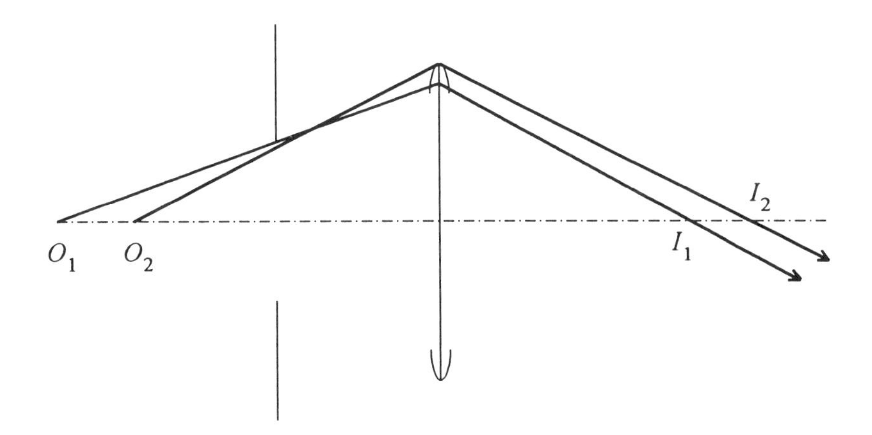

Figure 4.3 Dependence of the aperture stop on object position.

$$A = \min\left|\frac{R_j}{y_j}\right|, \tag{4.5}$$

where every surface j in the system is tested until the minimum is found.

Example 4.2

Determine which surface is the aperture stop for a system consisting of two thin lenses separated by a stop. The system data are h = 12.00, t_0 = 24.00, f'_1 = 13.33, R_1 = 3.50, t_1 = 5.00, R_2 = 2.50, t_2 = 5.00, f'_3 = 30.00, and R_3 = 2.50.

Set up a *y-nu* ray trace table as shown in Table 4.1. The clear radius at each surface is shown in parentheses on the same line with the reduced thicknesses, but in the surface column rather than the space column. Trace an arbitrary axial ray. The ratios of clear radii to ray height are also shown in parentheses on the same line with the reduced slope angles or the slope angles, but in the surface column rather than the space column.

The smallest ratio occurs at the second surface, therefore, surface 2 is the aperture stop. ■

4.3 ENTRANCE AND EXIT PUPILS

Quite frequently the aperture stop is buried deep within the optical system as in the example above. If two optical systems are to be used together efficiently, it is necessary to match the aperture stops of both systems optically, even though the aperture stop is not conveniently located. An example would be a telescope or other instrument used with the human eye. In this case, the pupil of the eye acts as the aperture stop of the eye. If it is not matched optically with the aperture stop of the instrument, then either too much light will be collected by the instrument and wasted on the eye, or not enough light will be collected. In the first case, image quality may be unnecessarily reduced and in the second case, the image brightness won't be as great as possible.

Table 4.1 Ray trace table for Example 4.2 showing an arbitrary axial ray.

	0		1		2		3		4
$-\phi$			-0.075		0.0		-0.033		
$(R)\ \tau$	(12)	24.0	(3.5)	5.0	(2.5)	5.0	(2.5)	12.0	(9.0)
y	0		2.4		2.0		1.6		0.0
α		0.1		-0.08		-0.08		-0.133	
u	(∞)	0.1	(1.46)	-0.08	(1.25)	-0.08	(1.56)	-0.133	

The problem of matching apertures can be overcome by remembering that an image is optically equivalent to the object which created it. If the aperture stop is used as an object for part of the optical system, then the image of the aperture stop can be made accessible. The image of the aperture stop projected into image space is called the *exit pupil*. It is designated by the symbol E'. The image of the aperture stop in object space is the *entrance pupil*, denoted by E.

It may be a little disconcerting to think of the aperture stop as being an object, and then finding images of this object through only part of the optical system. But remember that the stops and lens rims of an optical system are visible and that what the viewer is actually seeing are merely images.

For the thin lens system of Example 4.2, one way to find the pupils is to use the thin lens equations from Chapter 2. Some caution is required, however. The aperture is a real object. When finding the entrance pupil, the light travels from right to left rather than left to right as is the usual case. Therefore, to find the entrance pupil, let $s = -t_1$. Then the thin lens equations gives

$$\frac{1}{l_E} = \frac{1}{f_1} - \frac{1}{t_1}, \tag{4.6}$$

where the location of the entrance pupil measured from the first surface of the system is l_E. The size of the entrance pupil can be determined from the size of the aperture stop and the standard formula for magnification. In a similar way the location of the exit pupil can be found, but $s = -t_2$.

For more complicated systems or for thick lens systems, a better way to find the pupils, is to use the *y-nu* ray trace. An arbitrary ray which passes through the center of the aperture stop is traced forward into image space and backward into object space. Where this ray crosses the optical axis locates the positions of the exit and entrance pupils, respectively. This ray is called a *chief ray*, and it will be denoted as the second ray by putting a bar over the ray slopes and ray heights. The size of the pupils can be determined from the size of the aperture stop and the appropriate magnification. The magnification is found using the ratio of the reduced slope angles as introduced in the last chapter.

Example 4.3

Find the location and size of the pupils for the system given in Example 4.2.

First, perform a chief ray trace beginning at surface 2 (the aperture stop) and ending in both object and image space, as shown in Table 4.2.

From the ray trace data we can find the pupil sizes and positions in the same way that image size and location is determined from axial ray data:

$$l_E = -\frac{\bar{y}_1}{\bar{u}_0} = -\frac{(-0.5)}{(0.0625)} = 8.0 ,$$

Table 4.2 Chief ray trace for Example 4.3.

	0		1		2		3		4
y	-2.0		-0.5		0.0		0.5		1.5
α		0.0625		0.1		0.1		0.0833	
u	(6.0)		(7.0)		(∞)		(5.0)		(6.0)

$$R_E = R_A \frac{n_A \bar{u}_A}{n_0 \bar{u}_0} = (2.5)\frac{(1.0)(0.1)}{(1.0)(0.0625)} = 4.0 \; ,$$

$$l_{E'} = -\frac{\bar{y}_k}{\bar{u}_k} = -\frac{(0.5)}{(0.0833)} = -6.0 \; ,$$

$$R_{E'} = R_A \frac{n_A \bar{u}_A}{n_k \bar{u}_k} = (2.5)\frac{(1.0)(0.1)}{(1.0)(0.0833)} = 3.0 \; .$$

■

This discussion suggests a third method for finding the aperture stop. By definition, the entrance pupil is the image of the aperture stop in object space, and the aperture stop limits the rays which pass through the system. If the stop or lens rim which forms the aperture stop is unknown, then all of the stops and lens rims of the system can be imaged through whatever elements precede them back into object space. Each image formed then is potentially the entrance pupil. The real entrance pupil will be the image which limits the rays coming from the object. It will have the smallest slope angle for the ray which just passes through the image.

Example 4.4

Determine which surface of the system given in Example 4.2 is the aperture stop by first determining which surface images into the entrance pupil. (I know we already know that surface 2 is the aperture stop, but let's pretend we don't know this.)

In Example 4.3 we traced a ray from the second surface into object space to find the image of the stop. Now trace a ray from the third surface into image space, as shown in Table 4.3. From this ray trace, the size and location of the image of the third surface may be determined:

$$l_3 = -\frac{(1.0)}{(0.025)} = 40.0 \text{ cm} \; ,$$

Table 4.3 Third surface ray trace for Example 4.3.

	0		1		2		3		4
y	-1.6		-1.0		-0.5		0.0		1.2
α		0.025		0.1		0.1		0.1	
τ		0.025		0.1		0.1		0.1	

$$R_3' = (2.5)\frac{(1.0)(0.1)}{(1.0)(0.025)} = 10.0 \text{ cm} .$$

Finally, a scale drawing may me made as shown in Figure 4.4. The entrance pupil is determined by laying a straight edge on the object point on-axis and drawing a line to the edge of one of the images such that the line makes the smallest angle with the optical axis.

A mathematically equivalent procedure which does not require a scale drawing is to look at the ratios of the image sizes to the distances between the object and the surface image. The smallest ratio defines the image pupil:

$$\text{first surface } \angle = \frac{R_1}{t_1} = \frac{(3.5)}{(24.0)} = 0.146 \text{ rad} ,$$

$$\text{second surface } \angle = \frac{R_2'}{t_1 + l_2} = \frac{(4.0)}{(24.0) + (8.0)} = 0.125 \text{ rad} ,$$

$$\text{third surface } \angle = \frac{R_3'}{t_1 + l_3} = \frac{(10.0)}{(24.0) + (40.0)} = 0.156 \text{ rad} .$$

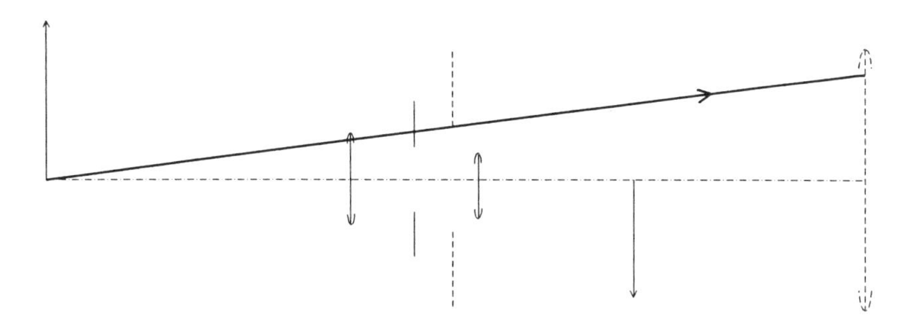

Figure 4.4 Scale drawing used to determine entrance pupil.

From the calculations shown above, we once again see that the second surface is the aperture stop. ■

Once the entrance and exit pupils of a system are determined, combining two optical systems efficiently only requires that the exit pupil of the first system is the same size and at the same location as the entrance pupil of the second system. When the pupils are matched, any ray which passes through one system will also pass through the other system.

We are now in a position to introduce an important characteristic of an optical system of which you may have heard. The quantity known as the *f*-number (written as $f/\#$) determines the aperture size relative to the focal length or overall size of the system. The $f/\#$ of an optical system is defined as the ratio of the effective focal length to the diameter of the entrance pupil for an object at infinity:

$$f/\# = \frac{f'}{D_E} = \frac{f'}{2R_E} . \tag{4.7}$$

The smaller the $f/\#$, the more light that can be collected by the system. If this is a camera lens, more light would translate into a shorter required exposure time. Thus, a lens with small $f/\#$ is said to be a "fast" lens.

4.4 THE FIELD STOP

If the chief ray described above has too large a slope angle at the aperture stop, it may not make it through the system. The stop or lens rim which limits the maximum slope angle of the chief ray is called the *field stop.*

Once the aperture stop has been determined, the field stop can be determined by tracing an arbitrary chief ray forward and backward through the system. The ratio of the stop or lens rim size to the ray height at each surface indicates which surface would limit the ray if it were scaled by a factor

$$B = \min\left|\frac{R_j}{\bar{y}_j}\right| . \tag{4.8}$$

Note that, unlike the aperture stop, both the object and image planes must be included in determining the minimum. The chief ray which just barely passes through the system will be called the *full-field chief ray*. This is in reference to the field of view of the system. The field stop tends to limit the field of view of the system. If the field stop is at either the object plane or image plane, then the field stop sharply cuts off how much of the object plane can be viewed. If the field stop is located elsewhere, then there will not be a sharp cutoff, but there will be a rapid decrease in the brightness of the image.

It is very important to note that because of the definitions of aperture stop and field stop, these two stops can never be located in the same place. They are absolutely complementary to each other.

The marginal ray which determines the aperture stop and the full-field chief ray which determines the field stop can be used as the two rays to calculate the optical invariant

$$\Lambda = \alpha_j \bar{y}_j - \bar{\alpha}_j y_j . \tag{4.9}$$

Because of the importance of the aperture stop and field stop for image quality and brightness, we will use the marginal axial ray and the full field chief ray when calculating optical invariants from now on. This choice of rays for calculating the optical invariant is a standard convention.

Just as the aperture stop of a system might be deep within the system, so might the field stop. Images of the field stop in object space and image space can be found just as with the entrance and exit pupils. The images of the field stop will help determine how much of object space can be seen through the system in image space.

The image of the field stop in object space is called the *entrance window* W. The image of the field stop in image space is the *exit window* W'. Again, since the images are optically equivalent to the object, if the entrance window lies in the object plane, then the field of view of the system is sharply cut off by the field stop. If this is the situation, then the exit window must lie in the image plane.

Since the aperture stop and the field stop cannot lie in the same plane, neither can the windows lie in the same plane as the pupils.

Example 4.5

Determine which surface is the field stop in the system given in Example 4.2. Also, find the location and size of the entrance and exit windows.

Back in Example 4.3, we traced an arbitrary chief ray. Looking at the ratios of the ray height to the clear radius, we see that surface 3 is the field stop. In Example 4.4, we traced a ray starting at surface 3 into object and image space. Using this ray trace we find the size and location of the entrance window:

$$l_W = -\frac{(-1.0)}{(0.025)} = 4.0 \text{ cm} ,$$

$$R_W = (2.5)\frac{(1.0)(0.1)}{(1.0)(0.025)} = 10.0 \text{ cm} .$$

Since the field stop is at the third surface which is already in image space, the exit window is also at the third surface. ■

4.5 VIGNETTING

As stated earlier, the field stop tends to limit tends to limit the field of view of the system. The field stop will cut off the chief rays if the object point moves far enough off axis. But the chief rays alone do not control the field of view. Even with the chief ray cutoff, some other rays may make it through the system and extend the field of view beyond the entrance window. Only if the entrance window is at the object will all rays be cut off at the same off-axis position. There is an additional complication. When viewed from an off-axis object position, the apparent stops and lens rims do not appear to be centered on a common axis any more. Therefore, surfaces other than the aperture stop or field stop may stop some rays. The effect of some rays being cut off while others pass through causes a gradual dimming of the image as distance from the optical axis increases. The effect is called *vignetting* after a type of photograph which gradually shades off at the edges.

Vignetting is a complicated phenomena, but a vignetting diagram can be drawn to clarify what is happening within the optical system. The intention of a vignetting diagram is to show the relative sizes and positions of all of the lens rims and stops in a system as seen from either the object or image. A vignetting diagram will show which lens rims or stops limit the rays passing through the system. In addition to the vignetting diagram, a vignetting factor may be calculated. The vignetting factor V is the ratio of the unvignetted area shown on the vignetting diagram to the area of the entrance or exit pupil. If $V = 1$, then there is no vignetting and just as much light passes through the system for an off-axis object point as for an on-axis object point. If $V = 0$, then no light makes it through the system at all. Just as with the determination of the aperture stop, vignetting is object-dependent. In this section, drawing a vignetting diagram will be discussed first, then calculating the vignetting factor.

The first step in drawing a vignetting diagram is to image all of the stops and lens rims into object space. Image space could be used equally well. As we have already seen, when the stops and lens rims are imaged into object space, we can treat the images equally. A ray which is drawn from an object point toward the edge of one of the images is guaranteed to actually pass through the edge of the corresponding physical stop or lens rim. As seen from an off-axis object point, the images will appear to be circles of various sizes whose centers are displaced by various amounts. The apparent displacements of the images can be determined by drawing a straight line from the off-axis object point to the center of each image. The apparent radii can be determined by drawing a straight line through the edge of each image. These lines are extended to the entrance pupil, thus projecting the images onto a common plane.

You can see the importance of projecting the images by looking through two rings of the same size. Hold one ring close to your eye and the other at arm's length. The further ring looks smaller and would cut off more rays if it were a lens rim or stop. If you look at the two rings from a location off of the axis through their centers, the centers seem to be displaced. As you move your eye upwards, the further ring appears to move downwards. When the two rings appear to overlap, some rays would be cut off by one ring and some rays would be cut off by the other ring. This is vignetting.

The next step in drawing a vignetting diagram is to measure the center displacements and the projected radii. The final step is to draw the diagram using the measurements. Because any ray outside a stop or lens rim is stopped at that surface, only rays that pass inside all of the images of the stops and lens rims make it through the system to the image. The unvignetted area is the area inside all of the circles on the vignetting diagram.

Measuring the displacements and radii is not very accurate, so let's take a minute to derive some formulas. Consider a system, as shown in Figure 4.5, that has an entrance pupil behind of the first surface a distance l_E. The image of a general surface j is located a distance l_j behind the first surface. By similar triangles, we can find the displacement of the center of the image of surface j on the entrance pupil as

$$\Delta_j = h\left(\frac{l_j - l_E}{t_0 + l_j}\right) . \tag{4.10}$$

The projected radius can also be found by using a different pair of similar triangles:

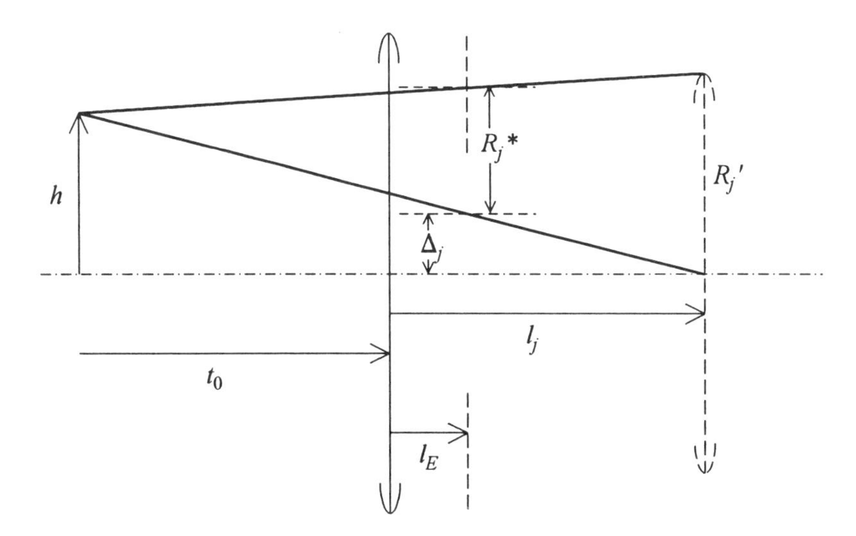

Figure 4.5 Projection of the image of a general surface j onto the entrance pupil.

$$R_j^* = R_j'\left(\frac{t_0 + l_E}{t_0 + l_j}\right) . \tag{4.11}$$

Example 4.6

Draw a vignetting diagram for the system given in Example 4.2, but change the object height to 8.0 cm instead of 12.0 cm.

When the object height is 12.0 for this system, the vignetting factor turns out to be zero. The object height is changed to get a nonzero value for V. This change will also change the field stop to the object, but as we shall see, there is still plenty of vignetting taking place.

As we have so throughly established, surface 2 is the aperture stop. Surface 1 and the image of surface 3 must be projected onto the entrance pupil, which is the image of surface 2. The projections are shown in Figure 4.6. Using the equations just derived, we obtain

$$\Delta_1 = (8.0)\left(\frac{(0.0) - (8.0)}{(24.0) + (0.0)}\right) = -2.67 ,$$

$$R_1^* = (3.5)\left(\frac{(24.0) + (8.0)}{(24.0) + (0.0)}\right) = 4.67 .$$

Similarly, for surface 3 we get

$$\Delta_3 = (8.0)\left(\frac{(40.0) - (8.0)}{(24.0) + (40.0)}\right) = 4.0 ,$$

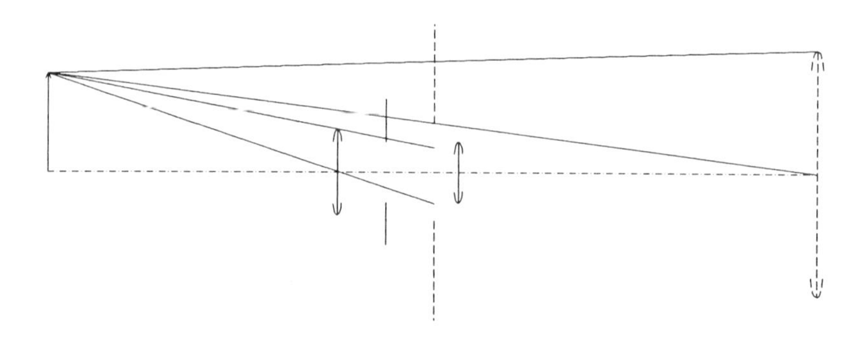

Figure 4.6 Projection of the images of the stops and lens rims onto the entrance pupil.

$$R_3^* = (10.0)\left(\frac{(24.0) + (8.0)}{(24.0) + (40.0)}\right) = 5.0 \ .$$

The vignetting diagram for this system is shown Figure 4.7. Notice that the entrance pupil is no longer limiting rays. Some rays are stopped by surfaces 1 and 3 before getting close to the edge of surface 2. ■

The final step in vignetting analysis is to determine the unvignetted area so that the reduction in image brightness can be calculated. Figure 4.8 shows how the unvignetted area is bounded by two circle segments. One segment has a radius of R_1^* and includes an angle of $2\theta_1$. The other segment has a radius R_3^* and includes an angle of $2\theta_3$. The area of any segment of a circle is given by the formula

$$A = \frac{1}{2}R^2(\theta - \sin\theta) \ , \tag{4.12}$$

where θ is the total angle included.

The problem now becomes to find formulae for the included angles θ_1 and θ_3. These angles are defined by the intersection points of two circles. To simplify the derivation we will change the location of the origin to the center of the lower of the two circles. We will also drop the asterisk from the symbol for projected radius. The equation for circle 1 becomes

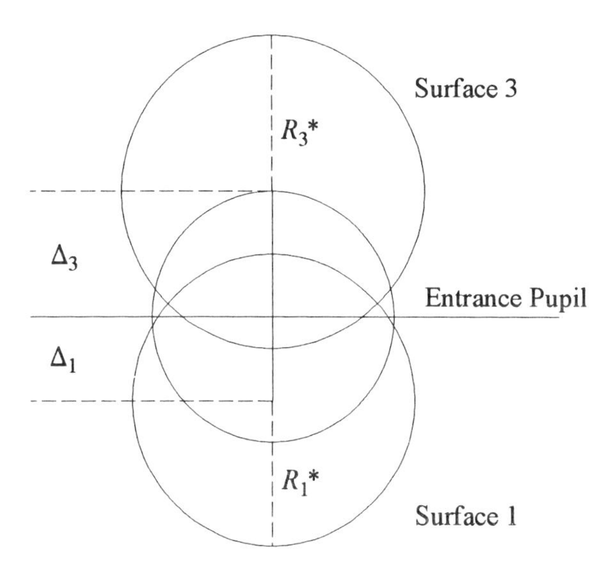

Figure 4.7 Vignetting diagram for Example 4.6.

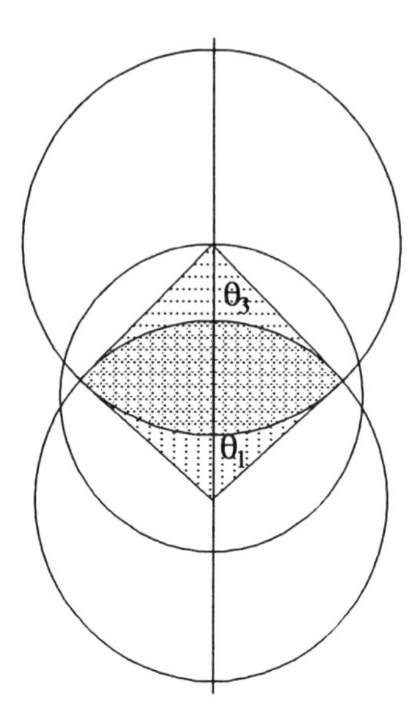

Figure 4.8 Overlapping circular segments form the unvignetted area.

$$x^2 + y^2 = R_1^2 , \tag{4.13}$$

and for circle 2 we have

$$x^2 + (y - d)^2 = R_3^2 , \tag{4.14}$$

where $d = \Delta_3 - \Delta_1$. If we eliminate x from these two equations, we can solve for the y coordinate:

$$y = \frac{R_3^2 - R_1^2 + d^2}{2d} . \tag{4.15}$$

Finally, we can see that

$$\cos\theta_1 = \frac{y}{R_1} \tag{4.16}$$

and

$$\cos\theta_3 = \frac{d - y}{R_3} . \tag{4.17}$$

Assuming that the angles are found in radians when taking the arccosines, we can now find the unvignetted area:

$$A = R_1^2 \theta_1 + R_3^2 \theta_3 - R_1 R_3 \sin(\theta_1 + \theta_3) . \tag{4.18}$$

The area calculated in the equation above should be compared with the area of the entrance pupil. This ratio indicates by how much the intensity is reduced due solely to vignetting.

Note that throughout this chapter we have been treating pupils as paraxial entities. Since the pupils are actual images of the aperture stop, they may be subject to aberrations just like any other image. Pupil aberrations are beyond the scope of this text, but they may change the vignetting factor significantly in some systems.

Example 4.7

Calculate the vignetting factor for the system described in Example 4.6.

From the previous example we know the radii and displacements of the circles in the plane of the entrance pupil. We only need to "plug and chug":

$$y = \frac{(5.0)^2 - (4.67)^2 + (6.67)^2}{2(6.67)(4.67)} = 3.575 ,$$

$$\theta_3 = \arccos\left(\frac{(6.67) - (3.575)}{(5.0)}\right) = 0.904 ,$$

$$\theta_1 = \arccos\left(\frac{(3.575)}{(4.67)}\right) = 0.698 .$$

And now the area is given by

$$A = (4.67)^2(0.698) + (5.0)^2(0.904)$$

$$- (4.67)(5.0)\sin(0.698 + 0.904) = 14.5 ,$$

and the vignetting factor is

$$V = \frac{A}{\pi R_E^2} = \frac{(14.5)}{\pi(4.0)^2} = 0.29 .$$

■

Calculating vignetting factors analytically can be done fairly easily for two surfaces as shown above. But what if there are several surfaces involved? An alternative approach is to use a computer to do a numerical integration. The first step in such an approach is to determine which projected radius is the smallest. This will set the limits on the integration:

$$0 \le x \le \min R_j \, . \tag{4.19}$$

Next a step size is established:

$$\delta x = \frac{\min R_j}{N} \tag{4.20}$$

where N is some fairly large number (say 50). Figure 4.9 shows a fairly typical vignetting diagram with three different circles forming the boundary of the unvignetted area. As you move out along the x axis, which circle forms the boundary changes. Therefore, for each step, you need to calculate the y coordinate for each circle. The upper y coordinate is given by

$$y_u = \Delta_j + \sqrt{R_j^2 - x^2} \tag{4.21}$$

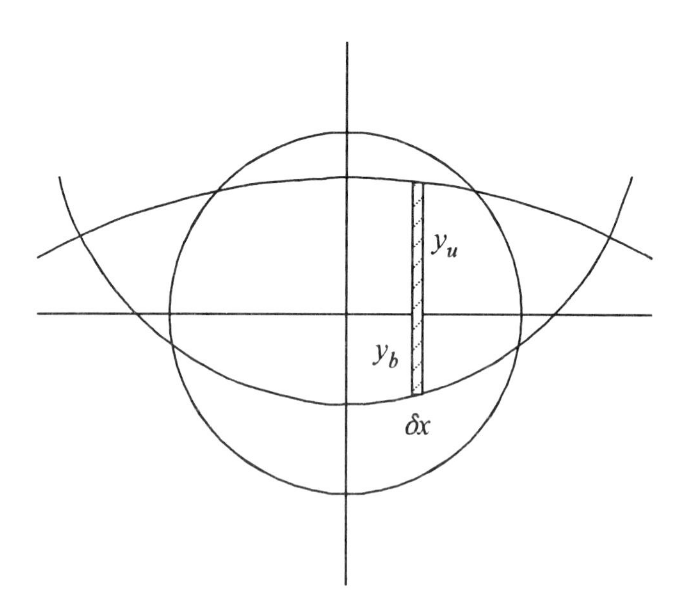

Figure 4.9 A vignetting diagram showing how to calculate the unvignetted area using numerical integration.

and the bottom y coordinate is

$$y_b = \Delta_j - \sqrt{R_j^2 - x^2} \; . \tag{4.22}$$

At each step, the minimum absolute value for y_u and y_b is determined. The unvignetted area is the sum of $(y_u - y_b) \times \delta x$ for each x.

Because the paraxial ray trace equations are all linear equations, there is a quick and easy way to check to see if there is no vignetting in a particular system without going to all of the trouble of drawing a vignetting diagram. The linear nature of paraxial optics means that any arbitrary paraxial ray may be thought of as a linear combination of the marginal axial ray and the full-field chief ray. There will be no vignetting if

$$R_j \geq |y_j| + |\bar{y}_j| \tag{4.23}$$

at each surface.

One final point remains to be covered. How do we draw a vignetting diagram when the object is at infinity (Figure 4.10)? The key here is to use the slope angle of the full-field chief ray $\bar{u}_0$ instead of the object height. Because the rays ar all parallel the projected circle is the same size as the corresponding image $R_j^* = R_j^{\prime}$. The displacement of the center of the circle is given by

$$\Delta_j = \bar{u}_0 (l_E - l_j) \; . \tag{4.24}$$

4.6 CHROMATIC ABERRATIONS

Recall that for a thick lens

$$\frac{1}{f^{\prime}} = (n - 1)\left[c_1 - c_2 + \left(1 - \frac{1}{n}\right) t_1 c_1 c_2\right] . \tag{4.25}$$

In general, the focal length of an optical system will depend on the indices of refraction used in the system. In addition, the back focal length, principal plane position, and most other first-order properties of the system will depend on the index. Because of dispersion the index of refraction of any material is a function of wave-length. We expect, therefore, that most of the first-order properties of a refracting system will depend on wavelength. Possible exceptions would typically include the determination of which surface is the aperture or field stop. Usually,

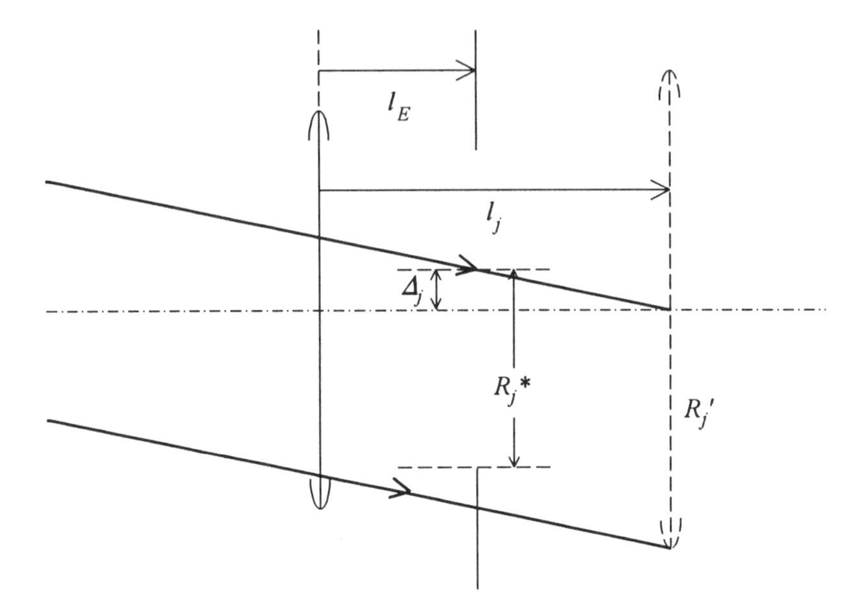

Figure 4.10 Projection scheme when the object is at infinity.

these do not change with wavelength, although the entrance and exit pupil and window sizes and positions may change slightly with wavelength.

The effect of dispersion on image quality is called *chromatic aberration*. The amount of chromatic aberration produced by a system can be determined to first order by tracing rays from the object using each of the different wavelengths.

Example 4.8

Determine the image position and image size for an object at infinity produced by a single lens made from LLF1 glass. The curvature of the first surface is 0.18244, the second surface is planar, and the lens is 1.8 cm thick. Repeat this calculation in F and C light.

Table 4.4 shows the calculations for F, d, and C light in that order. The index in F light is 1.55651 based on the Abbe value of 45.75 for LLF1 glass (see Chapter 1). The C light index is 1.54453. The ray slope angles and the incident angles are calculated for d light only.

The axial ray trace in F light yields the image position

$$t_2 = -\frac{(3.530351)}{(-0.40612)} = 8.692876 ,$$

while in C light we have

Table 4.4 Three-color paraxial ray trace for Example 4.8. The values for u and i are only given in d light.

	0		1		2		3
c			0.18244		0.0		
t		∞		1.8		8.837314	
				1.55651			
n		1.0		1.54814		1.0	
				1.54453			
			-0.10153				
$-\phi$			-0.10000		0.0		
			-0.09934				
				1.156433			
τ				1.162686			
				1.165403			
					3.530349		
y	4.0		4.0		3.534926		0.0
					3.536915		
				-0.40612		-0.40612	
α		0.0		-0.40000		-0.40000	
				-0.39736		-0.39736	
u		0.0		-0.25838			
i			0.729760		-0.25838		
					0.115643		0.999374
y			0.0		0.116269		1.000000
					0.116540		1.000271
α		0.1		0.1		0.1	
u		0.1		0.064594		0.1	
i					0.064594		

$$t_2 = -\frac{(3.536897)}{(-0.39736)} = 8.900989 .$$

I have purposefully kept too many digits in these answers because we will want to subtract them and look at the small difference between them. We can now define the longitudinal component of axial color as

$$\delta z'_{AC} = (t_2)_F - (t_2)_C , \tag{4.26}$$

from which we get

$$\delta z'_{AC} = (8.692872) - (8.901034) = -0.207468 .$$

The axial color is the longitudinal shift in image position as a result of dispersion.

Next, we trace full field chief rays in F and C light. The difference in where these rays cross the d light image plane is defined as the transverse component of lateral color:

$$\delta y'_{LC} = (\bar{y}_3)_F - (\bar{y}_3)_C . \tag{4.27}$$

In our example, we have

$$\delta y'_{LC} = (0.999374) - (1.000271) = -0.000897 .$$

Thus, lateral color is the change in image size or magnification due to change in wavelength. Of course, there is a change in image size due solely to the shift in image position caused by axial color. Lateral color is an additional effect beyond the change in image size due to axial color. ■

The example just presented defines and demonstrates axial and lateral color. Because paraxial optics is linear, the chromatic aberration of any paraxial ray may be determined from the chromatic aberration of the marginal axial ray and the full-field chief ray. The procedure outlined above is tedious and does not give any insight into how to reduce the chromatic aberration. Thus, we need to derive equations for axial and lateral color which will clearly show their functional dependence.

Our derivation makes use of the fact that the chromatic aberration effect is not due to translation through the media, since all light propagates in a straight line in a homogeneous medium. Rather, the effect is due to variations in the angle of refraction in Snell's Law. Thus an equation for longitudinal chromatic aberration can be derived based on the contributions due to all of the surfaces of the optical system. First, the variation in ray position due to refraction at one surface will be determined. This equation will then be applied to other surfaces to get the total aberration in a system.

Recall the equation for image and object distances for a single spherical refracting surface

$$\frac{n'}{s'} - \frac{n}{s} = \frac{n' - n}{r} . \tag{4.28}$$

This equation can be applied to the surface for different wavelengths by adding subscripts to the parameters which change with wavelength. The subscript F indicates blue light and the subscript C indicates red light.

Subtracting two versions of Eq. 4.28 gives

$$\left(\frac{n_C'}{s_C'} - \frac{n_F'}{s_F'} \right) - \left(\frac{n_C}{s_C} - \frac{n_F}{s_F} \right) = \frac{\delta n - \delta n'}{r} , \tag{4.29}$$

where $\delta n = n_F - n_C$ and $\delta n' = n_F' - n_C'$. Normally, the object distance would be the same for all colors of light, but the equation is written this way in case the object in question is actually an image produced by another refracting surface. This image will probably suffer from chromatic aberration which means that its position will depend on the wavelength. Equation 4.29 can be rearranged to give

$$\left(\frac{n_C'}{s_C'} - \frac{n_C' + \delta n'}{s_F'} \right) + \left(\frac{n_C}{s_C} - \frac{n_C + \delta n}{s_F} \right) = \frac{\delta n - \delta n'}{r} . \tag{4.30}$$

Rearranging terms gives

$$\frac{n_C'}{s_C' s_F'} (s_F' - s_C') - \frac{n_C}{s_C s_F} (s_F - s_C)$$

$$= \delta n \left(\frac{1}{r} - \frac{1}{s_F} \right) - \delta n' \left(\frac{1}{r} - \frac{1}{s_F'} \right) . \tag{4.31}$$

Now, let the difference in image position, $s_F' - s_C' = \delta z_{AC}'$, be defined as the longitudinal component of the axial chromatic aberration. We also need to define the difference in object position as the chromatic aberration of any preceding surfaces $\delta z_{AC} = s_F - s_C$. Since these equations are all based on the paraxial approximation anyway, small quantities such as δn, $\delta n'$, and $\delta z_{AC}'$ should appear only as first powers in Eq. 4.30. Thus, the difference between the various object and image distances in the denominators can be ignored. Also, the differences between n_C and

n_d can be ignored. This gives

$$\frac{n'}{(s')^2}\,\delta z'_{AC} - \frac{n}{s^2}\,\delta z = \delta n\left(\frac{1}{r} - \frac{1}{s}\right) - \delta n'\left(\frac{1}{r} - \frac{1}{s'}\right) . \tag{4.32}$$

In this equation, as before, quantities without subscripts are measured in d light. The previous equation can be put in a form which is compatible with the ray trace tables used in the previous chapter. Recall that the incidence angle i equals $u + y\,c$ and that the slope angle u equals $-y / s$ for a single surface. Similar equations apply to the refracted ray, and the incident and refracted rays are related by Snell's law $n'\,i' = n\,i$. Thus,

$$n'(u')^2\,\delta z'_{AC} - n\,u^2\,\delta z_{AC} = y\,i\,\delta n - y\,i'\,\delta n' . \tag{4.33}$$

Solving for the longitudinal axial color of the image gives

$$\delta z'_{AC} = \frac{n\,u^2}{n'\,u'^2}\,\delta z_{AC} + \frac{y\,n\,i}{n'\,u'^2}\left(\frac{\delta n}{n} - \frac{\delta n'}{n'}\right) . \tag{4.34}$$

As mentioned earlier, this equation applies to a single refracting surface. To get an equation for all of the surfaces of a system, this equation is applied sequentially to each surface. This isn't as difficult as it may sound. Remember $n' = n_j$, $n = n_{j-1}$, and so on. Using all of these identities for the first two surfaces gives

$$\left(\delta z'_{AC}\right)_1 = \frac{n_0\,u_0^2}{n_1\,u_1^2}\left(\delta z_{AC}\right)_1 + \frac{y_1\,n_0\,i_1}{n_1\,u_1^2}\left(\frac{\delta n_0}{n_0} - \frac{\delta n_1}{n_1}\right) \tag{4.35}$$

and

$$\left(\delta z'_{AC}\right)_2 = \frac{n_1\,u_1^2}{n_2\,u_2^2}\left(\delta z_{AC}\right)_2 + \frac{y_2\,n_1\,i_2}{n_2\,u_2^2}\left(\frac{\delta n_1}{n_1} - \frac{\delta n_2}{n_2}\right) . \tag{4.36}$$

Now, remember that at each surface, $\delta z'_{AC}$ becomes δz_{AC} for the next surface. We can now combine these two equations to get

$$\left(\delta z'_{AC}\right)_2 = \frac{n_0\,u_0^2}{n_2\,u_2^2}\left(\delta z_{AC}\right)_1$$

$$+\ \frac{1}{n_2\,u_2^2}\left[y_1\,n_0\,i_1\left(\frac{\delta n_0}{n_0} - \frac{\delta n_1}{n_1}\right) + y_2\,n_1\,i_2\left(\frac{\delta n_1}{n_1} - \frac{\delta n_2}{n_2}\right)\right] . \tag{4.37}$$

This equation can be generalized to

$$\delta z'_{AC} = \frac{n_0 u_0^2}{n_k u_k^2} \delta z_{AC} + \frac{1}{n_k u_k^2} \sum_{j=1}^{k} y_j n_{j-1} i_j \left(\frac{\delta n_{j-1}}{n_{j-1}} - \frac{\delta n_j}{n_j} \right) . \tag{4.38}$$

The term under the sum could be called the axial color coefficient or AC for short. Ignoring the object chromatic aberration we have

$$\delta z'_{AC} = \frac{AC}{n_k u_k^2} , \tag{4.39}$$

for the longitudinal component of axial color. The tangential component of axial forms one side of a right triangle, while the longitudinal component forms the other side. The ratio of these two sides is the tangent of the slope angle of the ray. Since we are dealing with paraxial rays, the tangent can be replaced by the slope angle. Thus, the tangential component of the axial color is

$$\delta y'_{AC} = -\frac{AC}{n_k u_k} . \tag{4.40}$$

This is the distance that separates F and C light marginal axial rays in the d light image plane.

The derivation for lateral color is somewhat more complex because we need to consider the location of the aperture stop. Figure 4.11 shows a single refracting surface with a separate aperture stop. To locate the images of an off-axis object point we first draw a ray through the center of curvature. For this ray $i = 0$, so the refracted ray is identical for all colors. Next, we draw rays for F and C light. Where these rays cross the ray through the center of curvature defines the off axis images in F and C light.

Recall that lateral color was defined as the transverse separation of the two chief rays in the d image plane. Since all of our calculations are assuming small angles, these two rays are nearly parallel in image space. It follows that for all practical purposes the lateral color could be calculated in the F or C image planes at our convenience. With this in mind, we can locate a couple of similar triangles in Figure 4.11 which yield

$$\frac{\delta y'_{LC}}{\delta z'_{AC}} = \frac{\bar{y} - \tilde{y}}{s'} . \tag{4.41}$$

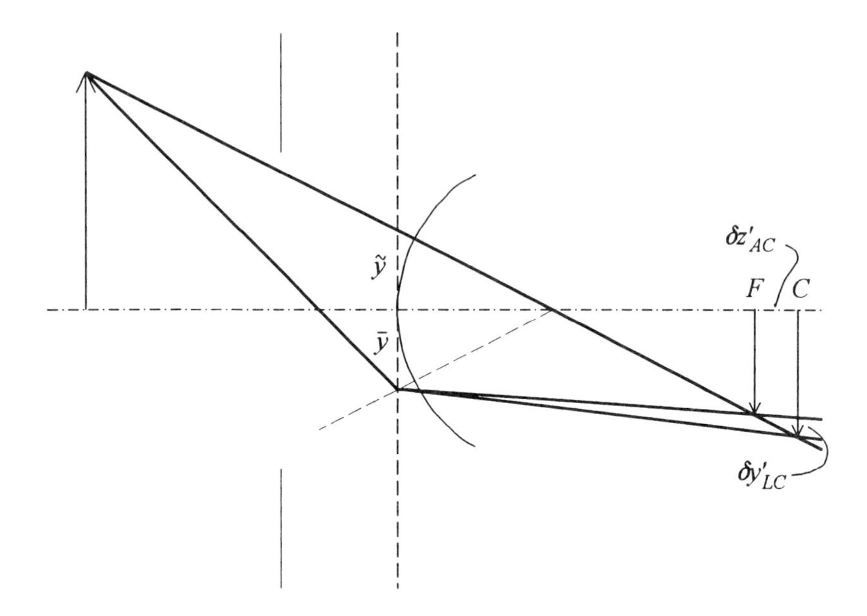

Figure 4.11 Paraxial ray drawing used to determine the lateral color.

The next step is to rewrite this equation in terms of *y-nu* ray trace parameters. Again, by similar triangles identified in Figure 4.11 we have

$$\frac{h'}{\tilde{y}} = \frac{s' - r}{r} . \tag{4.42}$$

Dividing the numerator and denominator of the right-hand side of this equation by $r\, s'$ gives

$$\frac{h'}{\tilde{y}} = \frac{\dfrac{1}{r} - \dfrac{1}{s'}}{\dfrac{1}{s'}} . \tag{4.43}$$

Next, multiplying the right-hand side numerator and denominator by y the height of the marginal axial ray (not shown in the figure) gives us some familiar ratios which can be replaced by *y-nu* ray trace angles. Rearranging terms gives

$$\tilde{y} = h' \frac{u'}{i'} . \tag{4.44}$$

Now we need to express h' in terms of y-nu ray trace parameters. If we multiply the top and bottom of the right-hand side of Eq. 4.44 by n' we get

$$\tilde{y} = \frac{n'u'h'}{n'i'} \tag{4.45}$$

which we recognize as the optical invariant divided by $n'\,i'$. As shown in the problems at the end of Chapter 3, there are a couple of different ways to write the optical invariant. For example,

$$\tilde{y} = \frac{n\bar{y}i - ny\bar{i}}{ni} = \bar{y} - y\frac{\bar{i}}{i} . \tag{4.46}$$

Substituting Eq. 4.46 into Eq. 4.41 gives

$$\delta y'_{LC} = -u'\frac{\bar{i}}{i}\delta z'_{AC} , \tag{4.47}$$

and

$$\delta y'_{LC} = -\frac{ny\bar{i}}{n'u'}\left(\frac{\delta n}{n} - \frac{\delta n'}{n'}\right) . \tag{4.48}$$

As with the axial color, we can sum over all the surfaces in a system and add a term for any lateral color in the object

$$\delta y'_{LC} = \frac{n_0 u_0}{n_k u_k}\delta y_{LC} - \frac{1}{n_k u_k}\sum_{j=1}^{k} n_{n-1} y_j \bar{i}_j \left(\frac{\delta n_{j-1}}{n_{j-1}} - \frac{\delta n_j}{n_j}\right) . \tag{4.49}$$

We can shorten this by calling LC the lateral color coefficient:

$$LC_j = n_{j-1} y_j \bar{i}_j \left(\frac{\delta n_{j-1}}{n_{j-1}} - \frac{\delta n_j}{n_j}\right) . \tag{4.50}$$

Example 4.9

Calculate the axial and lateral color coefficients and the amount of longitudinal axial color and lateral color for the previous example.

The d light ray trace information is given in the previous example. Here we just need to plug into the appropriate equations

$$AC = (4.0)(1.0)(0.729760)\left(\frac{(0.0)}{(1.0)} - \frac{(0.01198)}{(1.54814)}\right)$$

$$+ (3.534926)(1.54814)(-0.25838)\left(\frac{(0.01198)}{(1.54814)} - \frac{(0.0)}{(1.0)}\right)$$

$$= -0.03353 \; .$$

We can now calculate the longitudinal axial color

$$\delta z'_{AC} = \frac{(-0.03353)}{(1.0)(-0.4)^2} = -0.20956 \; ,$$

which compares favorably with previous results.

For lateral color we have

$$LC = (4.0)(1.0)(0.1)\left(-\frac{(0.01198)}{(1.54814)}\right)$$

$$+ (3.53492)(1.54814)(0.06459)\left(\frac{(0.01198)}{(1.54814)}\right)$$

$$= -0.00036 \; .$$

The lateral color is

$$\delta y'_{LC} = -\frac{(-0.00036)}{(1.0)(-0.4)} = -0.0009 \; .$$

Again, this compares favorably with previous results. ■

4.7 METHODS FOR REDUCING CHROMATIC ABERRATION

Equations 4.38 and 4.49 are complicated enough that it might seem easier to just do separate ray traces for the different indices of refraction. The real power of this equation is not its use for analysis but for design. By looking at the coefficients

$$AC = \sum_{j=1}^{k} y_j n_{j-1} i_j \left(\frac{\delta n_{j-1}}{n_{j-1}} - \frac{\delta n_j}{n_j} \right) \tag{4.51}$$

and

$$LC = - \sum_{j=1}^{k} n_{j-1} y_j \bar{i}_j \left(\frac{\delta n_{j-1}}{n_{j-1}} - \frac{\delta n_j}{n_j} \right) , \tag{4.52}$$

we can learn ways to reduce the amount of chromatic aberration in a system.

The simplest and most direct way to reduce chromatic aberration is to simply stop the system down. Reducing the aperture stop clear radius will reduce the maximum ray height y_j throughout the system. Unfortunately, the image brightness will also be reduced. This should be the method of last resort.

Reducing the ratio $\delta n / n$ will reduce the chromatic aberration. Unfortunately, glasses that have small dispersions (large Abbe value) also tend to have small indices. The ideal glass from the point of view of reducing chromatic aberration would have both large Abbe value and large d light index. Glass manufacturers are continually trying to formulate such glass. A secondary effect of large index is that the curvatures of the lens can be reduced for a given power. Smaller curvatures result in smaller incident angles.

The third and most useful method of reducing chromatic aberration is to try to balance the terms in the coefficient sum. This technique won't work in general for a single lens because the two incident angles tend to have opposite signs and the dispersion terms also change signs. Thus both terms in the coefficient sum for a thin lens will have the same sign. Balancing terms will work for more complicated lens systems, as we will see in a minute.

For lateral color, the chief ray incident angles tend to have the same signs in a singlet. In fact, if the lens is completely symmetric about a central aperture stop, including the object and image, then $LC = 0$. Also, for a single thin lens, the stop is at the lens so $\bar{y} = 0$ and the lateral color coefficient tends to be quite small.

We can simplify Eqs. 4.38 and 4.49 for systems of thin lenses. Let's begin by writing Eq. 4.38 for a single thin lens in air. In this case the object and image space dispersions are zero and the ray heights at the two surfaces are equal:

$$\delta z'_{AC} = \frac{y}{u_2^2} \frac{\delta n}{n} \left(- i_1 + n\, i_2 \right) . \tag{4.53}$$

Once again using $i = u + y\,c$ and substituting the ray trace equation for refraction at a surface we can show that

$$\delta z'_{AC} = \frac{y^2}{u_2^2} \delta n \left(c_2 - c_1 \right) . \tag{4.54}$$

Next, using the equation for the focal length of a thin lens and the definition of the Abbe value we get

$$\delta z'_{AC} = -\frac{y^2}{u_2^2}\frac{1}{f' V} . \tag{4.55}$$

Finally, for a series of thin lenses in contact, we have

$$\delta z'_{AC} = -\frac{1}{u_k^2}\sum_{j=1}^{k}\frac{y_j^2}{f'_j V_j} . \tag{4.56}$$

As an example of the use of this equation, consider the design of a cemented achromat, that is two thin lenses in contact. For this system, the ray heights are the same in both lenses. The overall effective focal length is given by

$$\frac{1}{f_T} = \frac{1}{f_1} + \frac{1}{f_2} . \tag{4.57}$$

By combining the previous two equations, the following equations are found which give the individual focal lengths necessary to achieve the overall focal length without chromatic aberration:

$$f'_1 = f'_T\left(\frac{V_1 - V_2}{V_1}\right) , \tag{4.58}$$

$$f'_2 = f'_T\left(\frac{V_2 - V_1}{V_2}\right) . \tag{4.59}$$

Example 4.10

Design a thin lens cemented achromat with focal length 10.0 using K10 and F8 glass.

We arbitrarily select the K10 glass as the glass for the first lens. Then

$$f'_1 = (10.0)\left(\frac{(56.41) - (39.18)}{(56.41)}\right) = 3.054 ,$$

and

$$f_2' = (10.0)\left(\frac{(39.18) - (56.41)}{(39.18)}\right) = -4.398 .$$

If we are given some additional information on the shape of the lens, then we can determine the individual curvatures of the lenses. For example, assume the positive element in this system is equiconvex ($c_1 = -c_2$), then

$$c_1 = \frac{1}{2(n_1 - 1)f_1'} = \frac{1}{2(0.50137)(3.054)} = 0.3265 .$$

And for the remaining surface we have

$$c_3 = c_2 - \frac{1}{(n_2 - 1)f_2'} ,$$

$$c_3 = (-0.3265) - \frac{1}{(0.5955)(-4.398)} = 0.0553 .$$

If appropriate thickness are now inserted to make a realistic lens, the lens will have some modest amount of chromatic aberration, but not nearly as much as in a singlet. ■

Finally, we can write a thin lens formula for lateral color. The derivation is identical to the derivation for the thin lens version of axial color, except that chief ray data are used in place of the axial ray data. It should be easy to see that

$$\delta y_{LC}' = \frac{1}{u_k} \sum_{j=1}^{k} \frac{y_j \bar{y}_j}{f_j V_j} . \tag{4.60}$$

Note that when the stop is in contact with the thin lens, the lateral color vanishes because $\bar{y} = 0$.

4.8 ANGLE AND HEIGHT SOLVES

We now have a method for finding the curvatures of an achromatic doublet (given some additional information regarding the shape of the lens) and a knowledge of how to add thickness to the thin lens to make a real lens. As mentioned earlier, adding reasonable thicknesses will cause a small amount of color to return to the system. This problem will be dealt with when we discuss optimization. Another problem arises which we can take care of now and that is that adding thickness to a lens will also change the effective focal length of the lens.

There are several ways to correct the focal length. One way is to scale the system by multiplying all lengths by the ratio of the desired focal length to the actual focal length. Curvatures must be multiplied by the reciprocal of the ratio. The only problem with this technique is that it may make a lens unnecessarily thick or unrealistically thin. An alternative approach is to solve for the curvature of the last surface of the system which will give the correct focal length.

Using an axial ray trace from an object at infinity, we know that the effective focal length of a lens is given by

$$f' = -\frac{y_1}{u_k} . \tag{4.61}$$

The first surface axial ray height can be set and the desired focal length is known, so the axial ray slope angle in image space is fixed. Next we use the refraction equation

$$n_k u_k = n_{k-1} u_{k-1} - c_k (n_k - n_{k-1}) y_k \tag{4.62}$$

and eliminate u_k. All of the quantities in the resulting equation are known, so we can solve for the curvature of the last surface

$$c_k = \frac{n_{k-1} u_{k-1} + n_k \dfrac{y_1}{f'}}{(n_k - n_{k-1}) y_k} . \tag{4.63}$$

For the problem we just examined, we needed a particular angle

$$u_k = -\frac{y_1}{f'} \tag{4.64}$$

to solve the problem. This type of problem is called an *angle solve*. The same technique may be used after any surface when we want u_j to have a certain value.

A height solve is used when we need a ray with a particular height at a given surface. This may happen when a stop or lens rim is constrained by some external consideration to have a certain size and we need a certain ray to pass through the edge (marginal ray) or center (chief ray for example) of the surface. We change ray heights by changing the separation between surfaces. Using the translation equation we see

$$t_{j-1} = \frac{y_j - y_{j-1}}{u_{j-1}} . \tag{4.65}$$

The point of all of this discussion is that the paraxial ray trace equations are linear and linear equations can be easily inverted. Ray trace equation inversion is the central theme of the chapter on y-$\bar{y}$ diagrams.

Example 4.11

Add realistic thickness (to the next largest mm) to the lens from Example 4.10 and correct the focal length to 10.0 cm. Assume that the system is intended to $f/5$.

The first step is to determine the clear radius:

$$R = \frac{f'}{2f/\#} = \frac{(10.0)}{2(5.0)} = 1.0 .$$

We will assume that this is the clear radius for all surfaces. Next, the first sag is calculated:

$$z_1 = (3.0628 \text{ cm}) - (3.0628 \text{ cm})\sqrt{1 - (0.3265 \text{ cm}^{-1})^2 (1.0)^2}$$
$$= 0.17 \text{ cm} .$$

Because the first lens is symmetrical, the second sag is equal to the negative of the first sag. The lens thickness is

$$t_1 = (0.17) + 0.1 - (-0.17) = 0.44 \text{ cm} .$$

The second lens is a negative lens. Its thickness is

$$t_2 = 0.12R = 0.12(1.0) = 0.12 \text{ cm} .$$

The ray trace table is shown in Table 4.5.

From the information in the ray trace table we can now calculate c_3:

Table 4.5 Paraxial ray trace used to find the focal length for Example 4.11.

	0	1	2	3
c		0.3265	−0.3265	0.0553
t		0.44	0.12	
n	1.0	1.50137	1.59551	
$-\phi$		−0.163697	0.030737	0.032932
τ		0.293066	0.075211	
y	1.0	1.0	0.952026	0.941915
α	0.0	−0.163697	−0.134435	

$$c_3 = \frac{(-0.134435) + (1.0)\dfrac{(1.0)}{(10.0)}}{((1.0) - (1.59551))(0.941915)}$$

$$= 0.06139 \text{ cm}^{-1} .$$

■

SUMMARY

Lens thickness is related to the clear radius of the lens and to the curvatures of the surfaces. The sag of a surface can be calculated with

$$z_j = r_j - r_j \sqrt{1 - \frac{R_j^2}{r_j^2}} . \tag{4.66}$$

The overall thickness of a positive lens is given by

$$t_j = z_j + t_e - z_{j+1} . \tag{4.67}$$

For a negative lens, the axial thickness should be no less than 12% of the clear radius.

The aperture stop of an optical system is the stop or lens rim which limits rays coming from an on-axis object point. The aperture stop is dependent on the object position. One method for finding the aperture stop is to trace an arbitrary axial ray through the system. This ray is also needed to locate the image position. At each surface, calculate the ratio of the clear radius to the height of the axial ray. The surface with the smallest magnitude for this ratio is the aperture stop. The arbitrary axial ray can be converted into a marginal axial ray by multiplying all of the ray data by the factor

$$A = \min\left|\frac{R_j}{y_j}\right| . \tag{4.68}$$

This will guarantee that the axial ray passes through the edge of the aperture stop and that u_0 is as large as possible.

The entrance pupil is defined as the image of the aperture stop in object space. The exit pupil is the image of the aperture stop in image space. To find the pupils a chief ray is traced. A chief ray is a ray which passes through the center of the aperture stop. A chief ray always has $\bar{y}_A = 0$ and is denoted with a bar over the ray parameters. The location of the entrance pupil is given by

$$l_E = -\frac{\bar{y}_1}{\bar{u}_0} . \tag{4.69}$$

The size of the entrance pupil is given by

$$R_E = R_A \frac{n_A \bar{u}_A}{n_0 \bar{u}_0} . \tag{4.70}$$

Similar equations, using image space chief ray data, are used to determine the size and location of the exit pupil.

The *f*-number of an optical system is given by

$$f/\# = \frac{f'}{2 R_E} . \tag{4.71}$$

It is assumed that the object is at infinity when determining the entrance pupil used in this equation.

The field stop is defined as the stop or lens rim or object or image height which limits the chief rays. The field stop can be determined by looking for the minimum absolute value for the ratio of the clear radius at each surface to the height of an arbitrary chief ray. The arbitrary chief ray may be made into a full-field chief ray by multiplying by this ratio. The entrance window is defined as the image of the field stop in object space and the exit window is the image of the field stop in image space. The windows are found in a manner identical to the way the pupils were found, but a ray through the center of the field stop is used instead of a chief ray.

Vignetting occurs when rays from an off-axis object point are cut off by a lens rim or stop other than the aperture stop. A vignetting diagram may be drawn by first imaging all of the stops and lens rims into object space. All of these images as seen from an off-axis object point are then projected onto the entrance pupil. The offset of the center of each image is given by

$$\Delta_j = h\left(\frac{l_j - l_E}{t_0 + l_j}\right), \tag{4.72}$$

where l_j is the location of the image of the jth surface measured from the first surface. The projected radius is given by

$$R_j^* = R_j'\left(\frac{t_0 + l_E}{t_0 + l_j}\right), \tag{4.73}$$

where R_j' is the radius of the image of the jth surface.

Axial color is the longitudinal shift in image location as a result of dispersion. It may be calculated, in general, by

$$\delta z'_{AC} = \frac{n_0 u_0^2}{n_k u_k^2}\delta z_{AC} + \frac{1}{n_k u_k^2}\sum_{j=1}^{k} y_j n_{j-1} i_j\left(\frac{\delta n_{j-1}}{n_{j-1}} - \frac{\delta n_j}{n_j}\right). \tag{4.74}$$

For thin lenses in air a simpler version is

$$\delta z'_{AC} = -\frac{1}{u_k^2}\sum_{j=1}^{k}\frac{y_j^2}{f_j' V_j}. \tag{4.75}$$

Lateral color is the shift in image size due to dispersion effects on the chief ray. Lateral color is calculated using

$$\delta y'_{LC} = \frac{n_0 u_0}{n_k u_k}\delta y_{LC} - \frac{1}{n_k u_k}\sum_{j=1}^{k} n_{n-1} y_j \bar{i}_j\left(\frac{\delta n_{j-1}}{n_{j-1}} - \frac{\delta n_j}{n_j}\right). \tag{4.76}$$

The thin lens version of this equation is

$$\delta y'_{LC} = \frac{1}{u_k} \sum_{j=1}^{k} \frac{y_j \bar{y}_j}{f'_j V_j} . \tag{4.77}$$

When the aperture stop is located at a thin lens, then the lateral color is zero.

An achromatic doublet can be designed by cementing a positive thin lens to a negative thin lens. The appropriate focal lengths are given by

$$f'_1 = f'_T \left(\frac{V_1 - V_2}{V_1} \right) , \tag{4.78}$$

$$f'_2 = f'_T \left(\frac{V_2 - V_1}{V_2} \right) . \tag{4.79}$$

An angle solve can be used to determine the curvature of the last surface of an optical system which results in a desired focal length. The curvature is given by

$$c_k = \frac{n_{k-1} u_{k-1} + n_k \dfrac{y_1}{f'}}{(n_k - n_{k-1}) y_k} . \tag{4.80}$$

A height solve is used to determine the separation between two surfaces which results in a desired ray height. The separation is given by

$$t_{j-1} = \frac{y_j - y_{j-1}}{u_{j-1}} . \tag{4.81}$$

REFERENCES

A. E. Conrady, *Applied Optics and Optical Design* (Dover Publications, Toronto, 1985).

R. Kingslake, *Lens Design Fundamentals* (Academic Press, Orlando, FL, 1978).

PROBLEMS

4.1 Define the following terms: (a) aperture stop, (b) entrance and exit pupils, (c) chief ray, (d) field stop, (e) entrance and exit windows, (f) marginal axial ray, (g) vignetting, (h) vignetting factor, (i) axial color, and (j) lateral color.

4.2 Describe the specific steps involved in determining the stops, pupils and windows of an optical system using the *y-nu* ray trace table. How many paraxial rays need to be traced?

4.3 Describe the steps involved in drawing a vignetting diagram.

4.4 A thick equiconvex lens has two surfaces either of which might serve as an aperture stop. For what range of object positions is the first surface the aperture stop? For what range of object positions is the second surface the aperture stop? You should not need to resort to calculations to answer this question.

4.5 Calculate the minimum thickness for each lens described below. Round your answers to the next higher millimeter. (a) $r_1 = 5.0$ cm, $r_2 = -2.0$ cm, $R = 1.0$ cm. (b) $r_1 = 6.0$ cm, $r_2 = -3.0$ cm, $R = 1.2$ cm. (c) $r_1 = -6.0$ cm, $r_2 = 4.0$ cm, $R = 1.5$ cm. (d) $r_1 = 8.0$ cm, $r_2 = 4.0$ cm, $R = 2.0$ cm.

4.6 An optical system consists of a 2.00-cm-diameter stop located 2.00 cm in front of a thin lens. The thin lens has a diameter of 4.00 cm and an effective focal length of 4.000 cm. An object is located 6.00 cm in front of the stop. (a) Make a one-to-one scale drawing of this system. (b) Use a straight edge to determine which surface is the aperture stop and draw a marginal axial ray in object space. (c) Draw three special rays from the aperture stop to find the size and location of the entrance and exit pupils. (d) Use the exit pupil to draw the chief ray in image space and locate the image of the object.

4.7 Find the aperture stop and entrance and exit pupils for the system given in Problem 4.6 using thin lens formulae.

4.8 Find the aperture stop and entrance and exit pupils for the system given in Problem 4.6 using the *y-nu* ray trace table.

4.9 Same system as in Problem 4.6, but now the object is only 1.00 cm in front of the stop. Determine the aperture stop and find the size and location of the entrance and exit pupils by any method. Find the object position where either the stop or the lens rim could be considered the aperture stop.

4.10 A 6.00-cm-focal-length thin lens and a stop located 1.00 cm to the right form an optical system. The diameters of the lens and stop are 2.00 cm and 1.60 cm, respectively. For an object located 18.00 cm in front of the lens, determine the aperture stop and find the size and location of the entrance and exit pupils.

4.11 A telephoto lens consists of two thin lenses and a stop. The first lens has a focal length of 4.00 cm and a diameter of 3.00 cm. The stop is located 2.00 cm behind the first lens, and it has a diameter of 1.125 cm. The second lens is 1.00 cm behind the stop. The second lens has a focal length of −1.25 cm and a diameter of 0.75 cm. For an object at infinity, determine the aperture stop and find the size and location of the entrance and exit pupils. What is the $f/\#$ for this system?

4.12 An astronomical telescope consists of a 100.0-cm-focal-length objective and a 1.00-cm-focal-length eyepiece. The diameter of the objective is 10.0 cm and the diameter of the eyepiece is 0.50 cm. Assuming thin lenses and final image at infinity, determine which lens is the aperture stop and which is the field stop. Find the size and location of the entrance and exit pupils and windows.

4.13 A compound microscope is made from two thin lenses and uses the standard tube length of 16.0 cm. The objective lens has a focal length of 0.75 cm and a diameter of 0.60 cm. The eyepiece has a focal length of 1.00 cm and a diameter of 0.50 cm. Find the aperture stop and field stop when the microscope is focused and the eye is relaxed. Find the size and location of the entrance and exit pupils and windows.

4.14 A terrestrial telescope (spy glass) is made from the astronomical telescope of Problem 4.12 by inserting an erecting lens between the objective and the eyepiece in such a way that the magnification and image position do not change. This lens is also treated as a thin lens. Its focal length is 5.00 cm and its diameter is 2.00 cm. Determine which lens is the aperture stop and which is the field stop. Find the size and location of the entrance and exit pupils and windows.

4.15 An achromatic doublet is to be made from SF57 and FK5 glasses. The overall focal length is 75 mm. The first lens is equiconvex. Find the focal lengths of the two thin lenses which form this achromat. Find the radii of curvature for all surfaces.

4.16 An achromatic doublet is to be made from LaK8 and F8 glasses. The overall focal length is 10 cm. The interface between the two elements is planar. Find the focal lengths of the two thin lenses which form this achromat. Find the radii of curvature for the first and last surfaces.

4.17 Find the curvature of the second surface of a single lens needed to give the lens a focal length of 25.0 cm. The curvature of the first surface is 0.04 cm^{-1} and the lens thickness is 0.7 cm. The lens is made of BaLF4 glass.

4.18 An achromatic doublet made from SF2 and SF57 glasses has $c_1 = 0.521583$ cm^{-1}, $c_2 = 0.0\ \text{cm}^{-1}$, and $c_3 = 0.280897\ \text{cm}^{-1}$. If the clear radius of this lens is to be 1.00 cm, find appropriate thicknesses for the lenses. With these thicknesses inserted, use an angle solve for the last curvature to correct the overall focal length of the lens back to 10.0 cm.

4.19 A single lens is made of SF5 glass with $c_1 = 0.159421\ \text{cm}^{-1}$, $c_2 = 0.0\ \text{cm}^{-1}$, $t_1 = 0.5$ cm, and $R = 2.0$ cm. An object 2.0 cm tall is located at $t_0 = 40.0$ cm. The first surface is the aperture stop and the object is the field stop. Trace a marginal axial ray and a full field chief ray, then calculate the axial and lateral color coefficients.

4.20 A Coddington magnifier is made by reshaping a single glass sphere (Figure 4.12). In this case, the radius of the sphere is 1.50 cm. The glass is BK1. The sides of the sphere are ground down to make a cylinder of diameter 2.00 cm. A notch is ground into the center of the sides of the cylinder so that its diameter is 1.20 cm. (a) Find the object position which gives a virtual image 25.00 cm in front of the last surface of the lens. (b) For this object position, determine the aperture stop and find the size and location of the entrance and exit pupils using the *y-nu* ray trace and the optical invariant.

4.21 Calculate the axial and lateral color coefficients for the Coddington magnifier described in the previous problem. Take the full field object height to be 0.01 cm.

4.22 Rework the previous two problems involving the Coddington magnifier without the notch in the middle of the lens. What surface is the new aperture stop? How are the color coefficients affected? Does this explain the existence of the notch?

4.23 An optical system consists of three thin lenses. The first lens has a focal length of 4 and a clear radius of 3. The second lens has a focal length of −6 and a clear radius of 2. The third lens has a focal length of 10 and a clear radius of 2. The distance between the first and second lens is 2 and between the second and third lens is 1.5. The object of interest is 10 units to the left of the first lens and two units high. Do a *y-nu* ray trace to confirm that the second lens is the aperture stop. Do a second *y-nu* ray trace to find the field stop and determine the size and position of the entrance pupil. Do a third ray

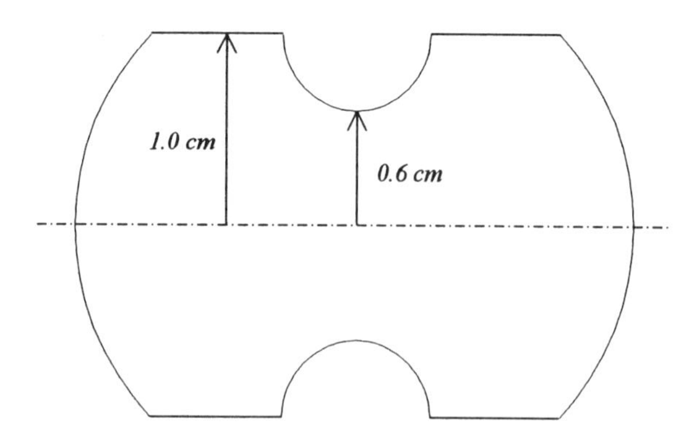

Figure 4.12 A Coddington type magnifier for Problems 4.20 through 4.22.

trace to find the size and location of the image of the third lens in object space. From all of the information provided, make a full-scale drawing of the system and draw a vignetting diagram.

4.24 Write a computer program or spreadsheet to calculate the curvatures and thicknesses for achromatic doublets. Inputs would be the index and Abbe value for two different glasses, the desired system focal length, and the clear radius of the system. Your program should be able to handle having either the positive or negative element as the first lens. Remember that an additional piece of shape information is needed. Options for the shape information include making the first or second element equiconvex or equiconcave or making any one of the three surfaces planar.

4.25 Write a spreadsheet or program to calculate the thickness of a single thin lens as a function of the lens shape. Assume the focal length for the lens is exactly 20.0 cm, the clear radius is 1.0 cm and the index is 1.500. The independent parameter should be the curvature of the first surface. Let this curvature vary in small steps (say 0.025 cm^{-1}) from -0.9 cm^{-1} to 0.975 cm^{-1}. Calculate the curvature of the second surface which gives the correct focal length using the thick lens formula. Since this curvature depends on the thickness of the lens, you will need to make an initial guess of the thickness, then iterate until you converge on a solution. Note that the thickness depends on both curvatures. As a check on your work, calculate the focal length using the thick lens formula. From your results, what is the shape which gives minimum thickness? Does thickness depend strongly or weakly on shape?

4.26 Design an achromatic doublet using BaSF10 and SK5 glasses. The overall focal length should be 40.0 cm and the clear radius should be 3.0 cm. Make the first lens equiconcave. Calculate and insert appropriate thicknesses. Correct the last surface with an angle solve to get back to the desired focal length.

4.27 Calculate the axial and lateral color coefficients for the achromat from Problem 4.15. Assume the object is 1.0 cm tall and 15.0 cm in front of the lens. Also, assume zero thickness between surfaces initially. Next, insert appropriate thicknesses for the two elements and recalculate the color coefficients.

4.28 Design an achromatic doublet using SF57 and FK5 glasses. Make the focal length 7.5 cm and the clear radius 1.0 cm. Let the interface between the two elements be planar. Calculate appropriate thicknesses. Assume the object is 1.0 cm tall and 15.0 cm in front of the lens, calculate the color coefficients.

4.29 Equation 4.56 gives the axial color for a system of thin lenses. In the text, this equation was used to derive equations for the focal lengths of the elements of an achromatic cemented doublet. Derive similar equations for an air-spaced doublet. The two focal lengths should be given in terms of the overall focal length, the two Abbe values and the ratio of the lens separation to the focal length of the first lens.

CHAPTER 5

MATRIX METHODS

Matrix methods provide an alternative way of solving paraxial optics problems. The same *y-nu* ray trace equations are used, but instead of a ray trace table being set up, a matrix is set up. Matrices are a mathematical tool designed to deal with linear equations like the ray trace equations, so it is natural to apply matrices to ray tracing.

This chapter introduces the use of matrices for paraxial ray tracing. Matrices are used to re-derive some of the equations discussed in previous chapters. Matrices will be used to find the first-order properties of lens systems. Hopefully, presenting this material from a different point of view will serve as a useful review. Matrix methods also offer additional insight into the meaning of the cardinal points, the definition of the optical invariant, and the importance of the marginal axial and full-field chief rays. In addition, matrices will be used in applications which would otherwise be cumbersome to solve. Finally, matrices will be used later in this book in the discussion of optimization and in other courses such as Physical Optics. Some practice now certainly will not hurt later.

5.1 PARAXIAL MATRIX EQUATIONS

A paraxial ray trace is accomplished by repeated application of the *y-nu* equations. These equations were derived in Section 3.1. First, for a translation we have

$$y_j = y_{j-1} + \tau_{j-1}\,\alpha_{j-1} \tag{5.1}$$

and

$$\alpha_{j-1} = \alpha_{j-1}\ . \tag{5.2}$$

As a reminder, in these equations, $\alpha = n\,u$ is the reduced ray angle and $\tau = t\,/\,n$ is the reduced separation. Even though it appears wasteful, we write the obvious Eq. 5.2 to complete the matrix as you will see shortly.

Next, for reflection or refraction we have

$$y_j = y_j \tag{5.3}$$

and

$$\alpha_j = \alpha_{j-1} - \phi_j y_j . \tag{5.4}$$

The power of a surface is

$$\phi_j = (n_j - n_{j-1})\, c_j , \tag{5.5}$$

or the inverse of the focal length for a thin lens. In the remainder of this chapter, the phrase "or reflection" will not be used, but you should remember that the refraction equations work for mirrors when $n_j = -n_{j-1}$.

Linear equations can be written in a shorthand notation using matrices. First, the ray parameters are written as a column vector (a 1×2 matrix)

$$\vec{y}_j = \begin{bmatrix} y_j \\ \alpha_{j-1} \end{bmatrix} \tag{5.6}$$

or

$$\vec{y}_j^{\,\prime} = \begin{bmatrix} y_j \\ \alpha_j \end{bmatrix} . \tag{5.7}$$

Notice that we need two different vectors. Both use the ray height at the surface given by the subscript, but one vector uses the ray slope immediately before the surface (unprimed) and the other uses the ray slope after the surface (primed).

To translate a ray from one surface to the next, we take the ray parameters immediately after the surface and apply the translation formulae to find the ray parameters just before the next surface

$$\begin{bmatrix} y_j \\ \alpha_{j-1} \end{bmatrix} = \begin{bmatrix} 1 & \tau_{j-1} \\ 0 & 1 \end{bmatrix} \begin{bmatrix} y_{j-1} \\ \alpha_{j-1} \end{bmatrix} , \tag{5.8}$$

or in matrix shorthand

$$\vec{y}_j = \vec{T}_{j-1}\, \vec{y}_{j-1}^{\,\prime} , \tag{5.9}$$

where the translation matrix is given by

$$\overleftrightarrow{T}_{j-1} = \begin{bmatrix} 1 & \tau_{j-1} \\ 0 & 1 \end{bmatrix}. \tag{5.10}$$

At this point, a word about matrix multiplication may be in order for anyone not familiar with the subject. Let

$$\overleftrightarrow{A} = \begin{bmatrix} a_{11} & a_{12} \\ a_{21} & a_{22} \end{bmatrix} \tag{5.11}$$

and

$$\vec{x} = \begin{bmatrix} x_1 \\ x_2 \end{bmatrix}. \tag{5.12}$$

To multiply a vector by a matrix, take the top row of the matrix and multiply the first element in that row with the first element in the vector and the second element in the row with the second element in the vector. Adding the two products gives the first element in the product vector. A similar procedure for the bottom row gives the second element of the result

$$\overleftrightarrow{A}\,\vec{x} = \begin{bmatrix} a_{11}x_1 + a_{12}x_2 \\ a_{21}x_1 + a_{22}x_2 \end{bmatrix}. \tag{5.13}$$

Refraction converts the ray parameters just before a surface into the ray parameters immediately after the surface

$$\begin{bmatrix} y_j \\ \alpha_j \end{bmatrix} = \begin{bmatrix} 1 & 0 \\ -\phi_j & 1 \end{bmatrix} \begin{bmatrix} y_j \\ \alpha_{j-1} \end{bmatrix}. \tag{5.14}$$

In matrix notation we have

$$\vec{y}_j^{\,\prime} = \overleftrightarrow{R}_j\,\vec{y}_j\,, \tag{5.15}$$

where

$$\overleftrightarrow{R}_j = \begin{bmatrix} 1 & 0 \\ -\phi_j & 1 \end{bmatrix}, \tag{5.16}$$

for refraction by a single surface. For a thin lens the optical power is equal to the inverse of the focal length of the lens. Therefore, the refraction matrix for a thin lens can be written as

$$\overleftrightarrow{R}_j = \begin{bmatrix} 1 & 0 \\ -\dfrac{1}{f_j'} & 1 \end{bmatrix} . \tag{5.17}$$

Next, we combine individual translation and refraction matrices to build up a single matrix which will take any ray in object space and give us the resulting ray in image space. To see how this is accomplished, start with a translation from the object surface to the first surface of the optical system

$$\vec{y}_1 = \overleftrightarrow{T}_0 \, \vec{y}_0^{\,\prime} . \tag{5.18}$$

We find the ray after the first surface by refraction, so

$$\vec{y}_1^{\,\prime} = \overleftrightarrow{R}_1 \, \vec{y}_1 . \tag{5.19}$$

Clearly we can write

$$y_1^{\prime} = \overleftrightarrow{R}_1 \, \overleftrightarrow{T}_0 \, \vec{y}_0^{\,\prime} . \tag{5.20}$$

By repeating this procedure, we eventually get to the ray coordinates at the image

$$\vec{y}_{k+1} = \overleftrightarrow{T}_k \, \overleftrightarrow{R}_k \cdots \overleftrightarrow{T}_j \, \overleftrightarrow{R}_j \cdots \overleftrightarrow{T}_1 \, \overleftrightarrow{R}_1 \, \overleftrightarrow{T}_0 \, \vec{y}_0^{\,\prime} . \tag{5.21}$$

All of the matrices can be multiplied together to get one 2×2 matrix. Since this matrix takes a ray from the object to the image, we call the total matrix a conjugate matrix:

$$\overleftrightarrow{C} = \overleftrightarrow{T}_k \, \overleftrightarrow{R}_k \cdots \overleftrightarrow{T}_j \, \overleftrightarrow{R}_j \cdots \overleftrightarrow{T}_1 \, \overleftrightarrow{R}_1 \, \overleftrightarrow{T}_0 . \tag{5.22}$$

Some texts identify this matrix as an image matrix. I think the name conjugate matrix better conveys the fact that the matrix connects an object and its image. As we shall see later, conjugate matrices have very special properties.

Another word about matrix multiplication is in order. Multiplying two matrices together is very similar to multiplying a vector by a matrix. Simply imagine that the second matrix is made of two side-by-side column vectors and proceed as before. To illustrate again let

$$\overleftrightarrow{A} = \begin{bmatrix} a_{11} & a_{12} \\ a_{21} & a_{22} \end{bmatrix} \tag{5.23}$$

and

$$\overleftrightarrow{B} = \begin{bmatrix} b_{11} & b_{12} \\ b_{21} & b_{22} \end{bmatrix} . \tag{5.24}$$

Then

$$\overleftrightarrow{A}\,\overleftrightarrow{B} = \begin{bmatrix} a_{11}b_{11} + a_{12}b_{21} & a_{11}b_{12} + a_{12}b_{22} \\ a_{21}b_{11} + a_{22}b_{21} & a_{21}b_{12} + a_{22}b_{22} \end{bmatrix} . \tag{5.25}$$

Note that in matrix multiplication, the order of the matrices is important. That is,

$$\overleftrightarrow{A}\,\overleftrightarrow{B} \neq \overleftrightarrow{B}\,\overleftrightarrow{A} . \tag{5.26}$$

On the other hand, the order in which the multiplication is performed is not important. Using parentheses to show which operation is performed first, we have

$$\overleftrightarrow{A}\,(\overleftrightarrow{B}\,\overleftrightarrow{C}) = (\overleftrightarrow{A}\,\overleftrightarrow{B})\,\overleftrightarrow{C} . \tag{5.27}$$

Example 5.1

Write out the three translation matrices and two refraction matrices for the single-lens problem described below. Multiply the matrices together to get the conjugate matrix.

The object is 40.0 cm from the first surface, and the lens is 1.20 cm thick and made of glass with $n = 1.50$. The curvature of the first surface is 0.300 cm^{-1} and the curvature of the second surface is 0.167 cm^{-1}. The image is 18.0 cm to the right of the second surface.

To save space I will write out the matrices as I multiply them together. Starting in object space and working toward image space we have

$$\overleftrightarrow{R}_1 \overleftrightarrow{T}_0 = \begin{bmatrix} 1 & 0 \\ -0.15 & 1 \end{bmatrix} \begin{bmatrix} 1 & 40 \\ 0 & 1 \end{bmatrix} = \begin{bmatrix} 1 & 40 \\ -0.15 & -5 \end{bmatrix},$$

$$\overleftrightarrow{T}_1 \overleftrightarrow{R}_1 \overleftrightarrow{T}_0 = \begin{bmatrix} 1 & 0.8 \\ 0 & 1 \end{bmatrix} \begin{bmatrix} 1 & 40 \\ -0.15 & -5 \end{bmatrix} = \begin{bmatrix} 0.88 & 36 \\ -0.15 & -5 \end{bmatrix},$$

$$\overleftrightarrow{R}_2 \overleftrightarrow{T}_1 \overleftrightarrow{R}_1 \overleftrightarrow{T}_0 = \begin{bmatrix} 1 & 0 \\ 0.0833 & 1 \end{bmatrix} \begin{bmatrix} 0.88 & 36 \\ -0.15 & -5 \end{bmatrix} = \begin{bmatrix} 0.88 & 36 \\ -0.0767 & -2 \end{bmatrix}.$$

And finally,

$$\overleftrightarrow{C} = \overleftrightarrow{T}_2 \overleftrightarrow{R}_2 \overleftrightarrow{T}_1 \overleftrightarrow{R}_1 \overleftrightarrow{T}_0 = \begin{bmatrix} 1 & 18 \\ 0 & 1 \end{bmatrix} \begin{bmatrix} 0.88 & 36 \\ -0.767 & -2 \end{bmatrix} = \begin{bmatrix} -0.5 & 0 \\ -0.767 & -2 \end{bmatrix}.$$

■

5.2 PROPERTIES OF THE CONJUGATE MATRIX

If you tried to work out the example on your own, chances are good you made a mistake at some point in the calculation. We need some checks on our work.

The most important check that can be made reflects the fact that the determinant of the matrices is always unity. The determinant of a 2 × 2 matrix is defined by

$$\| \overleftrightarrow{A} \| \equiv a_{11} b_{22} - a_{12} b_{21} . \tag{5.28}$$

Notice that the determinant of both the translation matrix and the refraction matrix is equal to one. This is a consequence of the way the equations were set up with the reduced separation and slope angle. Furthermore, the determinant of the product of two matrices is equal to the product of the determinants of those same two matrices:

$$\| \overleftrightarrow{A} \overleftrightarrow{B} \| = \| \overleftrightarrow{A} \| \| \overleftrightarrow{B} \| . \tag{5.29}$$

In the first example, your matrix multiplication can be checked at each stage by calculating the determinant. If you get something other than 1.0, you messed up.

While all combinations of refraction and translation matrices have the same determinant, the conjugate matrix has some very special properties. If we assume finite conjugates, then the matrix equation gives the ray height at the image in terms of the ray parameters at the object:

$$y_{k+1} = c_{11} y_0 + c_{12} \alpha_0 , \tag{5.30}$$

where the elements of the conjugate matrix are c_{11}, c_{12}, c_{21}, and c_{22}. With this notation, the elements of the conjugate matrix could be confused with the curvature of surfaces 11, 12, 21 and 22, but the context should make the meaning clear. Since we have an imaging situation, all of the rays from an object point must meet at the same image point regardless of their initial slope angle. For imaging to take place, then $c_{12} = 0$. In addition, because

$$y_{k+1} = c_{11} \, y_0 , \tag{5.31}$$

we can substitute the image and object heights

$$h' = c_{11} \, h \tag{5.32}$$

to show that $c_{11} = m$. Next, because the determinant must equal 1.0, $c_{22} = 1 / m$.

To find the remaining matrix element, imagine tracing a ray that starts from an object point and travels parallel to the optical axis in object space. From our matrix equations we have

$$\alpha_k \equiv n_k u_k = c_{21} h . \tag{5.33}$$

But, all rays with object space slope angle of zero must pass through the image space focal point (Figure 5.1). From the definition of the principal planes and focal points

$$f' = -\frac{y_1}{u_k} = -\frac{h}{u_k} . \tag{5.34}$$

Combining these two equations gives

$$c_{21} = -\frac{n_k}{f'} = -\phi , \tag{5.35}$$

the negative of the total power of the system.

Thus, for all finite conjugates, we obtain

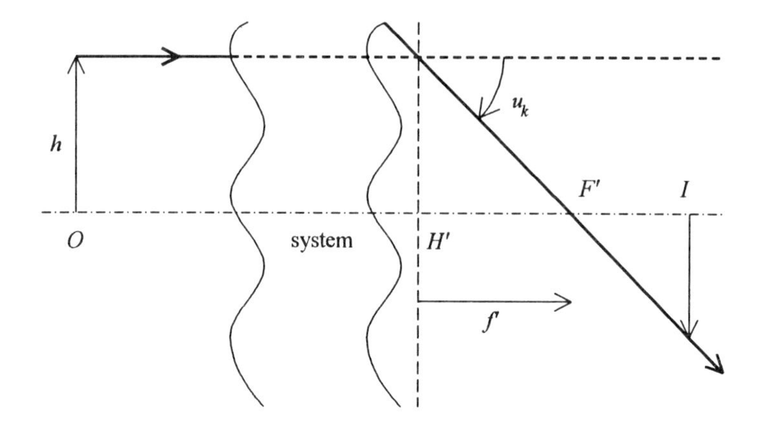

Figure 5.1 Determination of the optical power of a system.

$$\overleftrightarrow{C} = \begin{bmatrix} m & 0 \\ -\phi & \dfrac{1}{m} \end{bmatrix} . \tag{5.36}$$

Example 5.2

What is the effective focal length and magnification for the system in Example 5.1?

From the conjugate matrix we see immediately from the first element of the first row that the magnification is -0.5. Changing the sign and inverting the first element of the second row gives the focal length of $+13.04$ cm. ■

In the examples given above, the image distance is assumed to be known. However, the image distance is normally something we are trying to find out. There are a couple of ways to determine the image location using matrices. The most obvious is to construct a matrix from the object through the system to the last surface. Let's call this matrix $\overleftrightarrow{B}$ since it is not quite a full conjugate matrix.

$$\overleftrightarrow{B} = \overleftrightarrow{R}_k \cdots \overleftrightarrow{T}_j \overleftrightarrow{R}_j \cdots \overleftrightarrow{T}_1 \overleftrightarrow{R}_1 \overleftrightarrow{T}_0 , \tag{5.37}$$

$$\overleftrightarrow{C} = \overleftrightarrow{T}_k \overleftrightarrow{B} . \tag{5.38}$$

To find the image location we use this matrix to trace an axial ray from the object to the last surface:

$$\begin{bmatrix} y_k \\ \alpha_k \end{bmatrix} = \begin{bmatrix} b_{11} & b_{12} \\ b_{21} & b_{22} \end{bmatrix} \begin{bmatrix} 0 \\ \alpha_0 \end{bmatrix} . \tag{5.39}$$

Finally, use

$$t_k = -n_k \frac{y_k}{\alpha_k} \tag{5.40}$$

to determine the image location.

The procedure just shown is complete. But we can make use of Eq. 5.39 to show that

$$y_k = b_{12}\,\alpha_0 \tag{5.41}$$

and

$$\alpha_k = b_{22}\,\alpha_0 \; . \tag{5.42}$$

Therefore,

$$t_k = -n_k \frac{b_{12}}{b_{22}} \; . \tag{5.43}$$

This is a very useful shortcut.

Similarly, we don't need to find the conjugate matrix to determine the magnification. Since

$$m = \frac{\alpha_0}{\alpha_k} \, , \tag{5.44}$$

we can see that

$$m = \frac{1}{b_{22}} \; . \tag{5.45}$$

Example 5.3

What is the image location and magnification for the system in Example 5.1? Use the matrix from the object to the last surface found in Example 5.1.

$$t_k = -(1.0)\frac{(36.0)}{(-2.0)} = 18.0 ,$$

$$m = \frac{1}{(-2.0)} = -0.5 .$$

■

5.3 THE SYSTEM MATRIX

Obviously, the conjugate matrix is dependent on the object distance. We can define a system matrix which begins at the first surface of the system and ends at the last:

$$\overleftrightarrow{S} = \overleftrightarrow{R}_k \overleftrightarrow{T}_{k-1} \cdots \overleftrightarrow{T}_1 \overleftrightarrow{R}_1 . \tag{5.46}$$

The system matrix will be independent of the object distance so it can be used for different problems involving the same system. In its four elements the system matrix contains all of the important first-order properties of the lens. To simplify the notation a little, I will not use subscripts for the matrix elements. Instead, let

$$\overleftrightarrow{S} = \begin{bmatrix} a & b \\ c & d \end{bmatrix} . \tag{5.47}$$

The four elements are called *Gaussian constants*.

To see how to use the system matrix, let us begin by writing down a conjugate matrix

$$\overleftrightarrow{C} = \overleftrightarrow{T}_k \overleftrightarrow{S} \overleftrightarrow{T}_0 . \tag{5.48}$$

In expanded form, we can see all of the individual elements:

$$\begin{bmatrix} m & 0 \\ -\phi & \frac{1}{m} \end{bmatrix} = \begin{bmatrix} a + c\frac{t_k}{n_k} & a\frac{t_0}{n_0} + b + c\frac{t_0 t_k}{n_0 n_k} + d\frac{t_k}{n_k} \\ c & c\frac{t_0}{n_0} + d \end{bmatrix} . \tag{5.49}$$

There are two ways we can take advantage of the system matrix. First, if you know the object distance, then the magnification can be found by

$$m = \frac{1}{c\frac{t_0}{n_0} + d} . \tag{5.50}$$

With knowledge of the magnification, the image location can be found:

$$t_k = \frac{n_k}{c}(m - a) . \tag{5.51}$$

Example 5.4

Given the system matrix

$$\overleftrightarrow{S} = \begin{bmatrix} 2 & 46.6667 \\ -0.025 & -0.08333 \end{bmatrix} ,$$

and object location $t_0 = 10.0$, determine the magnification and image location. Assume the lens is surrounded by air.

First, we can find the magnification from

$$m = \frac{1}{(-0.025)(10.0) + (-0.08333)} = -3.0 .$$

Now, we can find the image location without tracing a ray using

$$t_k = \frac{(1.0)}{(-0.025)}((-3.0) - (2.0)) = 20.0 .$$

■

The other way that we can use Eq. 5.50 is if a certain magnification is desired. Then

$$t_0 = \frac{n_0}{c}\left(\frac{1}{m} - d\right) . \tag{5.52}$$

As mentioned earlier, the Gaussian constants contain all of the first-order information about the system. We should, therefore, be able to determine the location of the cardinal points in terms of the Gaussian constants.

First, recall the definition of principal planes as planes of unit magnification. The conjugate matrix will transfer rays from the object space principal plane to the image space principal plane. Also remember that t_0 is measured from the object to the first surface but l_H is measured from the first surface back to the object space principal plane. Therefore

$$l_H = -t_0 = \frac{n_0}{c}(d - 1) . \tag{5.53}$$

Similarly, for the image space principal plane we have

$$l_{H'} = t_k = \frac{n_k}{c}(1 - a) . \tag{5.54}$$

Equation 5.49 also shows us that

$$c = -\frac{n_k}{f'} . \tag{5.55}$$

Or stated another way

$$f' = -\frac{n_k}{c} . \tag{5.56}$$

From here it is fairly easy to show that

$$l_{F'} = af' \tag{5.57}$$

and

$$l_F = df . \tag{5.58}$$

The nodal points are also conjugates. They are defined by $u_k / u_0 = 1$. We can use this definition to find the magnification of the nodal planes:

$$m = \frac{n_0 u_0}{n_k u_k} = \frac{n_0}{n_k} . \quad (5.59)$$

As before, we can put the magnification into Eqs. 5.51 and 5.52 to get the locations of the nodal points

$$l_{N'} = f'\left(\frac{n_0}{n_k} - a \right) \quad (5.60)$$

and

$$l_N = f\left(\frac{n_k}{n_0} - d \right) . \quad (5.61)$$

As always, if $n_0 = n_k$, then the nodal planes and the principal planes are in the same locations.

Example 5.5

Given the system matrix

$$\overleftrightarrow{S} = \begin{bmatrix} 0.8 & 2.6 \\ -0.2 & 0.6 \end{bmatrix},$$

find the locations of the cardinal points of the system.

Assuming that the system is in air, we have

$$f' = -\frac{1}{c} = -\frac{1}{(-0.2)} = 5.0 ,$$

$$l_F = df = (0.6)(-5.0) = -3.0 ,$$

$$l_H = f(d-1) = (-5.0)((0.6)-1) = 2.0 ,$$

$$l_{F'} = af' = (0.8)(5.0) = 4.0 ,$$

$$l_{H'} = f'(a - 1) = (5.0)((0.8) - 1) = -1.0 .$$

■

The principal plane conjugate matrix can be used with another pair of translations to find a matrix equation for any pair of conjugates with the object and image distances measured from the principal planes instead of the first and last surfaces. This technique is simply breaking the translation from the object to the first surface of the system into two parts. The first part is a translation from the object to the first principal plane. The second part is a translation from the first principal plane to the first surface. The second translation is already included in the principal plane conjugate matrix. Similar arguments apply to the image translation. Therefore,

$$\begin{bmatrix} m & 0 \\ -\phi & \dfrac{1}{m} \end{bmatrix} = \begin{bmatrix} 1 & \dfrac{s'}{n_k} \\ 0 & 1 \end{bmatrix} \begin{bmatrix} 1 & 0 \\ -\phi & 1 \end{bmatrix} \begin{bmatrix} 1 & -\dfrac{s}{n_0} \\ 0 & 1 \end{bmatrix} . \tag{5.62}$$

When these matrices are multiplied together, some interesting equations result. For instance, the upper right element (which must be zero) becomes

$$-\frac{s}{n_0} + \frac{s s'}{n_0 n_k}\phi + \frac{s'}{n_k} = 0 . \tag{5.63}$$

From this, the Gaussian form of the imaging equation is easily found:

$$\frac{n_k}{s'} - \frac{n_0}{s} = \frac{n_k}{f'} = -\frac{n_0}{f} . \tag{5.64}$$

Using the upper left element and the magnification yields

$$m = \frac{n_0 s'}{n_k s} . \tag{5.65}$$

Plugging in $n_0 = n_k = 1$ gives the normal thin lens equations.

The third set of cardinal points, the focal points, are not conjugates, so conjugate matrices can't be set up. Still, it is possible to write a matrix equation for ray tracing from one focal plane to the other, it just won't be a conjugate equation. Since the two focal planes are located distances $-f$ and f' from the principal planes, a transfer matrix would look like

$$\begin{bmatrix} 1 - \frac{f'}{n_k}\phi & -\frac{f}{n_0} + \frac{ff'}{n_0 n_k}\phi + \frac{f'}{n_k} \\ -\phi & 1 + \frac{f}{n_0}\phi \end{bmatrix} . \tag{5.66}$$

This matrix can be simplified because

$$1 - \frac{f'}{n_k}\phi = 1 + \frac{f}{n_0}\phi = 0 . \tag{5.67}$$

We end up with

$$\begin{bmatrix} 0 & \frac{1}{\phi} \\ -\phi & 0 \end{bmatrix} \tag{5.68}$$

as the transfer matrix from the object space focal point to the image space focal point.

Back in Chapter 2, object and image distances measured from the focal points were called z and z'. Using these distances, a general conjugate matrix equation can be written using the focal point matrix

$$\begin{bmatrix} m & 0 \\ -\phi & \frac{1}{m} \end{bmatrix} = \begin{bmatrix} 1 & \frac{z'}{n_k} \\ 0 & 1 \end{bmatrix} \begin{bmatrix} 0 & \frac{1}{\phi} \\ -\phi & 0 \end{bmatrix} \begin{bmatrix} 1 & -\frac{z}{n_0} \\ 0 & 1 \end{bmatrix} . \tag{5.69}$$

After multiplication we have

$$\begin{bmatrix} m & 0 \\ -\phi & \frac{1}{m} \end{bmatrix} = \begin{bmatrix} -\phi\frac{z'}{n_k} & \phi\frac{zz'}{n_0 n_k} + \frac{1}{\phi} \\ -\phi & \phi\frac{z}{n_0} \end{bmatrix} . \tag{5.70}$$

The upper right-hand element yields

$$z z' = f f' . \tag{5.71}$$

The top left element gives

$$m = -\frac{z'}{f'} . \tag{5.72}$$

Similarly, the bottom right element yields

$$m = -\frac{f}{z} . \tag{5.73}$$

5.4 STOPS, PUPILS, WINDOWS, AND MATRICES

While the system matrix contains all of the longitudinal first-order properties of a system, it does not tell us directly such things as which surfaces are the aperture or field stops or where the pupils and windows are located. The clear radius information must be included somehow to answer such questions. Matrix methods can still be used, however, along with the clear radius data.

We begin by looking at how matrices can help determine which surface is the aperture stop. Recall that the aperture stop is the stop or lens rim which limits the light coming from an on axis object point. In other words, the aperture stop is responsible for the maximum object space slope angle for an axial ray which makes it through the system. Therefore, to find the aperture stop we need to trace an axial ray through the system and check how the ray height compares with the clear radius at each surface.

To trace a ray from the object to a particular surface j requires a transfer matrix of the form

$$\overleftrightarrow{M} = \overleftrightarrow{T}_{j-1} \overleftrightarrow{R}_{j-1} \cdots \overleftrightarrow{R}_1 \overleftrightarrow{T}_0 . \tag{5.74}$$

Note that this matrix is neither a complete system matrix nor a conjugate matrix, but simply a transfer matrix from the object to surface j. Let the elements of this matrix be m_{11}, m_{12}, m_{21}, and m_{22}. Since $y_0 = 0$ for an axial ray, the height of the ray at surface j is given by

$$y_j = m_{12} \alpha_0 . \tag{5.75}$$

If we want the ray to just barely pass through surface j, as shown in Figure 5.2, then $|y_j| = R_j$. We can make this substitution and rearrange Eq. 5.74 to find the object

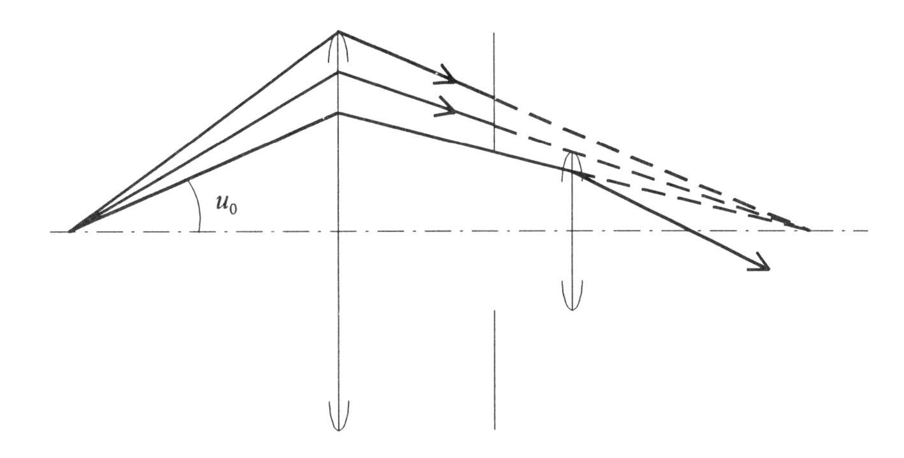

Figure 5.2 A lens system with various axial rays aimed at lens rims and the edge of the stop.

space reduced slope angle for a marginal ray:

$$\alpha_0 = \left| \frac{R_j}{m_{12}} \right| . \tag{5.76}$$

Now remember, there is a different transfer matrix for each surface as well as a different value for clear radius. The aperture stop is the surface with the smallest ratio.

Example 5.6

An optical system consists of three thin lenses. The first lens has a focal length of 64.0 cm and a clear radius of 5.0 cm. The second lens has $f' = 20.0$ cm, $R = 2.0$. The third lens also has $f' = 20.0$ but $R = 5.0$. Each lens is separated from the next by 40.0 cm. An object 8.0 cm high is located 320 cm in front of the first lens. Which lens is the aperture stop?

The first thing we need to do is find the transfer matrices from the object to each surface. For the first surface we have simply

$$\overleftrightarrow{M} = \overleftrightarrow{T}_0 = \begin{bmatrix} 1 & 320 \\ 0 & 1 \end{bmatrix} .$$

We can immediately calculate the desired ratio:

$$\left|\frac{R_1}{m_{12}}\right| = \frac{(5.0)}{(320)} = 0.015625 \; .$$

For the second surface, we must multiply the first translation matrix by the first refraction matrix and by the second translation matrix to get

$$\overleftrightarrow{M} = \overleftrightarrow{T}_1 \, \overleftrightarrow{R}_1 \, \overleftrightarrow{T}_0 = \begin{bmatrix} 0.375 & 160 \\ -0.015625 & -4 \end{bmatrix} .$$

Again the aperture stop ratio can be quickly determined:

$$\left|\frac{R_2}{m_{12}}\right| = \frac{(2.0)}{(160)} = 0.0125 \; .$$

And finally,

$$\overleftrightarrow{M} = \overleftrightarrow{T}_2 \, \overleftrightarrow{R}_2 \, \overleftrightarrow{T}_1 \, \overleftrightarrow{R}_1 \, \overleftrightarrow{T}_0 = \begin{bmatrix} -1 & -320 \\ -0.034375 & -12 \end{bmatrix} ,$$

from which we get

$$\left|\frac{R_3}{m_{12}}\right| = \left|\frac{(5.0)}{(-320)}\right| = 0.015625 \; .$$

You can see that surface 2 has the smallest value, and therefore surface 2 is the aperture stop. As always, the determinant of each matrix must be 1.0. If the matrix multiplications are done in the proper order, the results of each step can be used in the next step. ■

Now this looks like a lot more work, but remember that you end up finding each of the various transfer matrices in the process of finding the conjugate matrix. You just need to multiply your matrices in the proper order from object to image.

Once the aperture stop has been identified, the next step is to trace a chief ray to find the pupils. The entrance pupil, in general, will require a backward ray trace. We have not yet discussed how to do a backward ray trace with matrices, so we will do so now.

We could set up matrix equations like

$$\begin{bmatrix} y_{j-1} \\ \alpha_{j-1} \end{bmatrix} = \begin{bmatrix} 1 & -\tau \\ 0 & 1 \end{bmatrix} \begin{bmatrix} y_j \\ \alpha_{j-1} \end{bmatrix} \tag{5.77}$$

and

$$\begin{bmatrix} y_j \\ \alpha_{j-1} \end{bmatrix} = \begin{bmatrix} 1 & 0 \\ \phi_j & 1 \end{bmatrix} \begin{bmatrix} y_j \\ \alpha_j \end{bmatrix}. \tag{5.78}$$

This method requires setting up a new set of matrices and calculating their product. But we have already set up matrices to trace rays in the forward direction. There should be some way of using that information, After all, the rays must take the same path backwards that they did forward according to the principal of reversibility.

If we start with a matrix which describes a transfer from just in front of the first surface to just in front of some other surface j, our forward ray trace looks like

$$\begin{bmatrix} y_j \\ \alpha_{j-1} \end{bmatrix} = \overleftrightarrow{P} \begin{bmatrix} y_1 \\ \alpha_0 \end{bmatrix}. \tag{5.79}$$

Multiplying both sides of the previous equation by the inverse matrix gives

$$\overleftrightarrow{P}^{-1} \begin{bmatrix} y_j \\ \alpha_{j-1} \end{bmatrix} = \overleftrightarrow{P}^{-1} \overleftrightarrow{P} \begin{bmatrix} y_1 \\ \alpha_0 \end{bmatrix} = \begin{bmatrix} 1 & 0 \\ 0 & 1 \end{bmatrix} \begin{bmatrix} y_1 \\ \alpha_0 \end{bmatrix} = \begin{bmatrix} y_1 \\ \alpha_0 \end{bmatrix}. \tag{5.80}$$

For our matrices, the inverses are easy to find according to

$$\overleftrightarrow{P}^{-1} = \begin{bmatrix} p_{22} & -p_{12} \\ -p_{21} & p_{11} \end{bmatrix}. \tag{5.81}$$

With the inverse matrix available, it is a simple matter to trace a chief ray from the aperture stop back into image space. As shown in Figure 5.3, the entrance pupil location is given by

$$l_E = -\frac{\bar{y}_1}{\bar{u}_0}, \tag{5.82}$$

where

$$\begin{bmatrix} \bar{y}_1 \\ \bar{\alpha}_0 \end{bmatrix} = \begin{bmatrix} p_{22} & -p_{12} \\ -p_{21} & p_{11} \end{bmatrix} \begin{bmatrix} 0 \\ \bar{\alpha}_A \end{bmatrix}. \tag{5.83}$$

From this matrix equation, we can find the entrance pupil position more directly with

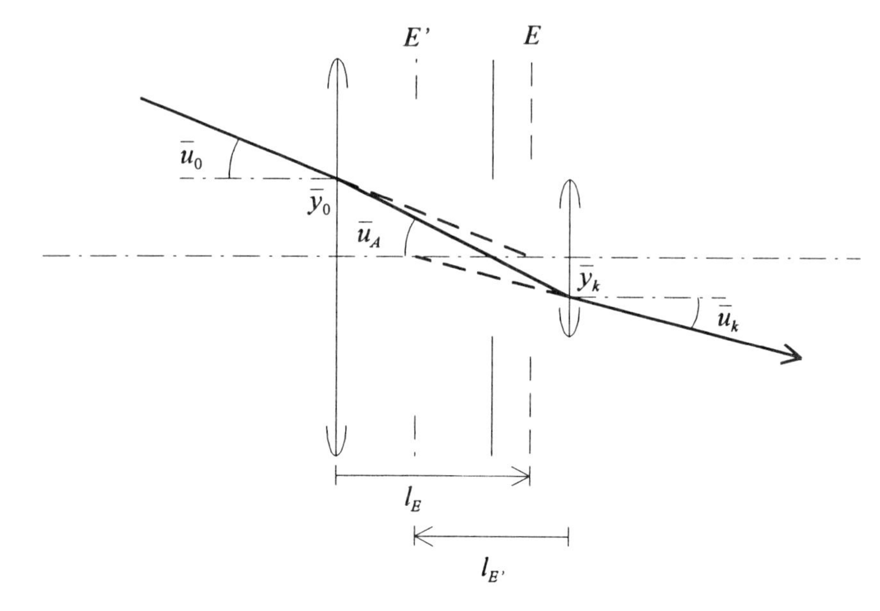

Figure 5.3 Entrance and exit pupil locations for a typical system.

$$l_E = n_0 \frac{p_{12}}{p_{11}} . \tag{5.84}$$

We can find the entrance pupil magnification by looking at the ratio of the reduced slope angles in the aperture space and object space

$$m_E = \frac{\bar{\alpha}_A}{\bar{\alpha}_0} = \frac{1}{p_{11}} . \tag{5.85}$$

Of course, the entrance pupil size follows directly.

The exit pupil is found by a similar procedure, but without the step of finding the inverse matrix. Once the aperture stop is determined, a matrix from the aperture stop to the last surface must be calculated:

$$\overleftrightarrow{P}' = \overleftrightarrow{R}_k \, \overleftrightarrow{T}_{k-1} \cdots \overleftrightarrow{T}_A . \tag{5.86}$$

Since the entrance pupil location is determined by

$$l_{E'} = -\frac{\bar{y}_k}{\bar{u}_k} , \tag{5.87}$$

it is easy to show that

$$l_{E'} = -n_k \frac{p_{12}'}{p_{22}'} \tag{5.88}$$

and

$$m_{E'} = \frac{\bar{\alpha}_A}{\bar{\alpha}_k} = \frac{1}{p_{22}'} . \tag{5.89}$$

Keep in mind that even though the equations for the entrance and exit pupils look somewhat similar, the matrix elements are entirely different.

Example 5.7

Find the entrance and exit pupils for the system described in Example 5.6.

We have already determined that surface 2 is the aperture stop. The next step is to find the matrix from the first surface to the aperture stop:

$$\overleftrightarrow{P} = \overleftrightarrow{T}_1 \overleftrightarrow{R}_1 = \begin{bmatrix} 0.375 & 40 \\ -0.015625 & 1 \end{bmatrix} .$$

Now we can find the position of the entrance pupil:

$$l_E = n_0 \frac{p_{12}}{p_{11}} = (1.0) \frac{(40)}{(0.375)} = 106.7 .$$

The entrance pupil size is

$$R_E = m_E R_A = \frac{1}{p_{11}} R_A = \frac{1}{(0.375)} (2.0) = 5.33 .$$

Now for the exit pupil, we must find a matrix to take us from the aperture stop to image space:

$$\overleftrightarrow{P}' = \overleftrightarrow{R}_3 \overleftrightarrow{T}_2 = \begin{bmatrix} 1 & 40 \\ -0.05 & -1 \end{bmatrix} .$$

From this matrix, we find

$$l_{E'} = -n_k \frac{p'_{12}}{p'_{22}} = -(1.0)\frac{(40)}{(-1.0)} = 40$$

and

$$R_{E'} = m_{E'} R_A = \frac{1}{p'_{22}} R_A = \frac{1}{(-1.0)}(2.0) = -2.0 \ .$$

■

Finding the field stop and windows can be done along similar lines. Forward and backward matrices must be set up between the aperture stop and each surface of the system including the object and image planes. These matrices are inspected along with the corresponding clear radii to find the minimum ratio just as we did with the aperture stop. More matrices are then set up to find the windows just as we did with the pupils.

5.5 FIRST-ORDER LAYOUT OF A ZOOM LENS

A zoom lens is defined as a lens with variable focal length but a fixed image location relative to the object. When object and image distances are measured from the principal planes, imaging requires that the distance obey the standard Gaussian lens equation. If the effective focal length changes, then the object and image distances must also change. Since the magnification is equal to the ratio of the image to object distances, we have the alternate definition of a zoom lens as a lens with variable magnification and fixed object and image positions. For an object at infinity, the linear magnification is zero. In this case we switch to using the angular magnification.

Our task in this section is to use matrices to try to determine how to complete the first-order layout of a zoom lens. We don't have to use matrices to solve this problem, but we choose to use them here as an example.

Obviously, a single thin lens cannot be a zoom lens, since its focal length is fixed. Even though the lens in the eye has a variable focal length and a fixed image position, it is not a zoom lens either. When the focal length of the eye changes, the object distance changes. It is a variable focus system, but not a zoom lens.

Similarly, a system consisting of two separated thin lenses can have a variable focal length by simply changing the spacing between the two lenses. The system matrix is found by combining two thin lens refraction equations and one translation equation:

$$\overleftrightarrow{S} = \begin{bmatrix} 1 & 0 \\ -\dfrac{1}{f_2'} & 1 \end{bmatrix} \begin{bmatrix} 1 & t_1 \\ 0 & 1 \end{bmatrix} \begin{bmatrix} 1 & 0 \\ -\dfrac{1}{f_1'} & 1 \end{bmatrix} . \tag{5.90}$$

After matrix multiplication we have

$$\overleftrightarrow{S} = \begin{bmatrix} 1 - \dfrac{t_1}{f_1'} & t_1 \\ \dfrac{t_1}{f_1' f_2'} - \dfrac{1}{f_1'} - \dfrac{1}{f_2'} & 1 - \dfrac{t_1}{f_2'} \end{bmatrix} . \tag{5.91}$$

The first element of the second row is the negative inverse of the focal length of the system:

$$f' = \frac{f_1' f_2'}{f_1' + f_2' - t_1} . \tag{5.92}$$

Thus, the first condition of a zoom system is satisfied. By moving the lenses relative to each other, the focal length can be changed. However, the second condition, that the image plane remains fixed, is not normally met. Assuming that the object is at infinity, then the image is at the back focal plane. This location is given by $l_{F'} = a\,f'$ or

$$l_{F'} = \frac{(f_1' - t_1)\, f_2'}{f_1' + f_2' - t_1} . \tag{5.93}$$

In other words, the entire lens must move to keep the object at infinity in focus.

A zoom lens requires a minimum of three thin lenses. With the substitution of the focal point separations for the lens separations, the three-lens system matrix is

$$a = 1 - \frac{t_1}{f_1'} - \frac{t_2}{f_1'} - \frac{t_2}{f_2'} + \frac{t_1 t_1}{f_1' f_2'} , \tag{5.94}$$

$$b = t_1 + t_2 - \frac{t_1 t_2}{f_2'} , \tag{5.95}$$

$$c = -\frac{1}{f_1'} - \frac{1}{f_2'} - \frac{1}{f_3'} + \frac{t_1}{f_1' f_2'} + \frac{t_1}{f_1' f_3'} + \frac{t_2}{f_1' f_3'} + \frac{t_2}{f_2' f_3'} + \frac{t_1 t_2}{f_1' f_2' f_3'} , \tag{5.96}$$

$$d = 1 - \frac{t_1}{f_2'} - \frac{t_1}{f_3'} - \frac{t_2}{f_3'} + \frac{t_1 t_2}{f_2' f_3'} . \tag{5.97}$$

Once again, we see that the focal length, given by

$$f' = \frac{f_1' f_2' f_3'}{f_1'(f_2' + f_3' - t_2) + f_2'(f_3' - t_1 - t_2) + t_1(f_3' - t_2)} , \tag{5.98}$$

changes as the separation between the lenses changes. The back focal length is given by

$$l_{F'} = \frac{(f_1'(f_2' - t_2) - f_2'(t_1 + t_2) - t_1 t_2) f_3'}{f_1'(f_2' + f_3' - t_2) + f_2'(f_3' - t_1 - t_2) + t_1(f_3' - t_2)} . \tag{5.99}$$

Clearly, the back focal length changes as we change the lens positions to get different focal lengths. But what is not obvious from Eq. 5.99, is that the back focal length does not change very much. For carefully chosen values of the focal lengths and lens separations, the numerator varies closely with the denominator.

If we move the first and third lenses together, while the second lens remains fixed, as shown in Figure 5.4, we have what is known as an optically compensated zoom lens. The requirement that the first and third lenses move together can be expressed mathematically as

$$t_1 + t_2 = k_1 , \tag{5.100}$$

where k_1 is a constant.

Since the third lens is moving, it is expected that the back focal length will vary. But because the second lens is fixed, the zoom requirement becomes

$$t_2 + l_{F'} = k_2 , \tag{5.101}$$

where k_2 is another constant. In actual practice, k_2 will not be exactly constant, but nearly so.

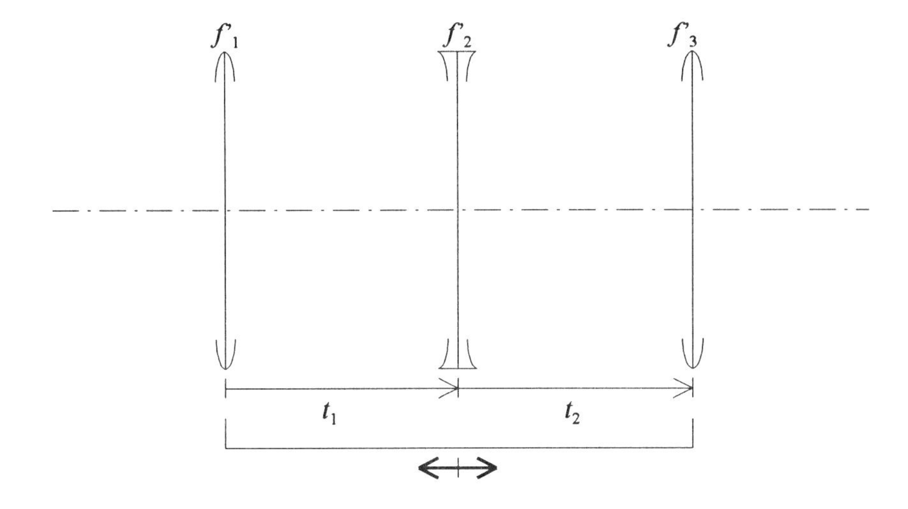

Figure 5.4 Layout of an optically compensated zoom lens. Lenses 1 and 3 move together.

Example 5.8

Design a zoom lens with a focal length which varies from 5.0 cm to 15.0 cm while keeping the back focal length at approximately 5.0 cm.

Our first step is to choose reasonable values for the constants k_1 and k_2. Obviously, k_2 must be larger than the back focal length. Therefore, choose

$$k_2 = 6.0 \ .$$

The other constant is chosen to allow space for the lenses to move:

$$k_1 = 7.0 \ .$$

The focal length of the system is now a function of the focal lengths of the three lenses and the separation between lens 1 and 2. If we assume that the minimum focal length is reached at a particular separation called t_{min} and the maximum focal length is reached at a separation called t_{max}, then we could write down two equations in the five unknowns.

$$\frac{f_1' f_2' f_3'}{f_1' f_2' - 7 f_1' - f_1' t_{min} + f_1' f_3' + 7 t_{min} - t_{min}^2 - f_3' t_{min} - 7 f_2' + f_2' f_3'} = 5.0$$

and

$$\frac{f_1' f_2' f_3'}{f_1' f_2' - 7 f_1' - f_1' t_{max} + f_1' f_3' + 7 t_{max} - t_{max}^2 - f_3' t_{max} - 7 f_2' + f_2' f_3'} = 15.0 \ .$$

To get three more equations, we use the zoom condition. If we set up the zoom condition for the minimum and maximum positions of the first lens and a point somewhere in between, then the value for k_2 will never be able to vary much from its assigned value.

Thus Eq. 5.101 can be used three times:

$$7 - t_{min} + \frac{(f_1' f_2' - 7 f_1' + f_1' t_{min} + 7 t_{min} - t_{min}^2 - 7 f_2') f_3'}{f_1' f_2' - 7 f_1' - f_1' t_{min} + f_1' f_3' + 7 t_{min} - t_{min}^2 - f_3' t_{min} - 7 f_2' + f_2' f_3'} = 6.0 \ ,$$

$$7 - t_{max} + \frac{(f_1' f_2' - 7 f_1' + f_1' t_{max} + 7 t_{max} - t_{max}^2 - 7 f_2') f_3'}{f_1' f_2' - 7 f_1' - f_1' t_{max} + f_1' f_3' + 7 t_{max} - t_{max}^2 - f_3' t_{max} - 7 f_2' + f_2' f_3'} = 6.0 \ ,$$

and

$$7 - t_{mid} + \frac{(f_1' f_2' - 7 f_1' + f_1' t_{mid} + 7 t_{mid} - t_{mid}^2 - 7 f_2') f_3'}{f_1' f_2' - 7 f_1' - f_1' t_{mid} + f_1' f_3' + 7 t_{mid} - t_{mid}^2 - f_3' t_{mid} - 7 f_2' + f_2' f_3'} = 6.0 \ ,$$

where

$$t_{mid} = \frac{t_{min} + t_{max}}{2} \ .$$

We now have five complicated equations in our five unknowns. A computer algebra program such as Maple can be used to solve the equations for unknowns. The results are $f_1 = 9.630$ cm, $f_2 = -1.463$ cm, $f_3 = 2.045$ cm, $t_{min} = 5.199$ cm, and $t_{max} =$ 6.555 cm. ■

5.6 THE *Y-NU* RAY TRACE AND MATRICES

The biggest problem with matrix methods is that the process of writing down the various refraction and translation matrices and multiplying them together is lengthy and prone to error. An attractive alternative is to use the standard *y-nu* ray trace table to trace two rays from the first surface to the last surface. As before, the second ray will be denoted by a bar over the ray parameters. Of course, the *y-nu* ray trace gives the same result that the matrix ray trace would give for these two rays. In addition, because the equations are linear, if we can trace any two linearly independent rays (nonzero optical invariant), then we have enough information to trace any ray. Put another way, this means that the information from the two *y-nu* ray traces can be used to determine the four elements of the system matrix or the four Gaussian constants.

Consider the system matrix used to trace a ray from the object space side of the first surface to the image space side of the last surface:

$$\begin{bmatrix} y_k \\ \alpha_k \end{bmatrix} = \begin{bmatrix} a & b \\ c & d \end{bmatrix} \begin{bmatrix} y_1 \\ \alpha_0 \end{bmatrix} . \tag{5.102}$$

When written out as a regular set of equations we get

$$y_k = a y_1 + b \alpha_0 , \tag{5.103}$$

$$\alpha_k = c y_1 + d \alpha_0 . \tag{5.104}$$

If we write out the equations again for a second ray we get two more equations:

$$\bar{y}_k = a \bar{y}_1 + b \bar{\alpha}_0 , \tag{5.105}$$

$$\bar{\alpha}_k = c \bar{y}_1 + d \bar{\alpha}_0 . \tag{5.106}$$

We now have four equations in the four Gaussian constants. Inverting gives

$$a = \frac{[\alpha_0 \bar{y}_k - \bar{\alpha}_0 y_k]}{\Lambda} , \tag{5.107}$$

$$b = \frac{[\bar{y}_1 y_k - y_1 \bar{y}_k]}{\Lambda} , \tag{5.108}$$

$$c = \frac{[\alpha_0 \bar{\alpha}_k - \bar{\alpha}_0 \alpha_k]}{\Lambda} , \tag{5.109}$$

and

$$d = \frac{[\alpha_k \bar{y}_1 - \bar{\alpha}_k y_1]}{\Lambda} . \tag{5.110}$$

In these equations, the optical invariant $\Lambda = \alpha_0 \bar{y}_1 - \bar{\alpha}_0 y_1$ is calculated for the two particular rays used in the *y-nu* ray trace. If the marginal ray and full-field chief ray are used, then all quantities have their normal meaning. It is not necessary to use these two rays, however. Any two rays which are linearly independent (nonzero optical invariant) will suffice.

As an example (Figure 5.5), select the first ray so that

$$\begin{bmatrix} y_1 \\ \alpha_0 \end{bmatrix} = \begin{bmatrix} 1.0 \\ 0.0 \end{bmatrix} . \tag{5.111}$$

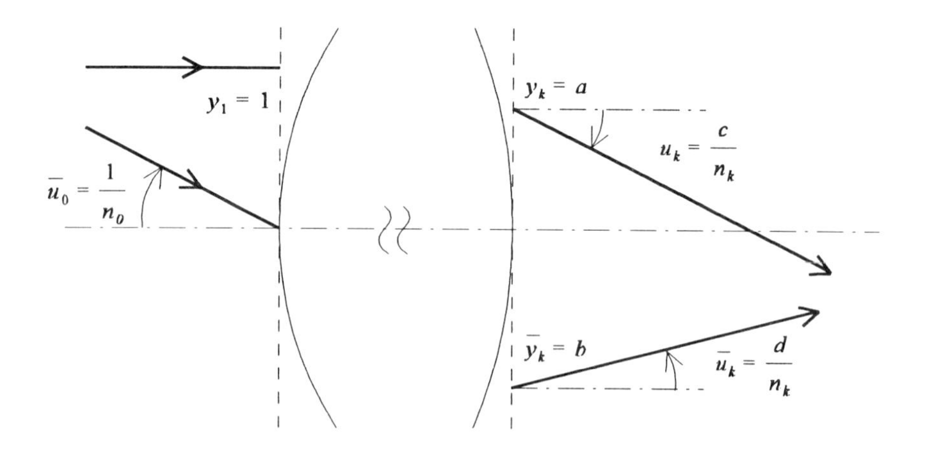

Figure 5.5 With a careful choice of input parameters, the results of a *y-nu* ray trace gives the Gaussian constants for the system.

Select the second ray

$$\begin{bmatrix} \bar{y}_1 \\ \bar{\alpha}_0 \end{bmatrix} = \begin{bmatrix} 0.0 \\ 1.0 \end{bmatrix} . \tag{5.112}$$

With these choices, evaluating the elements of the system matrix becomes very simple since

$$\begin{aligned} \Lambda &= -1.0 , \\ a &= y_k , \\ b &= \bar{y}_k , \\ c &= \alpha_k , \\ d &= \bar{\alpha}_k . \end{aligned} \tag{5.113}$$

Example 5.9

Determine the system matrix for the system described in Example 5.1 using *y-nu* ray tracing.

The *y-nu* ray trace for this system is shown in Table 5.1. From the ray trace we find that the system matrix is

$$\overleftrightarrow{S} = \begin{bmatrix} 0.88 & 0.8 \\ -0.07667 & 1.06667 \end{bmatrix} .$$

■

SUMMARY

All of the paraxial optics we have learned so far can be expressed in a matrix formulation where the ray parameters are represented by a two element column vector

Table 5.1 Ray trace table for Example 5.9.

	0	1	2
c		0.3	0.16667
t		1.2	
n	1.0	1.5	1.0
$-\phi$		-0.15	0.08333
τ		0.8	
y		1.0	0.88
α	0.0	-0.15	-0.07667
y		0.0	0.8
α	1.0	1.0	1.06667

$$\vec{y}_j = \begin{bmatrix} y_j \\ \alpha_{j-1} \end{bmatrix} . \tag{5.114}$$

The actual ray tracing is performed by means of translation matrices of the form

$$\overleftrightarrow{T}_j = \begin{bmatrix} 1 & \tau_j \\ 0 & 1 \end{bmatrix} \tag{5.115}$$

and refraction matrices of the form

$$\overleftrightarrow{R}_j = \begin{bmatrix} 1 & 0 \\ -\phi_j & 1 \end{bmatrix} . \tag{5.116}$$

A conjugate matrix is one which connects the object plane with the image plane. The conjugate matrix has the form

$$\overleftrightarrow{C} = \overleftrightarrow{T}_k \overleftrightarrow{R}_k \cdots \overleftrightarrow{R}_1 \overleftrightarrow{T}_0 = \begin{bmatrix} m & 0 \\ -\phi & \frac{1}{m} \end{bmatrix} . \tag{5.117}$$

To find the image size and location, we need to trace a ray from the object to the last surface of the system:

$$\overleftrightarrow{B} = \overleftrightarrow{R}_k \overleftrightarrow{T}_{k-1} \cdots \overleftrightarrow{T}_1 \overleftrightarrow{R}_1 \overleftrightarrow{T}_0 . \tag{5.118}$$

Then the image location is given by

$$t_k = -n_k \frac{b_{12}}{b_{22}} , \tag{5.119}$$

and the magnification is

$$m = \frac{1}{b_{22}} . \tag{5.120}$$

A system matrix contains information about the optical system only. No translation is included from the object or to the image.

$$\vec{\vec{S}} = \vec{\vec{R}}_k \vec{\vec{T}}_{k-1} \cdots \vec{\vec{T}}_1 \vec{\vec{R}}_1 = \begin{bmatrix} a & b \\ c & d \end{bmatrix} . \tag{5.121}$$

The elements of the system matrix have special meaning. For a system in air,

$$a = \frac{l_{F'}}{f'} , \tag{5.122}$$

$$b = f' + \frac{l_F l_{F'}}{f'} , \tag{5.123}$$

$$c = -\frac{1}{f'} , \tag{5.124}$$

$$d = \frac{l_F}{f'} . \tag{5.125}$$

The system matrix can be used with any object to find the image size and location. The magnification for a particular conjugate pair is given by

$$\frac{1}{m} = c \frac{t_0}{n_0} + d . \tag{5.126}$$

The image location is found with

$$t_k = \frac{n_k}{c}(m - a) . \tag{5.127}$$

Matrix methods can also be used to determine which surface is the aperture stop. For each surface j in the optical system, calculate the matrix:

$$\overleftrightarrow{M}(j) = \overleftrightarrow{T}_{j-1} \overleftrightarrow{R}_{j-1} \cdots \overleftrightarrow{R}_1 \overleftrightarrow{T}_0 . \tag{5.128}$$

Then compare the clear radius of each surface with the upper right element of each matrix:

$$\left| \frac{R_j}{m_{12}(j)} \right| . \tag{5.129}$$

The smallest value for this ratio defines the minimum object space slope angle of an axial ray and determines which surface is the aperture stop. Because pupils are images of the aperture stop, we can find the pupil size and location using matrices. For the entrance pupil, find the matrix:

$$\overleftrightarrow{P} = \overleftrightarrow{T}_{A-1} \overleftrightarrow{R}_{A-1} \cdots \overleftrightarrow{T}_1 \overleftrightarrow{R}_1 . \tag{5.130}$$

Well, actually we need the inverse of this matrix, but this matrix is available from the calculations for determining the aperture stop. Anyway, the entrance pupil location is given by

$$l_E = n_0 \frac{p_{12}}{p_{11}} , \tag{5.131}$$

and the entrance pupil magnification is

$$m_E = \frac{1}{p_{11}} . \tag{5.132}$$

To find the exit pupil location we need to determine the matrix which takes us from the aperture stop to image space

$$\overleftrightarrow{P}' = \overleftrightarrow{R}_k \overleftrightarrow{T}_{k-1} \cdots \overleftrightarrow{R}_{A+1} \overleftrightarrow{T}_A . \tag{5.133}$$

With this matrix, the exit pupil position is

$$l_{E'} = -n_k \frac{p'_{12}}{p'_{22}}, \tag{5.134}$$

and the exit pupil magnification is

$$m_{E'} = \frac{1}{p'_{22}}. \tag{5.135}$$

Finally, it should be noted that matrices can be used to derive complicated formulae by merely replacing numbers with symbols and performing the normal manipulations. An example of this procedure is presented for an optically corrected zoom lens.

REFERENCES

R. Kingslake, *Lens Design Fundamentals* (Academic Press, Orlando, FL, 1978).

A. Nussbaum and R. Phillips, *Contemporary Optics for Scientists and Engineers* (Prentice-Hall, Englewood Cliffs, NJ, 1976).

PROBLEMS

5.1 For 2×2 matrices show that $\overleftrightarrow{A}\overleftrightarrow{B} \neq \overleftrightarrow{B}\overleftrightarrow{A}$ in general.

5.2 For 2×2 matrices show that $\overleftrightarrow{A}(\overleftrightarrow{B}\overleftrightarrow{C}) = (\overleftrightarrow{A}\overleftrightarrow{B})\overleftrightarrow{C}$.

5.3 Find the general form of the inverse of a 2×2 matrix. Start with the equations (5.103) and (5.104). Solve these two equations for y_1 and α_0. Write the resulting pair of equations as a matrix equation.

5.4 Equations (5.107) through (5.110) can be used to calculate the determinant of the system matrix in terms of *y-nu* values. Given the formula for the optical invariant in object space and that the determinant must equal one, find the formula for the optical invariant in image space.

5.5 Write down in correct order, but do not multiply, the refraction and translation matrices for the system: $c_1 = 0.17900$, $t_1 = 0.90$, $n_1 = 1.620$, $c_2 = -0.15050$, $t_2 = 0.27$, $n_2 = 1.000$, $c_3 = -0.1790$, $t_3 = 0.25$, $n_3 = 1.720$, $c_4 = -0.02466$.

5.6 Write down in correct order, but do not multiply, the refraction and translation matrices for the system: $c_1 = 0.200$, $t_1 = 1.00$, $n_1 = 1.740$, $c_2 = -0.050$ (mirror). Remember that the light passes through the first surface again on the way out.

5.7 Multiply together the following refraction and translation matrices to find the system matrix.

$$\begin{bmatrix} 1 & 0 \\ -0.209 & 1 \end{bmatrix} \begin{bmatrix} 1 & 0.308 \\ 0 & 1 \end{bmatrix} \begin{bmatrix} 1 & 0 \\ -0.261 & 1 \end{bmatrix}.$$

5.8 Multiply togther the following refraction and translation matrices to find the system matrix.

$$\begin{bmatrix} 1 & 0 \\ -0.01 & 1 \end{bmatrix} \begin{bmatrix} 1 & 2 \\ 0 & 1 \end{bmatrix} \begin{bmatrix} 1 & 0 \\ 0.02 & 1 \end{bmatrix} \begin{bmatrix} 1 & 5 \\ 0 & 1 \end{bmatrix} \begin{bmatrix} 1 & 0 \\ 0.04 & 1 \end{bmatrix}.$$

5.9 Given the conjugate matrix

$$\begin{bmatrix} -2.0 & 0 \\ -0.1 & -0.5 \end{bmatrix},$$

calculate the determinant of the matrix, the focal length of the system, and the image magnification. Use the thin lens formula to find the object and image distances.

5.10 Given the conjugate matrix

$$\begin{bmatrix} -0.5 & 0 \\ -0.166667 & -2.0 \end{bmatrix},$$

calculate the determinant of the matrix, the focal length of the system, and the image magnification. Use the thin lens formula to find the object and image distances.

5.11 Given the system matrix

$$\begin{bmatrix} 0.8 & 1.0 \\ -0.04 & 1.2 \end{bmatrix},$$

calculate the effective focal length of the system, the front and back focal lengths, and the positions of the principal planes.

5.12 Given the system matrix

$$\begin{bmatrix} 0.5 & b \\ -0.02 & 2.0 \end{bmatrix},$$

what is the value for b? Calculate the effective focal length of the system, the front and back focal lengths, and the positions of the principal planes.

5.13 An object 2.00 cm high is located 45.00 cm in front of the system given in Problem 5.11. Find the image size and location.

5.14 An object 4.00 cm high is located 50.00 cm in front of the system given in Problem 5.12. Find the image size and location.

5.15 Derive formulae for the locations of the planes of -1 magnification in terms of the Gaussian constants. These planes could be called *inverted planes*.

5.16 Derive a conjugate formula with distances measured from the inverted planes.

5.17 Use matrices to find the effective focal length, front and back focal lengths, and principal planes for the following system. $c_1 = 2.32070$, $t_1 = 0.606$, $n_1 = 1.700$, $c_2 = 0.02910$, $t_2 = 0.1142$, $n_2 = 1.000$, $c_3 = -1.8727$, $t_3 = 0.0161$, $n_3 = 1.720$, $c_4 = 2.2173$, $t_4 = 0.1008$, $n_4 = 1.000$, $c_5 = 0.34190$, $t_5 = 0.0585$, $n_5 = 1.720$, $c_6 = -2.3753$.

5.18 Use matrices to find the aperture stop for the following system. $f'_1 = 30.0$, $R_1 = 3.00$, $t_1 = 5.00$, $f'_2 = -20.0$, $R_2 = 2.00$. Assume that the object is 100.0 in front of the first lens.

5.19 For the system described below, surface 2 is the aperture stop. Use matrices to find the size and location of the entrance and exit pupils. $f'_1 = 12.0$, $R_1 = 4.00$, $t_1 = 4.00$, $R_2 = 2.00$ (surface 2 is a stop), $t_2 = 4.00$, $f'_3 = 20.0$, $R_2 = 6.00$.

5.20 If an object 4.00 high is located 18.0 in front of the system described in Problem 5.19, determine the image size and location.

5.21 Draw a vignetting diagram for the system described in Problems 5.19 and 5.20.

5.22 Use matrices to determine the aperture stop, the entrance pupil, and the exit pupil for the system described below. $f'_1 = 10.0$, $R_1 = 5.00$, $t_1 = 5.00$, $f'_2 =$

-3.333, $R_2 = 3.00$, $t_2 = 2.00$, $f'_3 = 6.00$, $R_2 = 4.00$. Assume that the object is 10.0 high and 50.0 to the left of the first surface.

5.23 Find the cardinal points for the system described in Problem 5.22.

5.24 Find the image size and location for the system described in Problem 5.22.

5.25 Draw a vignetting diagram for the system described in Problem 5.22.

5.26 Use matrices to derive equations for a four-element zoom lens. The second and third elements should move together. How many exact solutions are there for the conjugate equation?

CHAPTER 6

EXACT RAY TRACING

Up to this point, we have studied only paraxial or first-order optics. Paraxial optics gives us a basic understanding of imaging through optical systems. In particular, we can find image location and size and determine the aperture stop, pupil sizes and locations, field stop, and window sizes and locations. But in all optical systems, the aperture stop is large enough that marginal rays do not follow paraxial ray paths. Also, the object or field of view is never vanishingly small so that full-field rays cannot be treated as strictly paraxial. In other words, the majority of light rays which actually pass through the system do not behave like paraxial rays. Thus, the rays which make up the image do not necessarily meet at the paraxial image point. The result is an image which is blurred or the wrong size or in the wrong position. An optical system may have all of the proper first-order properties, but be totally worthless if its image quality is poor. Providing good image quality is the real challenge of lens design, and the only way to evaluate image quality is to trace rays exactly.

In this chapter, we will discuss how to perform exact ray tracing, making no approximations in our calculations. The results of a ray trace calculation should be as accurate as the input data and the limitations of round-off error. After learning how to do an exact ray trace, we will look at various ways of displaying results to help evaluate image quality both quantitatively and qualitatively.

6.1 EXACT RAY TRACE EQUATIONS

There are three differences between paraxial ray tracing which we have done previously and exact ray tracing. The first is that we will now use Snell's law

$$n_{j-1} \sin i_j = n_j \sin i_j' , \tag{6.1}$$

instead of the paraxial version

$$n_{j-1} i_j = n_j i_j' , \tag{6.2}$$

in the refraction equations. The second difference is that translations will occur between the actual spherical surfaces rather than between vertex planes. Again, because we are now performing exact ray tracing, we want to eliminate all approximations. And the third difference is that the rays must be traced in three dimensions rather than just two. Two dimensions was satisfactory with paraxial rays because the rays remained close to the optical axis and changes in the plane of refraction were unimportant. That is no longer the case.

As with paraxial ray tracing, the equations will be derived in a computationally efficient form, but now the form is optimized for inclusion in a computer program rather than for hand calculation.

We will begin with a few definitions. A *meridional ray* is a ray which lies in a plane which contains the optical axis (the z axis in our coordinate system). Therefore, the intersection point of the ray and the first surface will also lie in a plane containing the optical axis. Because we are dealing only with axially symmetric systems, the centers of curvature for all of the surfaces lie along the optical axis. This means that the normal at the first surface will also be in a plane containing the optical axis. Recall that the law of refraction states that the refracted ray is in the same plane with the incident ray and the normal. This means that a meridional ray starts out in one particular plane and remains in that plane as it passes through the system. Once again, because of the axial symmetry, we could rotate our coordinate system, so that the meridional plane is the y–z plane. Paraxial rays are meridional rays which remain close to the optical axis.

Any other exact ray is called a *skew ray*. Skew rays do not lie in a plane containing the optical axis. Of course, the refracted ray is in the same plane as the incident ray and the normal at each surface, but for a skew ray the normals at each surface do not lie in one particular plane. A skew ray passing through a system will not remain in one single plane, but will refract from plane to plane.

A *sagittal ray* is a skew ray which intersects the entrance pupil in a plane perpendicular to the meridional plane. Since we have defined the y–z plane as the meridional plane, skew rays must cross the entrance pupil along the x axis.

Since there are two basic types of exact rays, you would expect two different sets of equations are needed to trace them. Indeed, you can derive equations for tracing meridional rays. But it turns out that meridional rays can be traced with skew ray trace equations with the proper choice of input data. There is no need for two lengthy derivations or for programming two different sets of equations.

6.1.1 Notation for Skew Ray Tracing

In specifying paraxial rays, we found that we needed two quantities, the ray height y and the ray slope angle u. For skew rays we need six quantities. We will use capital letters to indicate exact ray data. The ray position is given by the X, Y, Z coordinates of the point where the ray intersects a surface. Subscripts will indicate which surface. Ray direction is given by direction cosines L, M, and N. A direction cosine is the cosine of a direction angle. L is the cosine of the angle from the x axis to the ray, M is the cosine of the angle from the y axis to the ray, and N is the cosine of the angle from the z axis to the ray. Do not confuse the direction cosine N with the index n. A subscript will tell which surface precedes the space where the direction cosines are defined.

A meridional ray can be considered a special case of a skew ray with

$$L_0 = 0.0 \ , \tag{6.3}$$

$$M_0 = \cos(90^\circ - U_0) = \sin U_0 \ , \tag{6.4}$$

$$N_0 = \cos U_0 \ , \tag{6.5}$$

where U is the exact ray slope angle.

Technically, we do not need to specify all six parameters to specify a ray. The parameters X, Y, and Z are not completely independent of each other since the intersection point must lie on a sphere of known radius. Also, L, M, and N are not independent of each other because the sum of their squares must add to one. In practice, we will specify all six parameters, and use these facts to check our calculations.

6.1.2 Translation Equations

The first step in tracing an exact ray is translation between two surfaces. We will actually break this problem into two parts. First, we will find the point where the ray intersects the vertex plane of the next surface. We will use χ for the X coordinate of this point and ξ for the Y coordinate. Since this is a plane through the vertex of the surface, $Z = 0$. Also, because we will use χ and ξ only for intermediate calculations and will not normally output their values, we will not put subscripts on these variables. The second part of the translation problem is the find Δ the distance from the vertex plane intersection point to the surface intersection point. For a translation, we are given X_{j-1}, Y_{j-1}, Z_{j-1}, and L_{j-1}, M_{j-1}, N_{j-1}. Our final goal is to find X_j, Y_j, Z_j. Figure 6.1 shows an exact ray translation.

The vertex plane intersection point depends on the separation between the two surfaces t_{j-1}. More precisely, it depends on the length of the ray from the preceding surface to the vertex plane. If we call this distance T, then

$$\chi = X_{j-1} + TL_{j-1} \tag{6.6}$$

and

$$\xi = Y_{j-1} + TM_{j-1} \ , \tag{6.7}$$

where

$$T = \frac{t_{j-1} - Z_{j-1}}{N_{j-1}} . \tag{6.8}$$

Finding Δ is a lot more challenging. The point we are ultimately looking for is the intersection between the ray and the spherical surface. An equation for the spherical surface is

$$c_j (X_j^2 + Y_j^2 + Z_j^2) - 2 Z_j = 0 . \tag{6.9}$$

The intersection coordinates can be written in terms of the vertex plane point and the distance Δ:

$$X_j = \chi + \Delta L_{j-1} , \tag{6.10}$$

$$Y_j = \xi + \Delta M_{j-1} , \tag{6.11}$$

$$Z_j = \Delta N_{j-1} . \tag{6.12}$$

Substituting the previous three equations into Eq. 6.9 and rearranging gives a quadratic equation in the unknown distance:

$$\begin{aligned} c_j \Delta^2 - 2 \Delta [N_{j-1} - c_j (\chi L_{j-1} + \xi M_{j-1})] \\ + c_j [\chi^2 + \xi^2] = 0 . \end{aligned} \tag{6.13}$$

We can simplify this equation by introducing

$$G = N_{j-1} - c_j [\chi L_{j-1} + \xi M_{j-1}] \tag{6.14}$$

and

$$F = c_j [\chi^2 + \xi^2] . \tag{6.15}$$

The quadratic formula gives the solution as

$$\Delta = \frac{G \pm \sqrt{G^2 - c_j F}}{c_j} . \tag{6.16}$$

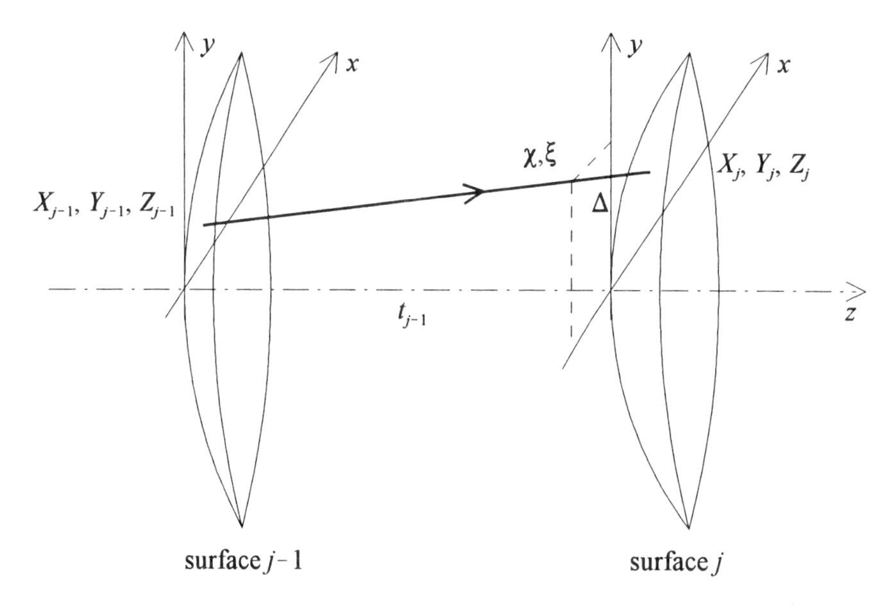

Figure 6.1 Exact ray translation.

We can use the result of Eq. 6.16 in Eqs. 6.10 to 6.12 to determine the intersection point.

There are three problems with Eq. 6.16, however. First, which sign should be used? Second, this formula becomes indeterminate as c_j approaches zero. Finally, if the product $c_j F$ is much smaller than G^2, which is not unusual, then the calculation will suffer a loss of numerical precision. The second and third problems can be addressed by rewriting the formula which we will do in a moment.

We can determine the correct sign to use by the following considerations. As the ray approaches the vertex (i.e., $\chi \to 0$ and $\xi \to 0$) then $\Delta \to 0$ also. From the definition of F, we see that F approaches zero and G approaches N_{j-1}. This leaves us with

$$\Delta = \frac{N_{j-1} \pm N_{j-1}}{c_j} . \tag{6.17}$$

Therefore, we need to use the negative sign in Eq. 6.16 to get the distance to go to zero.

Next, to rewrite Eq. 6.16 (with just the negative sign), multiply both the top and the bottom of the right-hand side by $G + \sqrt{G^2 - c_j F}$. This gives

$$\Delta = \frac{F}{G + \sqrt{G^2 - c_j F}} . \tag{6.18}$$

Equation 6.18 is in a usable form, but we shall modify it slightly after discussing the refraction equations.

6.1.3 Refraction Equations

The first step in deriving equations for refraction is to find an equation for the angle of incidence at the surface. Since refraction takes place in a plane containing the rays and the normal, we can represent the geometry for any skew ray in the plane defined by the incident ray and the normal as shown in Figure 6.2. Here the vertical axis extends from the point where the optical axis intercepts the refracting surface to the point where the incident ray intercepts the vertex plane. The horizontal axis is the optical axis. The distance from the center of curvature C to the intersection point of the incident ray and the vertex plane is D. The distance from the center of curvature to the surface intersection point is the radius of the surface r. The third side of this triangle is Δ, the distance from the vertex plane to the surface along the incident ray. We can use the law of cosines to write an equation in terms of these distances and the incidence angle:

$$D^2 = \Delta^2 + r_j^2 + 2\,\Delta\, r_j \cos I_j \ . \tag{6.19}$$

The distance D is also the hypotenuse of a right triangle which gives

$$D^2 = \chi^2 + \xi^2 + r_j^2 \ . \tag{6.20}$$

The previous two equations can be combined and solved for the cosine of the

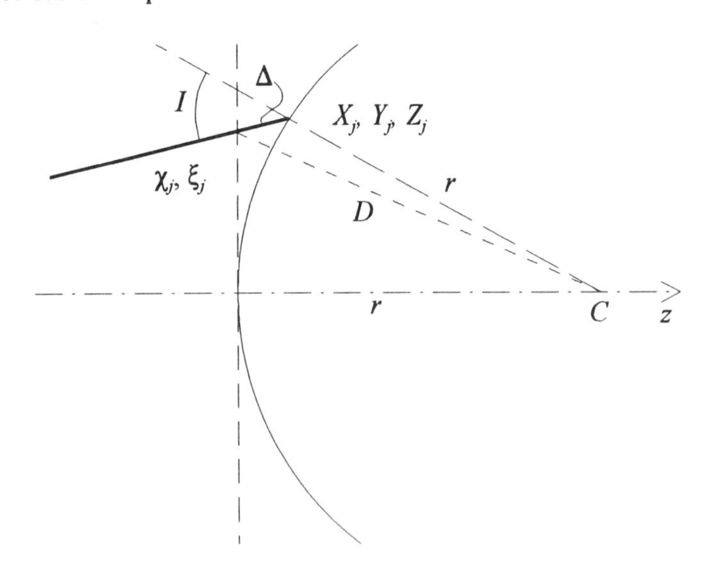

Figure 6.2 Refraction in a plane.

incidence angle. Recalling the definition of F we have

$$\cos I_j = \frac{F - c_j \Delta^2}{2\Delta} . \tag{6.21}$$

Substituting our original equation for Δ and simplifying finally gives

$$\cos I_j = \sqrt{G^2 - c_j F} . \tag{6.22}$$

The refracted angle can be found easily using Snell's law. Just take the arccosine of Eq. 6.22, then the sine, etc. Later we will have use for the cosines of the incidence and refracted angles, so there is really no need to find the angles themselves. If, we square Snell's law and replace the sine squared by one minus the cosine square, we can find the cosine of the refracted angle:

$$\cos I_j' = \sqrt{1 - \mu^2 (1 - \cos^2 I_j)} , \tag{6.23}$$

where

$$\mu = \frac{n_{j-1}}{n_j} . \tag{6.24}$$

Once again, μ is a temporary variable so a subscript is really not needed.

As an aside, notice that Eq. 6.23 will work for reflection as well as refraction. We have defined $\mu = -1$ for reflection. In this case, Eq. 6.23 says that the cosines are equal as they should be for reflection. None of the other equations we have used make reference to the type of surface. Therefore, our entire set of exact ray trace equations should work for systems with mirrors.

Finding the refracted angle was straightforward, but what we are really interested in is the direction of the refracted ray in terms of its direction cosines. The direction cosines are the components of a unit vector in the direction of the ray. We can use the notation

$$\hat{U}_{j-1} = L_j \hat{x} + M_j \hat{y} + N_j \hat{z} . \tag{6.25}$$

The direction of the refracted ray can be determined from the exact ray drawing technique we learned in Chapter 2. Figure 6.3 shows the relationship between the two direction vectors and the normal to the surface. From this figure we have

$$n_{j-1} \hat{U}_{j-1} = n_j \hat{U}_j + (n_j \cos I_j' - n_{j-1} \cos I_j) \hat{r}_j . \tag{6.26}$$

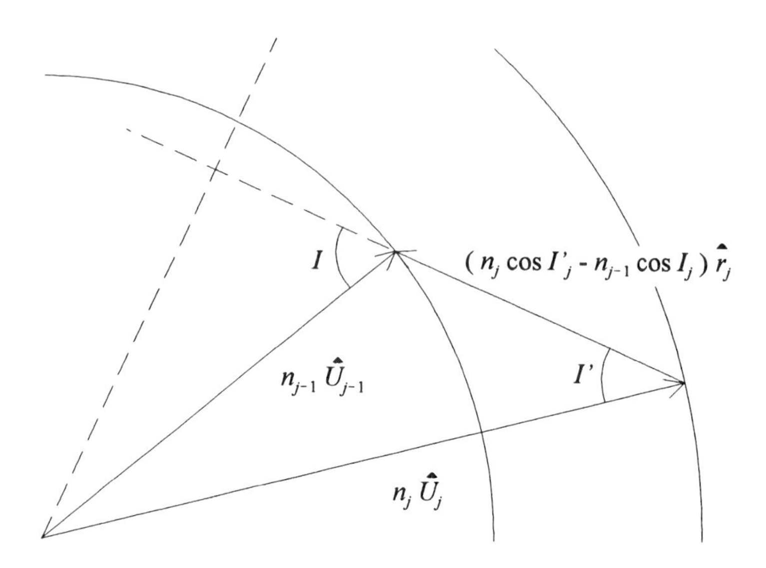

Figure 6.3 Exact ray drawing technique applied to the determination of the ray direction vector.

The next step is to figure out what the components of $\hat{r}_j$ are. Again, using vector notation we can write

$$\vec{X}_j = X_j\hat{x} + Y_j\hat{y} + Z_j\hat{z} \tag{6.27}$$

for the position of the ray intersection point on the surface. Figure 6.4 shows that

$$\vec{X}_j = r_j\hat{z} + r_j\hat{r}_j \,. \tag{6.28}$$

Solving for the unit vector in the direction of the normal gives

$$\hat{r}_j = (c_jX_j)\hat{x} + (c_jY_j)\hat{y} + (c_jZ_j - 1)\hat{z} \,. \tag{6.29}$$

Putting everything together, we get three equations for the three direction cosines of the refracted ray from the components

$$L_j = \mu L_{j-1} - \gamma\, c_jX_j \,, \tag{6.30}$$

$$M_j = \mu M_{j-1} - \gamma\, c_jY_j \,, \tag{6.31}$$

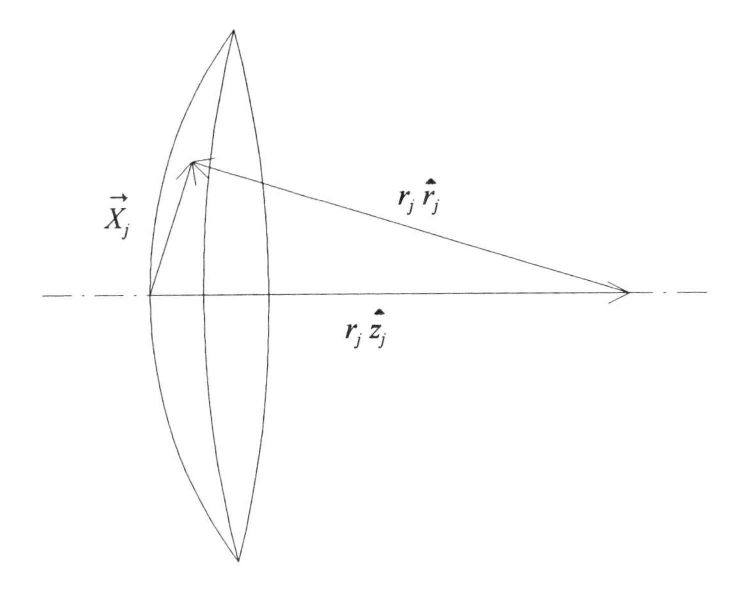

Figure 6.4 Vector representation of the ray intersection point on the surface.

$$N_j = \mu N_{j-1} - \gamma (c_j Z_j - 1) , \qquad (6.32)$$

where

$$\gamma = \cos I_j' - \mu \cos I_j . \qquad (6.33)$$

This completes the refraction equations. Tracing a ray requires alternately applying the translation and refraction equations for each surface of a system.

Example 6.1

Trace an exact ray given $X_0 = Z_0 = 0.0$, $Y_0 = 0.5$, $L_0 = M_0 = 0.07$, and $N_0 = 0.995088$ through a lens with $c_1 = 0.199453$, $c_2 = 0.0$, $t_1 = 0.5$, and $n_1 = 1.50137$. The object is 15.0 in front of the lens.

First, a paraxial ray is traced to locate the image plane. The result is $t_2 = 29.67$. Now we are ready to begin with a translation:

$$T = \frac{t_0 - Z_0}{N_0} ,$$

$$T = \frac{(15.0) - (0.0)}{(0.995088)} = 15.074045 ,$$

$$\chi = X_0 + T L_0 ,$$

$$\chi = (0.0) + (15.074045)(0.07) = 1.055183 ,$$

$$\xi = Y_0 + T M_0 ,$$

$$\xi = (0.5) + (15.074045)(0.07) = 1.555183 ,$$

$$G = N_0 - c_1 [\chi L_0 + \xi M_0] ,$$

$$G = (0.9902) - (0.199453) \times [(1.060392)(0.07) + (1.560392)(0.07)] ,$$

$$G = 0.958643 ,$$

$$F = c_1 [\chi^2 + \xi^2] ,$$

$$F = (0.199453)[(1.055183)^2 + (1.555183)^2] = 0.704469 ,$$

$$\cos I_1 = \sqrt{G^2 - c_1 F} ,$$

$$\cos I_1 = \sqrt{(0.958643)^2 - (0.199453)(0.704469)} = 0.882320 ,$$

$$\Delta = \frac{F}{G + \cos I_1} ,$$

$$\Delta = \frac{(0.704469)}{(0.958643) + (0.882320)} = 0.382663 ,$$

$$X_1 = \chi + \Delta L_0 ,$$

$$X_1 = (1.055183) + (0.382663)(0.07) = 1.081969 ,$$

$$Y_1 = \xi + \Delta M_0 ,$$

$$Y_1 = (1.555183) + (0.382663)(0.07) = 1.581969 ,$$

$$Z_1 = \Delta N_0 ,$$

$$Z_1 = (0.382663)(0.995088) = 0.380783 .$$

Now, begin the first refraction:

$$\mu = \frac{n_0}{n_1} = \frac{(1.0)}{(1.50137)} = 0.666058 ,$$

$$\cos I_1' = \sqrt{1 - \mu^2 (1 - \cos^2 I_1)} ,$$

$$\cos I_1' = \sqrt{1 - (0.666058)^2 (1 - (0.882320)^2)} = 0.949595 ,$$

$$\gamma = \cos I_1' - \mu \cos I_1 ,$$

$$\gamma = (0.949595) - (0.666058)(0.882320) = 0.361919 ,$$

$$L_1 = \mu L_0 - \gamma c_1 X_1 ,$$

$$L_1 = (0.666058)(0.07) - (0.361919)(0.199453)(1.081969) ,$$

$$L_1 = -0.031479 ,$$

$$M_1 = \mu M_0 - \gamma c_1 Y_1 ,$$

$$M_1 = (0.666058)(0.07) - (0.361919)(0.199453)(1.581969) ,$$

$$M_1 = -0.067572 \ ,$$

$$N_1 = \mu N_0 - \gamma (c_1 Z_1 - 1) \ ,$$

$$N_1 = (0.666058) (0.995088) - (0.361919) \times$$

$$[(0.199453) (0.380783) - 1] \ ,$$

$$N_1 = 0.997218 \ .$$

The next translation begins with

$$T = \frac{(0.5) - (0.380783)}{(0.997218)} = 0.119550 \ .$$

In the x direction we have

$$\chi = (1.081969) + (0.119550) (-0.031479) = 1.078206 \ .$$

And in the y direction we have

$$\xi = (1.581969) + (0.119550) (-0.067572) = 1.573891 \ .$$

The remaining translation equations are simplified because $c_2 = 0.0$, $G = 0.997218$, and $F = 0.0$.

$$\cos I_2 = 0.997218 \ ,$$

$$\Delta = 0.0 \ ,$$

and $X_2 = 1.078206$, $Y_2 = 1.573891$, and $Z_2 = 0.0$.

Refraction at the second surface begins with $\mu = 1.50137$

$$\cos I_2' = \sqrt{1 - (1.50137)^2 (1 - (0.997218)^2)} = 0.993718 \ ,$$

$$\gamma = (0.993718) - (1.50137) (0.997218) = -0.503475 \ ,$$

$$L_2 = (1.50137)(-0.031479) - (0.0) = -0.047262 ,$$

$$M_2 = (1.50137)(-0.067572) - (0.0) = -0.101451 ,$$

$$N_2 = (1.50137)(0.997218) - (-0.503475)[(0.0) - 1] ,$$

$$N_2 = 0.993718 .$$

Finally, one more translation, namely

$$T = \frac{(29.667195) - (0.0)}{(0.993718)} = 29.854748 ,$$

gives $X_3 = -0.332789$ and $Y_3 = -1.454903$. ■

6.1.4 Openings

In the previous example, we were given the initial values for X_0, Y_0, Z_0 and L_0, M_0, N_0. When you know these values, you are all set. This could be called an *absolute ray trace*. More typically, you will know that you want to trace a particular type of ray such as a full-field chief ray or a marginal axial ray. You then need to determine the initial ray values from the relative position of the ray in the entrance window and entrance pupil. This is called a *relative ray trace*.

Since we are dealing with axially symmetric systems, the system can always be rotated so that the object lies along the y axis. Also, we will assume that the object is planar. Therefore, the initial ray position can be described with just one parameter η, which is the height of the ray at the object relative to the height of the object H. In other words, if $\eta = 1$, then the ray is a full-field ray. If $\eta = 0$, then the ray is an axial ray.

The relative position of the ray in the entrance pupil is given by two parameters, ρ and θ. ρ measures how far from the optical axis the ray intersects the entrance pupil relative to the entrance pupil size. If $\rho = 1$, then the ray is a marginal ray. If $\rho = 0$, then the ray is a chief ray. θ is the angle between the y axis and the line connecting the center of the entrance pupil to the intersection point of the ray in the entrance pupil. If $\theta = 0°$, then the ray lies in the y–z plane and intersects the entrance pupil along the positive y axis. If $\theta = 180°$, the ray still lies in the y–z plane, but intersects the entrance pupil along the negative y axis. If $\theta = 90°$, then the ray is a sagittal ray and intersects the entrance pupil along the positive x axis. The coordinates of a ray in the entrance pupil can be expressed as

$$X_E = R_E \rho \sin\theta \tag{6.34}$$

and

$$Y_E = R_E \rho \cos\theta \;. \tag{6.35}$$

To perform a relative ray trace, specify η, ρ, and θ and determine the initial ray position and direction from the following equations:

$$X_0 = 0 \;, \tag{6.36}$$

$$Y_0 = \eta H \;, \tag{6.37}$$

$$Z_0 = 0 \;, \tag{6.38}$$

$$L_0 = \frac{X_E}{T} \;, \tag{6.39}$$

$$M_0 = \frac{Y_E - Y_0}{T} \;, \tag{6.40}$$

$$N_0 = \frac{t_0 + l_E}{T} \;, \tag{6.41}$$

where T is the length of the ray from the object point to the intersection of the ray with the entrance pupil as shown in Figure 6.5. The length of the ray can be determined from

$$T = \sqrt{(t_0 + l_E)^2 + (Y_E - Y_0)^2 + X_E^2} \;. \tag{6.42}$$

Example 6.2

Place a stop 5.00 cm in front of the lens of the previous problem. The clear radius of the stop is 1.00 cm and the clear radius of the lens is 2.00 cm. Find the initial ray parameters to trace a full-field chief ray.

Paraxial ray tracing will show that the stop is the aperture stop and the object is

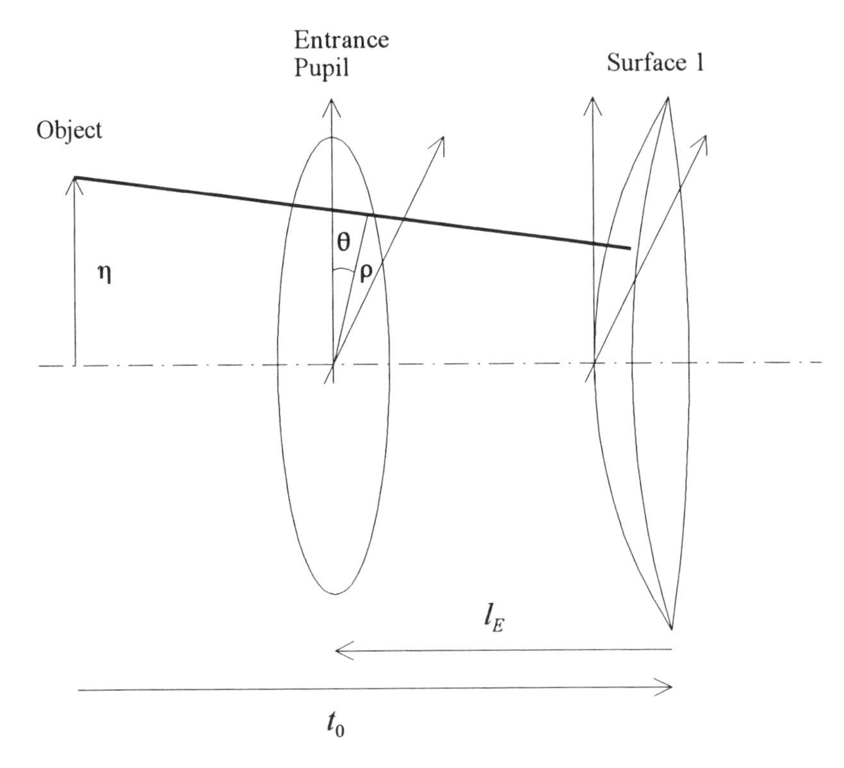

Figure 6.5 Relative ray coordinates for opening a relative ray trace.

the field stop. The object coordinates are $X_0 = Z_0 = 0.0$ and $Y_0 = 1.0$. For a chief ray $X_E = Y_E = 0.0$. The ray length is

$$T = \sqrt{(10.0 + 0.0)^2 + (0.0)^2 + (0.0 - 1.0)^2} = 10.049876 ,$$

$L_0 = 0.0$, and

$$M_0 = \frac{(0.0) - (1.0)}{(10.049876)} = -0.099504 ,$$

$$N_0 = \frac{(10.0) + (0.0)}{(10.049876)} = 0.995037 .$$

Now we can proceed with the ray trace in the normal way. ■

One additional complication is the situation where the ray is coming from infinity.

Equation 6.42 is no longer usable. Rays coming from a particular object point at infinity are all parallel. To specify such a ray, we only need its slope angle relative to the optical axis U_0 and ρ and θ. The entrance pupil coordinates can take the place of object plane coordinates in starting the ray trace and

$$L_0 = 0 , \tag{6.43}$$

$$M_0 = \sin U_0 , \tag{6.44}$$

$$N_0 = \cos U_0 . \tag{6.45}$$

Example 6.3

Same problem as in Example 6.2, except that the object is now located at infinity. The full-field ray slope angle is 0.1 radians.

The stop is still the aperture stop. From the definitions of the direction cosines we have $L_0 = 0.0$, $M_0 = 0.099833$, and $N_0 = 0.995004$. We take $X_0 = Y_0 = Z_0 = 0.0$ for the chief ray and set $t_0 = 0.0$. ■

6.1.5 Checks

It is always good, especially with calculations as complicated as exact ray tracing, to have some independent check on your work. As mentioned earlier, the ray intercept coordinates and the ray direction cosines are not independent. At each surface, you should check to see that

$$c_j [X_j^2 + Y_j^2 + Z_j^2] - 2 Z_j = 0 . \tag{6.46}$$

Example 6.4

Apply this check to the ray trace from Example 6.1.

For the first surface

$$(0.188453)[(1.081969)^2 + (1.581969)^2 + (0.380783)^2] - 2(0.380883)$$
$$= 0.000001 ,$$

which is close enough considering round-off errors. ■

After each surface the slope angles should be checked using that the direction cosines are related by

$$L_j^2 + M_j^2 + N_j^2 = 1 \ . \tag{6.47}$$

Example 6.5

Check the direction cosines after the first surface in Example 6.1:

$$(-0.031479)^2 + (-0.067572)^2 + (0.997218)^2 = 1.000001 \ .$$

■

There are two equations with square roots in the exact ray tracing sequence. The quantity under the square root can become negative in some situations. If you write a program to perform exact ray tracing, then these quantities should be checked for sign before attempting to take the square root.

If

$$G^2 - c_j F < 0 \ , \tag{6.48}$$

then the cosine of the incident angle is imaginary. Physically, this means that the ray entirely misses the surface as shown in Figure 6.6.

Example 6.6

As an example of what happens when a ray misses a surface, consider a glass sphere of radius 2.0 cm. The index of the glass is 1.5. Now try to trace a ray which is parallel to the optical axis and 2.5 cm above it.

We can begin the ray trace at the first surface by setting $X_0 = Z_0 = 0$, $t_0 = L_0 = M_0 = 0.0$, $Y_0 = 2.5$, $N_0 = 1.0$, $\chi = 0.0$, and $\xi = 2.5$:

$$G = 1.0 - (0.5)[(0.0)(0.0) + (2.5)(0.0)] = 1.0 \ ,$$

$$F = (0.5)[(0.0)^2 + (2.5)^2] = 3.125 \ ,$$

$$\cos I_1 = \sqrt{(1.0)^2 - (0.5)(3.125)} = \sqrt{(-0.5625)} \ .$$

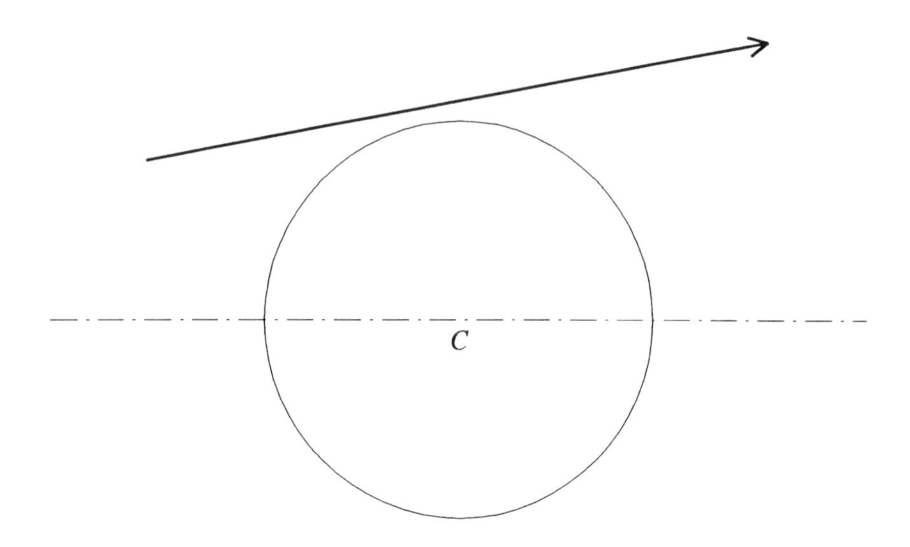

Figure 6.6 A ray which never intersects a surface cannot be traced.

The imaginary result for the cosine of the incident angle means that the ray does not intersect the surface. ■

If

$$1 - \mu^2 (1 - \cos^2 I_j) < 0 , \tag{6.49}$$

then the ray is totally internally reflected. In either of these cases, the ray trace must be terminated.

Example 6.7

As an example of what happens when we have total internal reflection, consider a surface with $c_1 = 0.0$, $n_0 = 1.5$, and $n_1 = 1.0$. A ray is incident on the vertex surface at $y = -1.0$ and has a slope angle of $20°$.

The starting coordinates for this ray are setting $X_0 = Z_0 = t_0 = 0.0$, $L_0 = 0.0$, $Y_0 = -1.0$, $M_0 = 0.342020$, and $N_0 = 0.939693$.

Here, $\chi = 0.0$ and $\xi = -1.0$:

$$G = (0.939693) - (-0.5)[(0.0)(0.0) + (-1.0)(0.342020)]$$
$$= 0.768683 ,$$

$$F = (-0.5)[(0.0)^2 + (-1.0)^2] = -0.5 ,$$

$$\cos I_1 = \sqrt{(0.768683)^2 - (-0.5)(-0.5)} = 0.583844 ,$$

$$\Delta = \frac{(-0.5)}{(0.768683) + (0.583844)} = -0.369678 .$$

Now, $X_1 = 0.0$ and

$$Y_1 = (-1.0) + (-0.369678)(0.342020) = -1.126437 ,$$

$$Z_1 = (-0.369678)(0.939693) = -0.347384 .$$

For refraction we have $\mu = 1.5$:

$$\cos I_1' = \sqrt{1 - (1.5)^2(1 - (0.583844)^2} = \sqrt{(-0.483034)} .$$

The imaginary value for the refracted angle cosine indicates total internal refraction. ■

6.2 IMAGE EVALUATION

Now that we know how to trace skew rays, how do we use this knowledge to evaluate image quality? In this section a variety of different diagrams and plots will be described. Each plot is useful for both qualitative and quantitative evaluation of image quality. Some of the diagrams are useful for specific aspects of image quality, while others are more general.

6.2.1 Spot Diagrams

Perhaps the most obvious thing to do is to try to simulate image formation by tracing a large number of rays to see where they end up on the paraxial image plane. The result is a spot diagram. If we trace rays from a particular object point in such a way that the rays are spaced evenly in the entrance pupil, then each ray will represent a certain fraction of the total energy being passed to the image. The spot diagram will indicate how this energy is distributed in the image plane. For good imaging, the ray traces will form a small tight spot. Poor imaging results in a large or asymmetric distribution of rays.

How should the rays be distributed in the entrance pupil? The simplest arrangement is a square grid as shown in Figure 6.7. A square grid makes finding the entrance pupil coordinates for the ray trace simple. A square grid is also applicable no matter what the shape of the entrance pupil. Unfortunately, for a circular entrance pupil, which is what we expect for axially symmetric systems, the rays near the edge of the entrance pupil do not have even weights. Some rays near the margin must represent smaller or larger areas than rays in the interior of the pupil. Also, the resulting spot diagram is not as aesthetically pleasing as a diagram based on a circular grid of rays. These limitations of a square grid can be overcome by making the grid spacing very small and tracing a large number of rays.

All of the systems we deal with are axially symmetric so the entrance pupil will always be a circle. We can space rays equally along circles as shown in Figure 6.8 to get uniform ray spacing and a nice-looking spot diagram. To determine the entrance pupil coordinates, we first divide the entrance pupil into $2Q$ circles, where Q is the number of rings of rays we will eventually have. The radius of each ring of rays is an odd multiple of R_E divided by $2Q$, hence

$$\rho = \frac{(2q - 1)}{2Q}, \tag{6.50}$$

where q is an integer between 1 and Q. Next, the θ angle for each ray is determined by dividing the semicircle into $2(2q - 1)$ segments. The angle for each ray is an odd multiple of $\pi / 2(2q - 1)$,

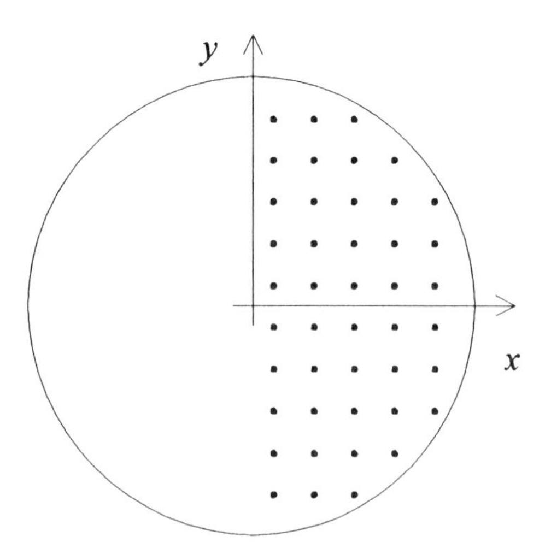

Figure 6.7 Square ray grid in the entrance pupil.

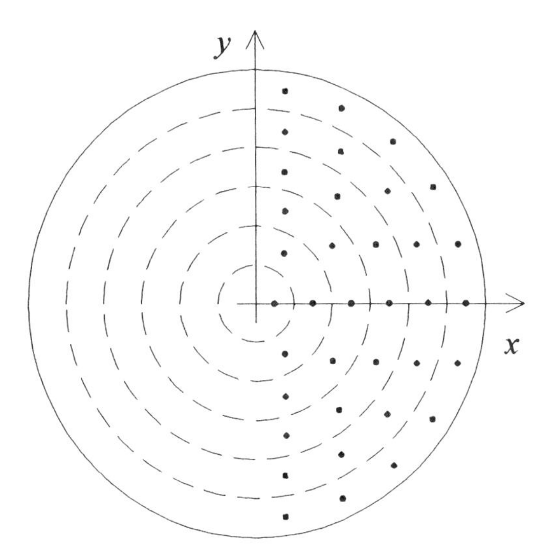

Figure 6.8 Circular ray grid in the entrance pupil.

$$\theta = \frac{(2p - 1)\pi}{2(2q - 1)}, \tag{6.51}$$

where p is an integer between 1 and $2q - 1$. The entrance pupil coordinates are given by

$$X_E = \frac{(2q - 1)R_E}{2Q} \sin\left[\frac{(2p - 1)\pi}{2(2q - 1)}\right], \tag{6.52}$$

$$Y_E = \frac{(2q - 1)R_E}{2Q} \cos\left[\frac{(2p - 1)\pi}{2(2q - 1)}\right]. \tag{6.53}$$

These formulae can be used with the relative ray trace openings to trace a set of rays for a spot diagram. Because of symmetry, only half of the entrance pupil needs to be used for an off-axis object point. For an on-axis object, only one-fourth of the entrance pupil is needed.

The more rays that are traced, the better the representation of the spot diagram is, but the longer it takes to do the calculations. Notice that the total number of rays traced through half the pupil is Q^2. Figure 6.9 shows spot diagrams for on axis and off axis for the system from Example 6.1.

Several important pieces of information can be gleaned from the spot diagram.

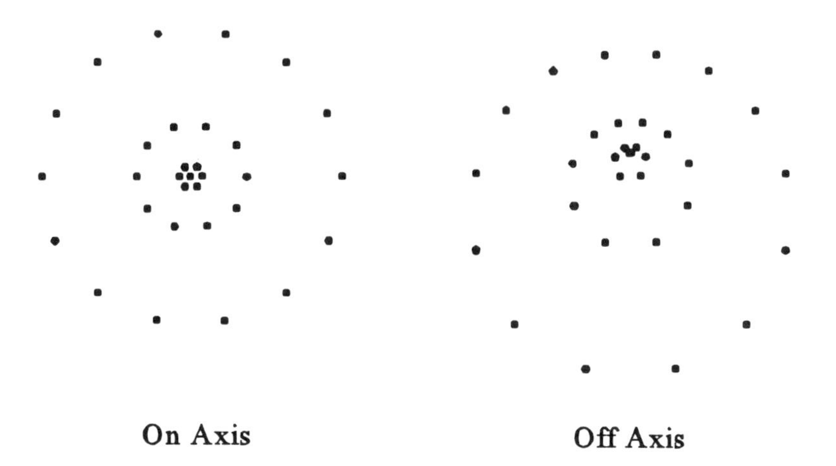

Figure 6.9 Spot diagrams for the system described in Example 6.1.

First, the center of the spot can be determined. The spot center may or may not be located at the same position as the paraxial image. It is obviously important to determine if the paraxial image and the exact image correspond. Because of the symmetry of our systems about the y–z plane, the x coordinate of the center of the spot must be zero. The y coordinate of the center of the exact spot can be determined from

$$Y_{center} = \frac{\sum_{q=1}^{Q} \sum_{p=1}^{2q-1} Y_{q,p}}{Q^2} . \tag{6.54}$$

The next useful piece of information is the overall size of the spot. The overall x dimension can be determined by doubling the maximum spot coordinate in the x direction. The overall y dimension is determined by finding the maximum spot coordinate in the x direction and subtracting the minimum spot coordinate. Usually, a better measure of the spot size is the root mean square spot size. This is calculated by squaring the coordinates of each ray relative to the center of the spot, finding the average, and taking the square root:

$$R_{rms} = \frac{\sqrt{\sum_{q=1}^{Q} \sum_{p=1}^{2q-1} \left[X_{q,p}^{\,2} + (Y_{center} - Y_{q,p})^2\right]}}{Q^2} . \tag{6.55}$$

This formula can be rewritten as

$$R_{rms} = \sqrt{\frac{\sum_{q=1}^{Q}\sum_{p=1}^{2q-1}(X_{q,p}^2 + Y_{q,p}^2)}{Q^2} - Y_{center}^2}\,, \tag{6.56}$$

which is much easier to compute in a program.

Spot diagrams also give qualitative information on the quality of the image. A spot diagram may consist of a very small nucleus and a large diffuse halo or a medium-size blob with uniform spot distribution. These different spot diagrams could have similar values for R_{rms}, but very different image qualities.

Unfortunately, because information on which ray creates which spot on the diagram is not included, spot diagrams do not lend insight into what is going wrong with the image. Several other types of plots have been developed to help demonstrate what may be wrong with an image as well as how wrong it is.

6.2.2 Longitudinal Spherical Aberration Plots

The first type of plot we will discuss is used for an on axis object. Several axial rays are traced from the object through the system with various initial slope angles. Because of symmetry, only positive slopes need to be traced. In image space, the ray will usually cross the optical axis at a point other than the paraxial image point. The distance from the paraxial image point to the crossing point is the longitudinal spherical aberration (*LSA*). Typically, *LSA* is a function of the initial ray slope or the height of the ray at the entrance pupil (ρ). A typical plot of *LSA* as a function of ρ is shown in Figure 6.10. Notice that *LSA* runs along the horizontal axis as if this axis were the optical axis. The location of the vertical axis corresponds to the paraxial image plane.

If a value is not specified to identify the ray, then the marginal ray is assumed. The *LSA* can be calculated from

$$LSA(\rho) = -\frac{N_k}{M_k} Y_{k+1}\,. \tag{6.57}$$

The negative sign is needed to indicate direction from the paraxial image plane.

The *LSA* can be determined in several wavelengths. The difference between the marginal *LSA* in *F* light and in *C* light is called *spherochromatism*. The difference in *LSA* for paraxial rays as a function of wavelength is just the longitudinal component of the axial color. To determine the color differences in *LSA* for paraxial rays, very small values of ρ are used because we cannot accurately trace exact rays from an on-axis object in the limit that ρ goes to zero. The particular value for ρ used in this calculation is not important because as ρ approaches zero, the value for *LSA* approaches a constant.

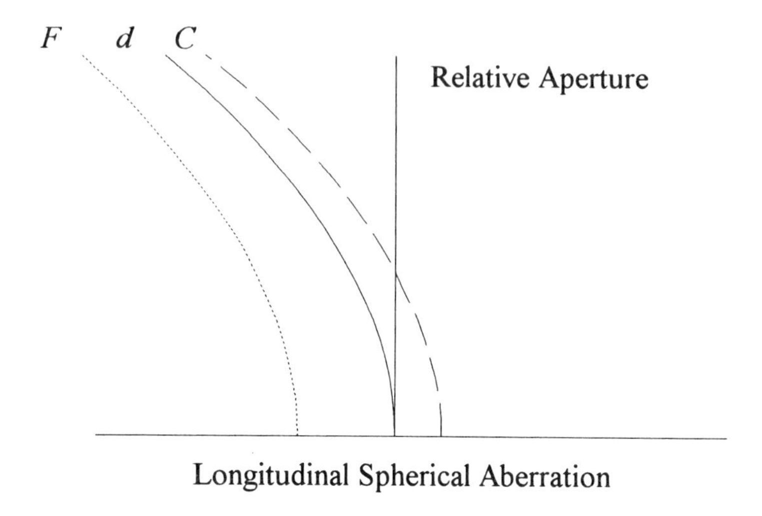

Figure 6.10 LSA plot for the system described in Example 6.1.

6.2.3 Ray Intercept Plots

In the previous section, we traced ray fans from an on axis object and plotted the longitudinal component of the aberration to create an LSA plot. We can also trace ray fans from off-axis objects. But now the symmetry is lost. We expect meridional and sagittal ray fans to give different results. Typically, in this type of plot the transverse aberrations are plotted as a function of the relative ray height ρ in the entrance pupil. For meridional rays, $\epsilon_y = Y_{k+1} - y_{k+1}$ is plotted versus ρ; and for sagittal rays, $\epsilon_x = X_{k+1} - x_{k+1}$ is plotted versus ρ. Not only are the sagittal and meridional not equivalent, the positive and negative branches of the meridional plots are not symmetric. As we shall discuss in the next chapter, the shape of these ray intercept curves can tell us which type of aberration is dominant. The ideal image would be a horizontal line along the ρ axis. The magnitude of variations from this line tells us how much aberration we have off axis. Figure 6.11 shows an example ray intercept plot.

6.2.4 Field Curvature Plots

As an object moves further off axis, it actually moves further from the lens. From the simple thin lens formula for image position, it is easy to see that as the object distance changes, the image position must change also. Thus a plane object will result in a curved image. This effect is called *field curvature*. As you might expect, this effect depends on ray direction so that meridional and sagittal rays form images at different distances from the plane paraxial image. This additional effect is called *astigmatism*. We can make a plot to show field curvature and astigmatism by tracing

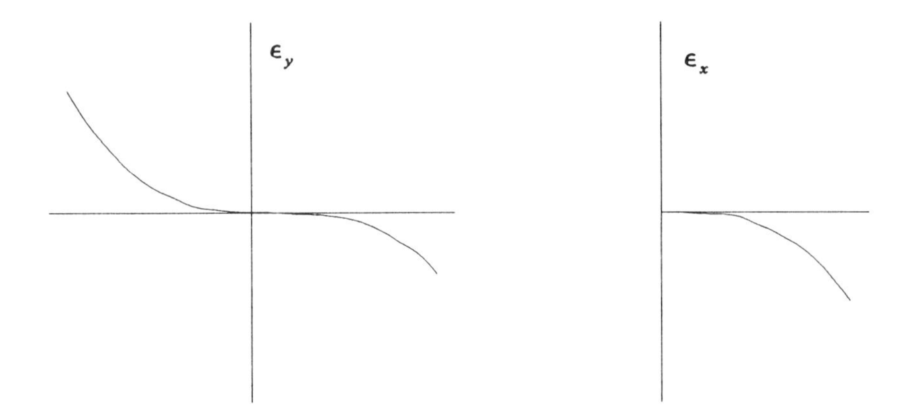

Figure 6.11 Tangential and sagittal ray intercept plots for the system from Example 6.1.

a chief ray and nearby sagittal and meridional rays. The intersection of the chief ray with the meridional ray locates the tangential image and the intersection of the sagittal ray with the chief ray locates the sagittal image.

We can write an equation for the height of the chief ray in image space as a function of the distance of the ray from the paraxial image plane:

$$\bar{y} = (\tan \bar{U})\,\bar{z} + \bar{Y}_{k+1}, \tag{6.58}$$

where $\bar{z}$ is the distance from the paraxial image plane. A similar equation, but without the bar, would hold for a nearby meridional ray. These two rays will cross when their y and z values are the same. We call the z distance at this point the *tangential astigmatism* or *TAst*:

$$TAst = -\frac{\bar{Y}_{k+1} - Y_{k+1}}{\tan \bar{U} - \tan U}. \tag{6.59}$$

In this equation, remember that

$$\tan U = \frac{M_k}{N_k}. \tag{6.60}$$

For sagittal astigmatism, a sagittal ray close to the chief ray is traced. Since the chief ray is strictly in the y–z plane, sagittal astigmatism (*SAst*) is determined when the close sagittal ray crosses the y–z plane. That is,

$$SAst = \frac{X_{k+1}}{\left(\dfrac{L_k}{N_k}\right)} . \tag{6.61}$$

Both the tangential and sagittal astigmatism calculated above are functions of the relative object height η or field angle. An astigmatism or field curvature plot can be made by plotting the *Ast* values on a horizontal axis and η on a vertical axis. Symmetry requires that only positive values of η need to be used. If the tangential and sagittal plots coincide, there is no astigmatism, only field curvature. If the lines are vertical, then there is no field curvature. An example of an astigmatism plot is shown in Figure 6.12.

6.2.5 Distortion Plots

The final type of plot indicates how far from the paraxial image point the chief ray misses. This aberration is called *distortion*. Distortion is usually measured in %:

$$Dist = \left[\frac{\bar{Y}_{k+1}}{\bar{y}_{k+1}} - 1\right] \times 100\% . \tag{6.62}$$

As with astigmatism and field curvature, distortion is a function of the relative field height η or field angle. Distortion can be plotted along the horizontal axis and η

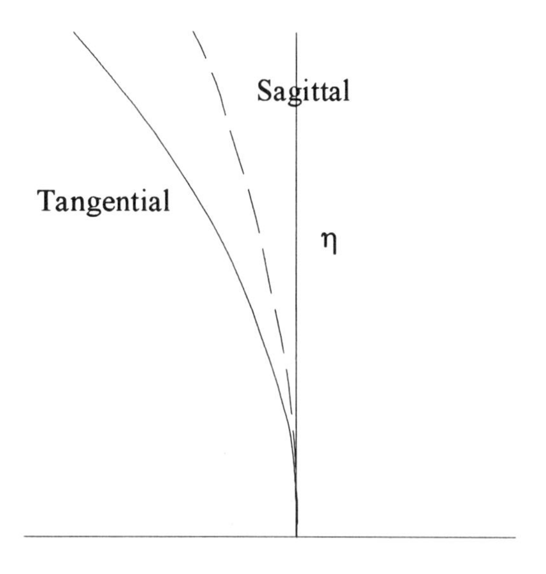

Figure 6.12 Tangential and sagittal astigmatism for Example 6.1.

along the vertical axis. A vertical line is the ideal distortion plot. Figure 6.13 shows an example of a distortion plot.

SUMMARY

While image size and location can be determined using paraxial optics, the quality of the image can be evaluated fully only by means of exact ray tracing.

Exact ray tracing begins by identifying the ray to be traced. If the ray starts at a finite object, then we must define the starting coordinates of the ray. Because of axial symmetry, only one dimension needs to be nonzero:

$$X_0 = 0 \ , \tag{6.63}$$

$$Y_0 = \eta H \ , \tag{6.64}$$

$$Z_0 = 0 \ . \tag{6.65}$$

Next, the ray direction needs to be set. A good way to do this is to specify a point in the entrance pupil where you would like to aim the ray. Then

$$L_0 = \frac{X_E}{T} \ , \tag{6.66}$$

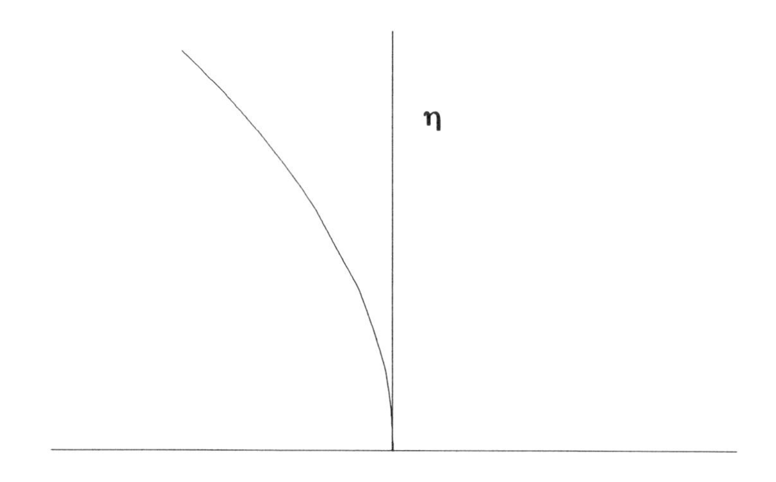

Figure 6.13 Distortion plot for the system described in Example 6.1.

$$M_0 = \frac{Y_E - Y_0}{T} , \tag{6.67}$$

$$N_0 = \frac{t_0 + l_E}{T} , \tag{6.68}$$

where

$$T = \sqrt{(t_0 + l_E)^2 + (Y_E - Y_0)^2 + X_E^2} . \tag{6.69}$$

If the object is at infinity, then the ray direction can be defined by

$$L_0 = 0 , \tag{6.70}$$

$$M_0 = \sin U_0 , \tag{6.71}$$

$$N_0 = \cos U_0 . \tag{6.72}$$

The ray location is defined by the desired intersection point of the ray with the entrance pupil.

Exact ray tracing consists of two basic steps, translation and refraction, just as with paraxial ray tracing. Unfortunately, the equations for these two steps are more complicated. First a translation distance is calculated:

$$T = \frac{t_{j-1} - Z_{j-1}}{N_{j-1}} . \tag{6.73}$$

Then, the location of the ray in the vertex plane is found using

$$\chi = X_{j-1} + T L_{j-1} \tag{6.74}$$

and

$$\xi = Y_{j-1} + T M_{j-1} . \tag{6.75}$$

Next, a couple of special values are found:

$$G = N_{j-1} - c_j [\chi L_{j-1} + \xi M_{j-1}] \tag{6.76}$$

and

$$F = c_j [\chi^2 + \xi^2] . \tag{6.77}$$

Even though we are discussing translation and its a little out of sequence, we can calculate the cosine of the incidence angle by

$$\cos I_j = \sqrt{G^2 - c_j F} . \tag{6.78}$$

Now, we can determine the distance from the vertex plane to the refracting surface measured along the ray by using

$$\Delta = \frac{F}{G + \cos I} . \tag{6.79}$$

Now the location of the intersection point between the ray and the surface may be found:

$$X_j = \chi + \Delta L_{j-1} , \tag{6.80}$$

$$Y_j = \xi + \Delta M_{j-1} , \tag{6.81}$$

$$Z_j = \Delta N_{j-1} . \tag{6.82}$$

The second phase of exact ray tracing is refraction which begins with finding the ratio of the indices

$$\mu = \frac{n_{j-1}}{n_j} . \tag{6.83}$$

Then, the cosine of the refracted angle is given by

$$\cos I_j' = \sqrt{1 - \mu^2 (1 - \cos^2 I_j)} . \tag{6.84}$$

Note that the sines of the angles do not need to be found. In fact, the angles themselves are not needed.

At this point another intermediate value is needed

$$\gamma = \cos I_j' - \mu \cos I_j \ . \tag{6.85}$$

Finally, we can find the new direction cosines

$$L_j = \mu L_{j-1} - \gamma c_j X_j \ , \tag{6.86}$$

$$M_j = \mu M_{j-1} - \gamma c_j Y_j \ , \tag{6.87}$$

$$N_j = \mu N_{j-1} - \gamma (c_j Z_j - 1) \ . \tag{6.88}$$

The steps outlined above can be repeated for each surface. As we carry through these calculations, we also need to check our work. At each surface the ray coordinates can be checked with

$$c_j [X_j^2 + Y_j^2 + Z_j^2] - 2 Z_j = 0 \ . \tag{6.89}$$

And after each surface the direction cosines can be checked with

$$L_j^2 + M_j^2 + N_j^2 = 1 \ . \tag{6.90}$$

The ray trace can be terminated with one final translation to the image plane.

A variety of diagrams and plots derived from exact ray traces are utilized to evaluate image quality. Perhaps the most obvious is a spot diagram which represents the appearance of a point object in image space. A spot diagram is made by tracing a large number of rays from an object point and noting where they intersect the image plane. One way to define the set of rays is with

$$X_E = \frac{(2q - 1) R_E}{2Q} \sin \left[\frac{(2p - 1)\pi}{2(2q - 1)} \right] , \tag{6.91}$$

$$Y_E = \frac{(2q - 1) R_E}{2Q} \cos \left[\frac{(2p - 1)\pi}{2(2q - 1)} \right] . \tag{6.92}$$

The center of the spot diagram for an off-axis object point is not necessarily located where the paraxial image is located. To find the center of the spot we use

$$Y_{center} = \frac{\sum_{q=1}^{Q} \sum_{p=1}^{2q-1} Y_{q,p}}{Q^2} . \tag{6.93}$$

The size of the spot is given by

$$R_{rms} = \sqrt{\frac{\sum_{q=1}^{Q} \sum_{p=1}^{2q-1} (X_{q,p}^2 + Y_{q,p}^2)}{Q^2} - Y_{center}^2} . \tag{6.94}$$

For an on-axis object a plot of the longitudinal spherical aberration is useful since this is the only applicable aberration. An *LSA* plot is made by tracing axial rays for various values of ρ. To calculate the *LSA* values, we use the results of an exact ray:

$$LSA(\rho) = -\frac{N_k}{M_k} Y_{k+1} . \tag{6.95}$$

A ray intercept plot is created by tracing ray fans in the tangential and sagittal planes. The aberrations ϵ_x and ϵ_y are plotted against the position of the ray in the entrance pupil ρ.

A tangential astigmatism plot is made by tracing a chief ray (denoted with a bar over the ray parameters) and a nearby meridional ray. The intersection of these two rays locates the tangential image. The distance of the tangential image from the paraxial image is the tangential astigmatism given by

$$TAst = -\frac{\bar{Y}_{k+1} - Y_{k+1}}{\tan \bar{U} - \tan U} . \tag{6.96}$$

In this equation, the ray angles are determined by

$$\tan U = \frac{M_k}{N_k} . \tag{6.97}$$

For sagittal astigmatism, a sagittal ray close to the chief ray is traced. Since the chief ray is strictly in the y–z plane, sagittal astigmatism *SAst* is determined when the close sagittal ray crosses the y–z plane. That is,

$$SAst = \frac{X_{k+1}}{\left(\dfrac{L_k}{N_k}\right)} . \tag{6.98}$$

Both the sagittal and tangential astigmatism are functions of the field position η.

Finally, a distortion plot can be made by tracing a chief ray and comparing its intersection point in the paraxial plane with the corresponding paraxial image location. Distortion is usually expressed as a percentage:

$$Dist = \left[\frac{\bar{Y}_{k+1}}{\bar{y}_{k+1}} - 1\right] \times 100\% . \tag{6.99}$$

REFERENCE

R. Kingslake, *Lens Design Fundamentals* (Academic Press, Orlando, FL, 1978).

PROBLEMS

6.1 Define the following types of rays: meridional, skew, sagittal.

6.2 Define spherical aberration.

6.3 Define astigmatism and field curvature.

6.4 Define distortion.

6.5 Describe the appearance of perfect imaging in each type of plot or diagram mentioned in this chapter (spot diagram, *LSA* plot, ray intercept plot, astigmatism plot, distortion plot).

6.6 When performing an exact ray trace, does a ray which passes through the center of the entrance pupil necessarily pass through the center of the aperture stop or the center of the entrance pupil? Explain.

Throughout this problem set we will use the term *minimal ray pattern*. A minimal ray pattern is defined as a set of five rays. One ray is the marginal axial ray ($\theta = 0°$). The second ray is the full-field chief ray. The remaining three rays are full-field marginal rays with $\theta = 0°$, $90°$, and $180°$.

6.7 Determine the initial values for X, Y, Z and L, M, N for all five rays in a minimal ray pattern for an object which is 2.00 cm tall and 25.0 cm in front of the first surface of a system. The entrance pupil is 5.00 cm in front of the first surface and has a clear radius of 4.00 cm.

6.8 Determine the initial values for X, Y, Z and L, M, N for all five rays in a minimal ray pattern for an object which is 1.50 cm tall, and 50.0 cm in front of the first surface of a system. The entrance pupil is 10.00 cm behind the first surface and has a clear radius of 3.00 cm.

6.9 Determine the initial values for X, Y, Z and L, M, N for all five rays in a minimal ray pattern for an object at infinity with field angle of 0.25 radians. The entrance pupil is 5.00 cm behind the first surface and has a clear radius of 2.00 cm.

6.10 Determine the initial values for X, Y, Z and L, M, N for all five rays in a minimal ray pattern for an object at infinity with field angle of 0.300 radians. The entrance pupil is 5.00 cm in front of the first surface and has a clear radius of 3.00 cm.

6.11 Show that $X = 1.263951$ cm, $Y = 1.144730$ cm, $Z = -0.104321$ cm is a valid ray intersection point on a surface with radius $r = -13.989809$ cm.

6.12 Show that $L = 0.087156$, $M = 0.173648$, $N = 0.980944$ is a valid set of direction cosines.

6.13 A thick meniscus lens is made from FK5 glass. The curvature of the first surface is -0.08 cm^{-1} and the curvature of the second surface is -0.12 cm^{-1}. The lens is 0.5 cm thick. If the object is located 100 cm in front of this lens, where is the paraxial image plane? Trace a single exact ray from $X_0 = 0$, $Y_0 =$ 2.00 cm, $Z_0 = 0$, $L_0 = 0.01$, $M_0 = 0.005$, and $N_0 = 0.999937$ to the paraxial image plane. What are the values for X_3 and Y_3?

6.14 For the lens in the previous problem the paraxial focal plane is located 49.98943 cm behind the lens. Trace a ray from infinity with $U = -0.2$ radians and $X_1 = 0.5$, $Y_1 = 0.990006$, and $Z_1 = 0.01318$ to find X and Y in the focal plane.

6.15 Consider a mirror with $r = -50.0$ cm. If the object is located 30.0 cm in front of this mirror, use the mirror formula to determine the image location and magnification. Trace a single exact ray from $X_0 = 0$, $Y_0 = 2.00$ cm, $Z_0 = 0$, $L_0 = 0.01$, $M_0 = 0.02$, and $N_0 = 0.99975$ to the paraxial image plane. What are the values for X_2 and Y_2?

6.16 For the mirror in the previous problem, trace a ray from infinity with $U = 0.2$ radians and $X_1 = -0.994987$, $Y_1 = -3.00$, and $Z_1 = -0.100$ to find X_2 and Y_2 in the focal plane.

6.17 Given the following data for a certain lens, trace four rings of rays to draw a spot diagram for an on-axis object point.

c	t	n	V
1.5400	0.110	1.617	55.0
-2.2300	0.025	1.720	29.3
-0.2975			

Complete the diagram using symmetry.

6.18 Given the spot diagram data for the previous problem, calculate the coordinates of the center of the spot and the rms size of the spot.

6.19 Given the system in Problem 6.17, make an *LSA* plot. Assume the clear radius of each surface is 0.20 cm.

6.20 Given the system in Problem 6.17, make sagittal and tangential ray intercept plots for an off-axis object located at infinity at a field angle of 1°.

6.21 Given the following data for a certain lens, make a distortion plot up to 10°. All surfaces have a clear radius of 0.15 cm.

c	t	n	V
2.8752	0.0756	1.623	60.3
-2.3804	0.0252	1.699	30.3
0.2012	0.0806	1.000	0.0
4.2662	0.0867	1.547	45.8
-1.5871	0.0302	1.613	57.6
6.9589			

6.22 Given the system in the previous problem, make an *LSA* plot.

6.23 Derive Eq. 6.22 from Eqs. 6.18 and 6.21.

6.24 Write a computer program or spreadsheet to trace absolute rays through any system with two refracting surfaces. Inputs should be the lens prescription and the initial values for X, Y, Z, L, M, and N. Output should include the paraxial image position and X and Y in the paraxial image plane.

6.25 Trace five meridional rays from an object at infinity on the optical axis separated by a perpendicular distance of 0.5 cm through the lens given in Problem 6.13. Use your results to make an *LSA* plot.

6.26 Repeat Problem 6.25, but with the lens reversed (i.e., $c_1 = 0.12\ \text{cm}^{-1}$, $c_2 = 0.08\ \text{cm}^{-1}$). Does spherical aberration depend strongly on shape of the lens?

6.27 Repeat Problem 6.25, but with SF57 glass substituted for the FK5 glass.

6.28 Repeat Problem 6.26 with SF57 glass substituted for the FK5 glass.

CHAPTER 7

THIRD-ORDER OPTICS

In the previous chapter we saw how to trace real rays through an optical system. We also saw how to use sets of rays to make spot diagrams and various types of plots to help analyze and evaluate optical system performance. With the exception of diffraction, which we will continue to ignore, real ray tracing provides the ultimate in image analysis. In Chapter 9, we will use real ray tracing to optimize a system to automatically reduce aberrations. This method is really a fancy version of trial-and-error. It lends no insight into the workings of imaging systems. And it certainly does not give you a good idea of what type of system to start with.

Back in the dark ages, before computers and pocket calculators, exact ray tracing (especially skew ray tracing) was avoided at all costs. The time involved in tracing even one ray through a moderately complicated system was enormous, as was the possibility of error. Using the sine function required extrapolating values from a table. Instead, an elegant theory of approximations to the exact calculations was developed. This theory results in two aberration polynomials. The purpose of the aberration polynomials is to express the transverse ray aberrations ϵ_x' and ϵ_y' as functions of the ray parameters. While paraxial optics completely ignores aberrations, the aberration polynomial uses paraxial ray tracing to determine an approximate value for aberrations. No real ray tracing is required.

Now that we have computers and software to do exact ray tracing for us, time and accuracy is of very little concern. We can quickly and easily trace as many exact rays as we want to trace through even the most complicated systems. We discuss the aberration polynomial because the equations involved will tell us which surface is the largest contributor to the total aberration. The equations will also indicate what changes in the system parameters will be effective in reducing the aberrations. The aberration polynomial will also allow us to classify aberrations and leads to better understanding of the image forming process. The aberration polynomial will help us select initial systems for further analysis and optimization. The chief drawback to the aberration polynomial is that it is not exact. The version presented here is only to third order, although the theory has been extended to fifth and higher orders (see, for example, Conrady's book).

The derivation of the aberration polynomial, even to just third order, is quite tedious, so it is presented in Appendix D for those brave and diligent souls who insist on going through it. In this chapter, the third-order aberration polynomials are presented in their entirety first. Then they are broken down into pieces and each piece is discussed in detail. For each piece, the functional dependence is discussed first and then formulae are given for the coefficient of that piece. It is hoped that, in this way the student will get the big picture first and then learn the details as needed.

7.1 THIRD-ORDER ABERRATION POLYNOMIAL

As mentioned previously, an aberration is the difference in position between an exact ray and the corresponding paraxial ray. We can define the transverse components of the aberration by first finding the paraxial image. But, before we find the image, we need to begin by defining the object position. Because our optical systems are all assumed to be axially symmetric, it does not matter where in the object plane the object is located. We can always rotate the coordinate system so that the object lies on the y-axis. Thus, $x_0 = 0$ and $y_0 = h$. After tracing the paraxial ray using *y-nu* ray tracing we find $x_{j+1} = 0$ and $y_{j+1} = h'$.

The next step is to trace an exact ray. We start at the same object point, but now we use capital letters to indicate exact quantities: $X_0 = 0$, $Y_0 = H = h$. The position of this ray in the paraxial image plane is given by X_{j+1} and Y_{j+1}. The transverse ray aberration components are defined by

$$\epsilon_x' = X_{j+1} \tag{7.1}$$

and

$$\epsilon_y' = Y_{j+1} - y_{j+1} \ . \tag{7.2}$$

From our previous experience with exact ray tracing, we know that rays traced from different object points will have different values for the transverse aberration components. Also, rays traced in different directions from the same object point will have different amounts of aberration. Therefore, we can say that ϵ_x' and ϵ_y' are functions of ray position and direction η, ρ, and θ which are defined below.

We can conveniently define ray position in terms of the fractional object height

$$\eta = \frac{Y_0}{H} , \tag{7.3}$$

where Y_0 is the actual starting position of the particular ray and H represents the full field of view. A similar definition may be used for objects at infinity:

$$\eta = \frac{U_0}{U_{HFOV}} , \tag{7.4}$$

where U_0 is the slope angle of the ray in object space and U_{HFOV} is the half-field-of-view angle. A full-field ray has $\eta = 1$ and an axial ray has $\eta = 0$.

Ray direction can be conveniently defined in terms of where the ray crosses the entrance pupil ($X_E = 0$): Y_E and Z_E. Actually, we will be using polar coordinates. Define

$$\rho = \frac{\sqrt{X_E^2 + Y_E^2}}{R_E} , \tag{7.5}$$

where R_E is the radius of the entrance pupil. Note that the angle θ is measured from the y axis. Therefore,

$$\theta = \arctan\left[\frac{X_E}{Y_E}\right] . \tag{7.6}$$

Figure 7.1 shows a general skew ray in object space. A chief ray is given by $\rho = 0$ and a marginal ray has $\rho = 1$. Rays passing through the positive x axis have $\theta = 90°$. The negative y axis is $\theta = 180°$ and the negative x axis is $\theta = 270°$. For a chief ray, θ is undefined.

Next, we need to discuss why we call this third-order optics. In the derivation of

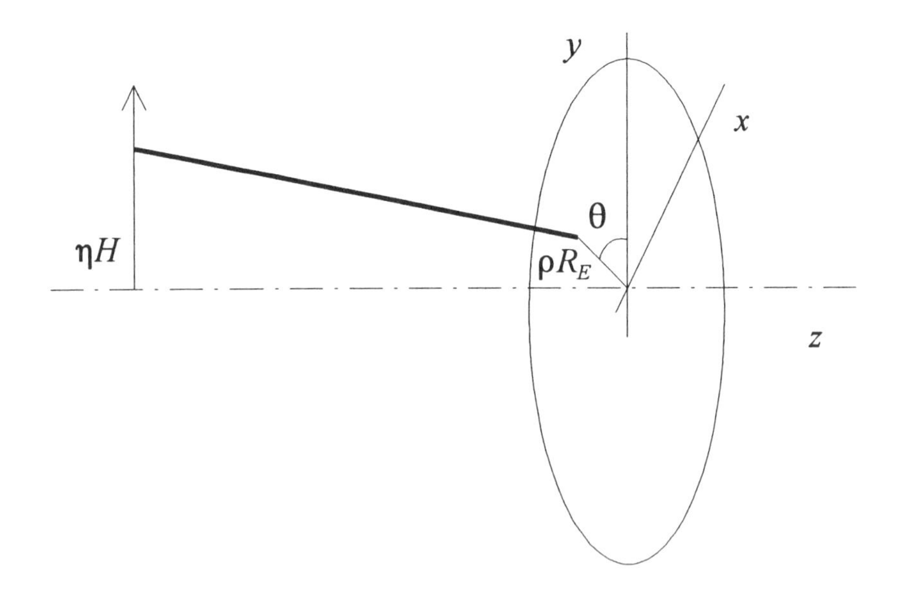

Figure 7.1 Ray tracing through a generalized optical system shown the relative object height and entrance pupil coordinates.

the aberration polynomial, skew rays are approximated by a linear combination of two paraxial rays, the marginal axial ray, and the full-field chief ray. The fraction of the marginal axial ray included in the linear combination is given by ρ and the fraction of the full-field ray is η, while θ defines the plane of the rays. In paraxial optics, the ray heights and slope angles are small quantities which appear only linearly in any equation. In third-order optics, these small quantities will appear as the third power. The ray parameters appear in the coefficients of the polynomial, so the third-order nature of the equations will be hidden initially.

After this lengthy introduction we are ready to present the aberration polynomials

$$\epsilon_x' = -\frac{1}{2 n_k u_k}\left[S \rho^3 \sin\theta + C \eta \rho^2 \sin 2\theta + (A + P\Lambda^2)\eta^2 \rho \sin\theta\right] \quad (7.7)$$

and

$$\epsilon_y' = -\frac{1}{2 n_k u_k}\left[S \rho^3 \cos\theta + C \eta \rho^2 (2 + \cos 2\theta) + (3A + P\Lambda^2)\eta^2 \rho \cos\theta + D\eta^3\right]. \quad (7.8)$$

The five coefficients S, C, A, P, and D are called *Seidel aberration coefficients.* They depend on the system parameters and the paraxial ray parameters. Each coefficient will be discussed in detail in the following sections. Also appearing in these equations are the index of refraction in image space n_k, the slope angle in image space of a marginal paraxial ray u_k, and the optical invariant Λ. The optical invariant is based on the marginal axial ray and the full-field chief ray.

Example 7.1

Given the aberration polynomial coefficients $S = -0.0926$, $C = 0.0112$, $A = -0.0018$, $P = -0.0666$, and $D = 0.0000$, the reduced slope angle in image space, $\alpha_k = -0.0667$, and the optical invariant, $\Lambda = 0.1333$, for the lens described in Example 6.1, calculate the transverse aberrations for a ray defined by $\eta = 0.3$, $\rho = 0.6$, and $\theta = 35°$.

The solution may be found by substituting the values given above into Eqs. 7.7 and 7.8:

$$\epsilon_x' = -\frac{1}{2(-0.0667)}[(-0.0926)(0.6)^3 \sin(35°) + (0.0112)(0.3)(0.6)^2 \sin(2 \times 35°) + ((-0.0018) + (-0.0666)(0.1333)^2) \times (0.3)^2 (0.6) \sin(35°)],$$

$$\epsilon_x' = -8.600 \times 10^{-2} + 8.521 \times 10^{-3} - 6.927 \times 10^{-4} ,$$

$$\epsilon_x' = -0.0782 ,$$

$$\epsilon_y' = -\frac{1}{2(-0.0667)} [(-0.0926)(0.6)^3 \cos(35°)$$

$$+ (0.0112)(0.3)(0.6)^2 (2 + \cos(2 \times 35°))$$

$$+ (3(-0.0018) + (-0.0666)(0.1333)^2)$$

$$\times (0.3)^2 (0.6) \cos(35°) + (0.0000)(0.3)^3] ,$$

$$\epsilon_y' = -0.1228 + 2.124 \times 10^{-2} - 2.183 \times 10^{-3} ,$$

$$\epsilon_y' = -0.1037 .$$

From this calculation, we can see that the S term is by far the largest contributor to the aberrations of this particular system. If a way could be found to reduce S, then the overall aberrations would be reduced.

We could also use the aberration polynomial to produce a spot diagram, or any of the other aberration plots shown in Chapter 6. The results are virtually the same as the real ray plots given in the previous chapter because the aberrations produced by this simple lens are primarily third-order. ■

Each term in the aberration polynomial indicates a different type of behavior. In other words, a group of actual rays which travel through an optical system will behave (at least in the third-order approximation) in a certain way depending on where the rays come from and what their initial direction is. Each of these terms, and each type of behavior, is given a name. The names are mostly historical in nature, but some of them are also descriptive of the type of behavior. Each will be discussed in detail in the following sections. It is very important to note that even though we discuss each aberration separately, in general they all occur together. Only in a couple of very special cases can we isolate each term individually in any real system.

7.2 SPHERICAL ABERRATION

If the object is on the optical axis, then all of the rays traced from that object are axial rays. Therefore, η is zero. All of the terms in the aberration polynomial except for the term with the S coefficient depend on η, which means that the only nonzero aberration in this situation is due to the S term. The S term is spherical aberration which was described in the previous chapter. It is the only aberration present for an on-axis object. And despite the warning in the last paragraph of the preceding section, this is a physically realizable situation. Spherical aberration can be measured separately from the other aberrations.

On axis, the aberration polynomial looks like

$$\epsilon_x' = -\frac{1}{2\,n_k u_k} S\,\rho^3 \sin\theta \tag{7.9}$$

and

$$\epsilon_y' = -\frac{1}{2\,n_k u_k} S\,\rho^3 \cos\theta \ , \tag{7.10}$$

where S, n_k, and u_k are all constants which depend on the system prescription and the object location. We can see that as ρ changes, the magnitude of the third-order spherical aberration varies as the cube (Figure 7.2). For a fixed value of ρ, the only remaining variable is θ. As θ varies, we generate a cone of rays from the object point as shown in Figure 7.3. When θ goes through its complete range of values from 0 to 360°, the intersection points of the rays with the paraxial image plane goes through the same range forming a circle with constant radius. The maximum radius of this circle is

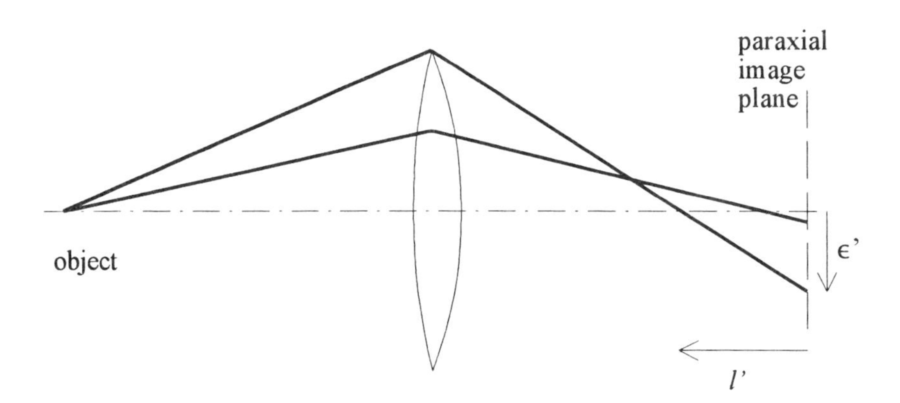

Figure 7.2 Spherical aberration in an idealized optical system. The third-order spherical aberration varies as the cube of the height of the ray in the entrance pupil.

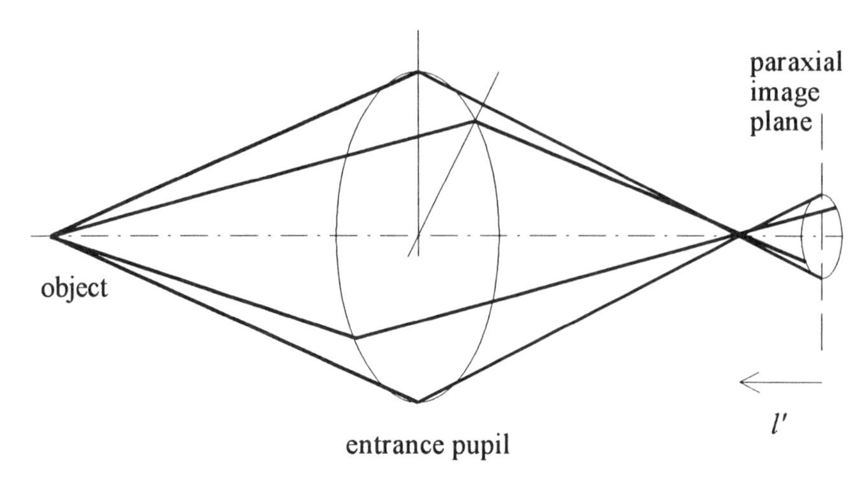

Figure 7.3 Third-order spherical aberration produced by an idealized optical system. Here, the entire system has been replaced by its entrance pupil.

$$\epsilon' = -\frac{1}{2 n_k u_k} S \,. \tag{7.11}$$

The geometry described here requires all of the rays with a certain value for ρ to meet at a point on the optical axis. Therefore, spherical aberration could be described as a shift in image location with aperture size. The size of the longitudinal shift is given approximately as

$$l' = -\frac{\epsilon'}{u_k} \,. \tag{7.12}$$

The radius of the aberration circle described above depends on S and ρ^3. Since the amount of light passing through a narrow range of ρ values depends linearly on ρ, the aberrated image of a point object will have a bright central core which fades rapidly toward the edges. The image of an extended object will be fuzzy and lack detail.

The most obvious way to reduce the size of the spherical aberration blur circle would be to reduce ρ by reducing the size of the aperture stop. Unfortunately, the size of the aperture stop is usually specified by the design requirements of the system and cannot be reduced. We really need to reduce S to improve the image quality. The coefficient is a summation of terms, one for each surface in the system

$$S = \sum_{j=1}^{k} W_j i_j^2 \,. \tag{7.13}$$

Each term is calculated according to the formula

$$W_j = y_j n_{j-1} \left[\frac{n_{j-1}}{n_j} - 1 \right] (u_j + i_j) . \tag{7.14}$$

The ray parameters, y_j, u_j, and i_j, all come from a paraxial, marginal axial ray trace.

Example 7.2

Determine the spherical aberration coefficient for the lens system described in Example 6.1.

The first step is to set up and trace a marginal axial ray as shown in Table 7.1. We now have the data necessary to calculate S:

$$S_1 = (2.0)(1.0)(0.532239)^2 \left[\frac{(1.0)}{(1.50137)} - 1 \right]$$

$$\times [(-0.044404) + (0.532239)]$$

$$= -0.092297 ,$$

Table 7.1 Ray trace table for Example 7.2.

	0		1		2	
c			0.199453		0.0	
t		15		0.5		
n		1.0		1.50137		1.0
$-\phi$			-0.1		0.0	
τ		15		0.33303		29.67
y	0.0		2.0		1.977798	
α		0.133333		-0.066667		-0.066667
u		0.133333		-0.044404		-0.066667
i			0.532239		-0.044404	

$$S_2 = (1.977798)(1.50137)(-0.044404)^2 \left[\frac{(1.50137)}{(1.0)} - 1 \right]$$
$$\times [(-0.066667) + (-0.044404)]$$
$$= -0.000326 ,$$

$$S = (-0.092297) + (-0.000326) = -0.092623 ,$$

$$\epsilon' = -\frac{1}{2(1.0)(-0.066667)}(-0.092623) = -0.694668 .$$

■

Equation 7.14 indicates ways to reduce the amount of spherical aberration in a system. Although it is not obvious, the signs of terms in this sum may be different. Combining lenses with negative and positive contributions can lead to cancellation. Positive thin lenses have negative or undercorrected spherical aberration ($LSA < 0$), while negative lenses have positive or overcorrected spherical ($LSA > 0$). This aberration reduction technique will be discussed fully in Chapter 10.

Another way to reduce the sum is to reduce the individual terms by making the incidence angle as small as possible at each surface. The ray height and slope angles are fixed by the object location and aperture size, but the incidence angle can be changed in a couple of ways without changing the focal length (or the image size and position) of the system. One thing that can be done is to increase the index of refraction of the glass in the lens. A higher-index glass requires a smaller curvature for a given focal length. The smaller curvature reduces the incidence angle for the marginal axial ray. A second approach is to change the shape of the lens to minimize the incidence angle. As we have frequently seen, lenses with different shapes can have the same imaging characteristics. The process of changing a lens' shape to reduce aberrations is called *lens bending*.

The affect of lens bending can be seen clearly by looking at the spherical aberration coefficient for a thin lens in air. From Eq. D.113 we have

$$S_t = -y^4(n-1)c\left[\frac{(n+2)}{n}c_1^2 - (2n+1)cc_1 - \frac{4(n+1)}{n}vc_1 + n^2c^2 + (3n+1)cv + \frac{(3n+2)}{n}v^2\right]. \tag{7.15}$$

Figure 7.4 shows the affect of lens bending and choice of index of refraction on the thin lens spherical aberration coefficient.

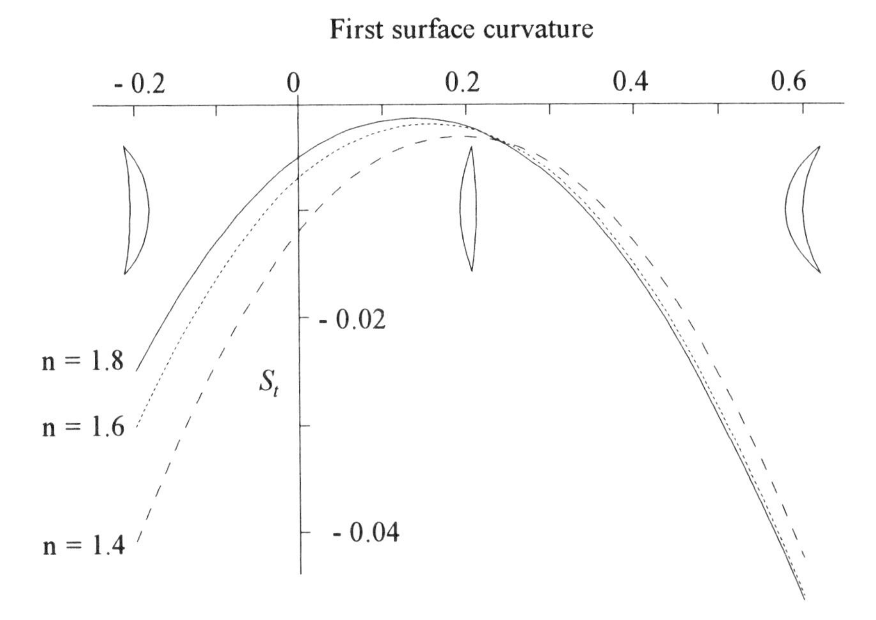

Figure 7.4 Spherical aberration coefficient as a function of first surface curvature (lens shape or bending) for various indices of refraction.

The previous equation represents just one thin lens. For multiple thin lenses, a similar equation must be determined for each lens and added together to get the overall S_t coefficient (see Section 10.1). As usual, c_1 is the curvature of the first surface. The net curvature is given by $c = c_1 - c_2$. Since

$$\frac{1}{f'} = (n - 1)(c_1 - c_2) = (n - 1)c \,, \tag{7.16}$$

for a thin lens the focal length of the lens remains fixed if c_1 and c_2 change by the same amount. This is lens bending. The remaining parameter, v, is the vergence:

$$v = \frac{1}{s} \,. \tag{7.17}$$

For real objects, v is negative.

The only free parameter in Eq. 7.15 is the curvature of the first surface. This will be our bending parameter. A minimum can be found by differentiating Eq. 7.15 with respect to c_1 and setting the result equal to zero:

$$\frac{\partial S_t}{\partial c_1} = -y^4(n - 1)c\left[\frac{2(n + 2)}{n}c_1 - (2n + 1)c - \frac{4(n + 1)}{n}v\right] \cdot \tag{7.18}$$

From this we can determine the thin lens shape which gives the minimum amount of spherical aberration:

$$c_1 = \frac{(2n + 1)nc + 4(n + 1)v}{2(n + 2)} . \tag{7.19}$$

Complicated systems of thick lenses may also be bent in a similar fashion, but this is best accomplished with a computer program.

Example 7.3

Determine the thin lens shape which gives the minimum amount of spherical aberration. The lens is to be made from LaK10 glass and should have an effective focal length of 10.0 cm. Assume the object is at infinity ($v = 0$).

First calculate the net curvature of the lens:

$$c = \frac{1}{(1.72 - 1)(10.0)} = 0.138889 .$$

From equation (7.19) we find the optimum lens shape:

$$c_1 = \frac{(2(1.72) + 1)(1.72)(0.138889)}{2(3.72)} = 0.142563 .$$

■

A special application of lens bending is another technique for reducing spherical aberration called lens doubling. The basic idea behind lens doubling is to reduce the *S* coefficient by reducing the incidence angle. This reduction is accomplished by doubling the number of surfaces in the system in such a way that each surface only needs to do half of the ray bending. The lens doubling leads to a theoretical decrease in spherical aberration by a factor of approximately 4.

To illustrate lens doubling, consider a single thin lens. This lens has a particular focal length and $f/\#$ which our final lens system must also have. The first step in lens doubling is to double all of the lengths in the lens. Since spherical aberration is a distance, it should also double in this process. This can be confirmed by looking at Eq. 7.15. The coefficient will double because y will double and it is raised to the fourth power, but c, c_1, and v will each be cut if half. Also included in the aberration calculation is the marginal axial ray slope angle in image space, but this angle does not change when all of the distances are double.

Now our lens has twice its intended focal length. To get back to the correct focal length we add a second thin lens with the same focal length and place it in contact with the first lens. Adding the second lens approximately doubles the amount of spherical aberration because we now have twice as many surfaces. We do not have exactly twice the aberration because the vergence for the second lens will be different from the first lens. What we need to do is to bend the second lens to its optimum shape for the vergence produced by the first lens. The shape for minimum spherical aberration and the amount of spherical aberration will be different than for the first lens. It turns out that adding the second lens adds only a small amount of spherical aberration when the second lens is optimized.

Now, our system has the proper focal length, but not the proper $f/\#$. To get back to the proper $f/\#$, we reduce the size of the aperture stop by a factor of 2. This causes a reduction in the S coefficient by a factor of 16. The new aperture stop also deceases the marginal axial ray slope back to its original value. The net result is an overall decrease in spherical aberration of approximately a factor of 4. The "approximately" part is due to the optimized second lens not quite doubling the amount of spherical aberration.

Example 7.4

Use lens doubling to reduce the spherical aberration caused by an $f/5$ thin lens with 10.0-cm effective focal length. Assume the object is at infinity and the index of refraction is 1.72.

The first step is to calculate the amount of spherical aberration. Using Eq. 7.15 and the net curvature and optimum shape determined in Example 7.3 we can find the S_t coefficient:

$$S_t = -(1.0)(0.72)(0.138889)\left[\frac{(3.72)}{(1.72)}(0.142563)^2 - (2(1.72)+1)(0.138889)(0.142563) + (1.72)^2(0.138889)^2\right]$$

$$= -0.001311 \ .$$

Note that several terms vanish because $v = 0$. Finally,

$$\epsilon' = -\frac{1}{2(1.0)(-0.1)}(-0.001311) = -0.006556 \ .$$

Next, we double all distances to get an $f' = 20.0$ cm, $f/5$ lens. For this lens $c = 0.069444$, $c_1 = 0.071281$, and $y = 2.0$. A calculation shows that the S_t has doubled, as expected. The amount of spherical aberration has also doubled since the slope angle of the marginal axial ray in image space remains the same. This lens becomes lens a with $S_{ta} = -0.002622$. We now add a second thin lens with $f' = 20.0$.

The second lens actually has a larger coefficient than the first lens ($S_{tb} > S_{ta}$) because the vergence is no longer zero. But, this lens can be bent to minimize its spherical aberration. For the second lens (lens b),

$$v_b = \frac{1}{(20.0)} = 0.05 .$$

The optimum shape of this lens is

$$c_{1b} = \frac{(2(1.72) + 1)(1.72)(0.069444) + 4(2.72)(0.05)}{2(3.72)}$$

$$= 0.144399 .$$

The spherical aberration contribution for the second lens is

$$S_{tb} = -(2.0)^4(0.72)(0.069444)\Bigg[\frac{(3.72)}{(1.72)}(0.144400)^2$$

$$-(2(1.72 + 1)(0.069444) - \frac{4(2.72)}{(1.72)}(0.05)(0.144400)$$

$$+(1.72)^2(0.069444)^2 + (3(1.72) + 1)(0.069444)(0.05)$$

$$+\frac{(3(1.72) + 1)}{(1.72)}(0.05)^2\Bigg] = -0.000773 .$$

The combined S coefficient is

$$S_t = (-0.002622) + (-0.000773) = -0.003395$$

and the aberration is

$$\epsilon' = -\frac{1}{2(1.0)(-0.2)}(-0.003395) = -0.008488 .$$

This is only about 1.3 times larger than our original aberration value. But, this system is $f/2.5$. When we stop the system down to get back to $f/5$, we gain a 16 times decrease in the S_t coefficient:

$$S_t = -0.000212 .$$

Finally, for the magnitude of the aberration we get

$$\epsilon' = -\frac{1}{2(1.0)(-0.1)}(-0.000212) = -0.001061 ,$$

which is smaller than our single-lens aberration value by a factor of more than 6.

■

One final comment on spherical aberration. Just because it is the only aberration of an on-axis object doesn't mean that it goes away for off-axis objects. As the aberration polynomial clearly shows, spherical aberration is also present for off-axis objects.

7.3 COMA

If we could isolate somehow on the C term of the aberration polynomial, we could observe and measure pure coma. The aberrations would look like

$$\epsilon_x' = -\frac{1}{2\, n_k u_k} C \eta \rho^2 \sin 2\theta \tag{7.20}$$

and

$$\epsilon_y' = -\frac{1}{2\, n_k u_k} C \eta \rho^2 (2 + \cos 2\theta) , \tag{7.21}$$

where C is the coma coefficient. As you can see, coma has a significantly more complicated form than spherical aberration. Coma depends linearly on η. If we double the object size or the field angle, coma will double. Thus, coma can become a problem for any system if we attempt to make the field of view too large. Coma also depends on aperture size, as ρ^2. Finally, the θ dependence is unique. If we fix η and ρ and vary θ, the cone of rays in object space will create a circle of intersection points in the paraxial image plane. Actually, because of the 2θ dependence, two circles are formed every time θ completes one circle in object space. That is, rays that are $180°$ apart in object space meet at the same point in image space as shown in Figure 7.5.

The maximum radius of a coma circle is

$$\epsilon' = -\frac{1}{2\, n_k u_k} C . \tag{7.22}$$

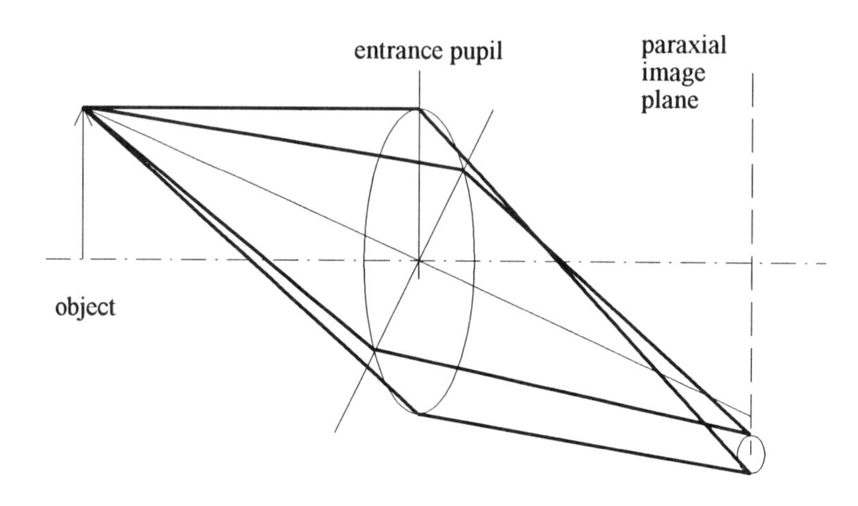

Figure 7.5 Coma produced by an idealized optical system.

Since this value represents the maximum width of the coma blur, it is called *sagittal coma*. Remember that the sagittal plane is perpendicular to the tangential or meridional plane. Figure 7.6 shows that the center of the circle is offset from the paraxial image point by twice this radius. The paraxial image point actually lies outside the largest aberration circle. The maximum height of the coma blur is three times the radius of the coma circle. This is called *tangential coma*. As ρ decreases, the radius of the blur circle decreases as ρ^2 and so does the offset. The net effect is that the comatic image of a point object looks something like a snow cone. Of course,

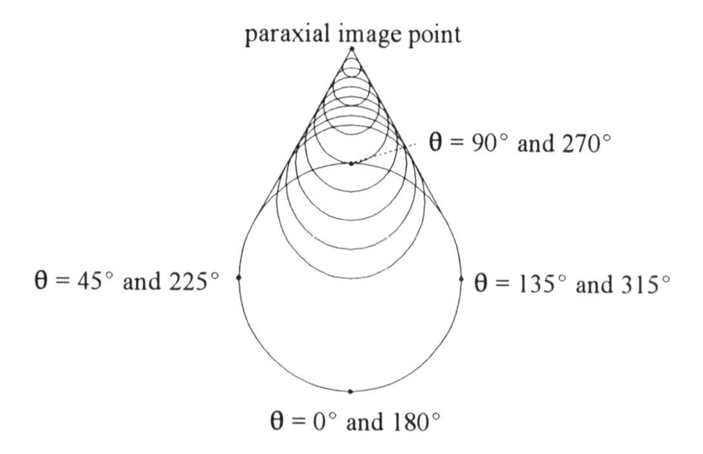

Figure 7.6 Details of the coma pattern produced in the paraxial image plane. Each circle represents light coming from a different zone of the entrance pupil.

before snow cones were invented, people thought the pattern looked like a comet with a small, bright nucleus and an extended tail which fades gradually away from the nucleus. The name coma comes from this appearance. The asymmetric appearance of coma makes it a particularly obnoxious aberration.

Coma may be reduced by limiting the field of view or by stopping down the system. Neither method can normally be used because the field of view and the entrance pupil size are almost always specified for our design. Coma can also be reduced by reducing the coefficient C. As with spherical aberration, C consists of summation of terms, one for each surface in the system. Each individual term looks like

$$C_j = y_j n_{j-1} i_j \bar{i}_j \left[\frac{n_{j-1}}{n_j} - 1 \right] (u_j + i_j) . \tag{7.23}$$

Because of the similarity between S_j and C_j, a shortcut for doing the calculation is

$$C = \sum_{j=1}^{k} W_j i_j \bar{i}_j . \tag{7.24}$$

Example 7.5

Determine the coma coefficient for the lens system described in Example 6.1.

We already set up the ray trace table in Example 7.1, so all we need to do here is trace a full-field chief ray (Table 7.2).

We are now ready to begin our coma calculations

$$C_1 = (-0.325817)(0.532239)(-0.066667) = 0.011561 .$$

For the second surface we have

Table 7.2 Full-field chief ray trace for Example 7.5.

	0		1		2	
y	1.0		0.0		-0.022202	
α		-0.066667		-0.066667		-0.066667
u		-0.066667		-0.044404		-0.066667
i			-0.066667		-0.044404	

$$C_2 = (-0.165359)(-0.044404)(-0.044404) = -0.000326 .$$

All together

$$C = (0.011561) + (-0.000326) = 0.011235 ,$$

$$\epsilon' = -\frac{1}{2(1.0)(-0.066667)}(0.011235) ,$$

$$\epsilon' = 0.084262 .$$

■

The thin lens version of the coma coefficient was derived in Appendix D:

$$C_t = C_t^* + \frac{\bar{y}}{y} S_t . \tag{7.25}$$

It consists of two parts. The first part is the coma contribution from the lens assuming that the stop is at the lens:

$$C_t^* = y^3 (n - 1) h v \left[n c^2 - \left(\frac{n + 1}{n} \right) c c_1 + \left(\frac{2n + 1}{n} \right) v c \right] . \tag{7.26}$$

Note that the coma is a linear function of the shape of the lens. As shown in Figure 7.7 a lens can be bent so that there is no coma contribution from the lens. The lens shape for zero coma is

$$c_1 = \frac{n^2 c + (2n + 1) v}{(n + 1)} . \tag{7.27}$$

The second term in Eq. 7.25 is the coma contribution if the stop is shifted away from the surface. This is our first example of a stop shift formula.

It is important to note that if the object is at infinity, then the substitution

$$h v = u_{HFOV} \tag{7.28}$$

should be made.

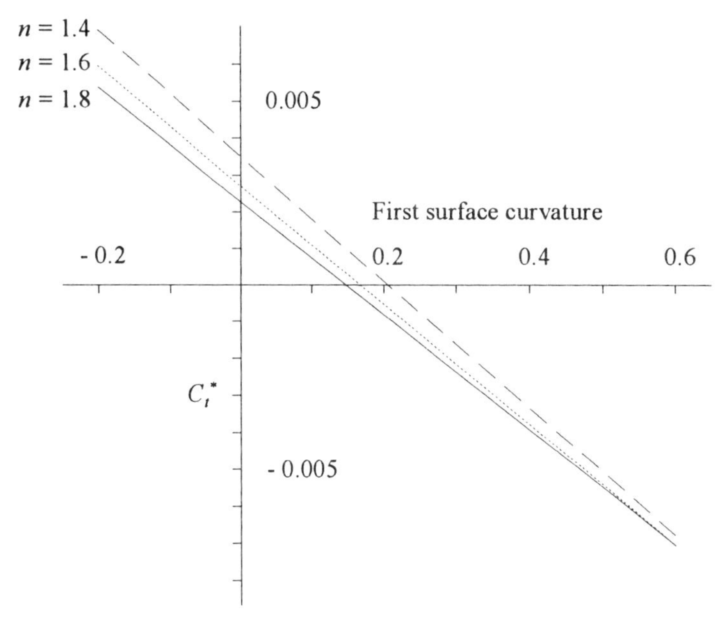

Figure 7.7 The coma coefficient as a function of lens shape and index of refraction.

Example 7.6

Calculate the lens shape which would give zero coma for the thin lens originally described in Example 7.3.

From Eq. 7.26 we calculate the lens shape which gives zero coma

$$c_1 = \frac{(1.72)^2(0.138889)}{(2.72)} = 0.151062 .$$

Note that this lens shape is very close to the lens shape which gives the minimum amount of spherical aberration. Since spherical aberration is quadratic in c_1, we would not expect S_t to differ very much from the minimum value. Using the zero coma lens shape, we can show that $S_t = -0.001327$. ■

One final method for reducing coma can be gleaned from Eq. 7.25. If the stop can be moved away from the lens, then stop position can be used to balance the two terms of Eq. 7.25 to get zero coma. This example illustrates the general technique of balancing one aberration against another. This general technique is illustrated in Chapter 10 for one of the lens design examples.

7.4 ASTIGMATISM

The next aberration is called *astigmatism*. You may have heard of astigmatism before if your eye doctor tested your eye sight and said that you had some astigmatism. Astigmatism in your vision is a result of your cornea not being spherical in shape, but rather more like a football. The astigmatism we are discussing arises even when the lens surfaces are spherical. The affect on images is similar for the two types of astigmatism. They both tend to create line images rather than point images.

In keeping with our nomenclature, astigmatism is, of course, given by the A terms in the aberration polynomials

$$\epsilon_x = -\frac{1}{2\, n_k u_k} A\, \eta^2 \rho \sin\theta \tag{7.29}$$

and

$$\epsilon_y = -\frac{3}{2\, n_k u_k} A\, \eta^2 \rho \cos\theta \ . \tag{7.30}$$

Note that like coma, the tangential astigmatism is three times larger than the sagittal astigmatism. The blur created by astigmatism in the paraxial image plane is not a circle, but an ellipse with long axis in the y direction as shown in Figure 7.8. The size of this ellipse depends linearly on the aperture size but quadratically on the object size.

An astigmatic image is formed when tangential rays come to a focus at a distance which is three times further away from the paraxial image plane than the focus of sagittal rays. Taking all rays together, we have a sagittal line image formed at the tangential focus and a tangential line formed at the sagittal focus. The effect on images can be seen by imagining a wagon wheel as an object (Figure 7.9). With the axle of the wheel on the optical axis, then by symmetry each spoke forms a tangential line. The spokes will appear in focus at the sagittal image. Conversely, the rim of the wheel can be viewed as a series of short sagittal lines. It will be in focus at the tangential image. Astigmatism is defined as a longitudinal shift in image position with ray direction.

You may have noticed that I scrupulously avoided using the words "image plane" in the previous paragraph. Both the tangential and sagittal images lie on curved surfaces as shown in Figure 7.10.

Because astigmatism depends on η^2, astigmatism is usually more of a problem at large field angles than is coma.

The A coefficient may be calculated from

$$A = \sum_{j=1}^{k} W_j \bar{i}_j^{\,2} \ . \tag{7.31}$$

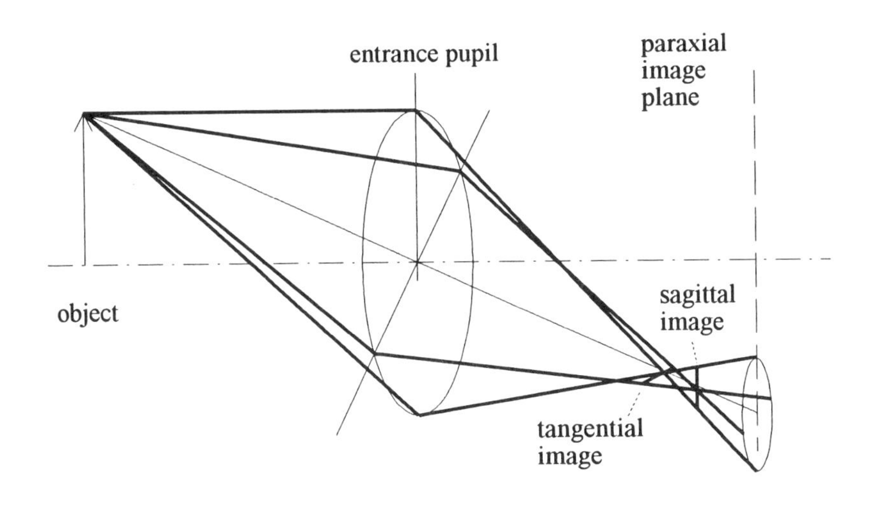

Figure 7.8 Astigmatism produced by an idealized optical system.

Example 7.6

Determine the astigmatism coefficient for our example lens.

All of the pieces are already in place from previous examples:

$$A_1 = (-0.325817)(-0.066667)^2 = -0.001448 ,$$

$$A_2 = (-0.165359)(-0.044404)^2 = -0.000326 ,$$

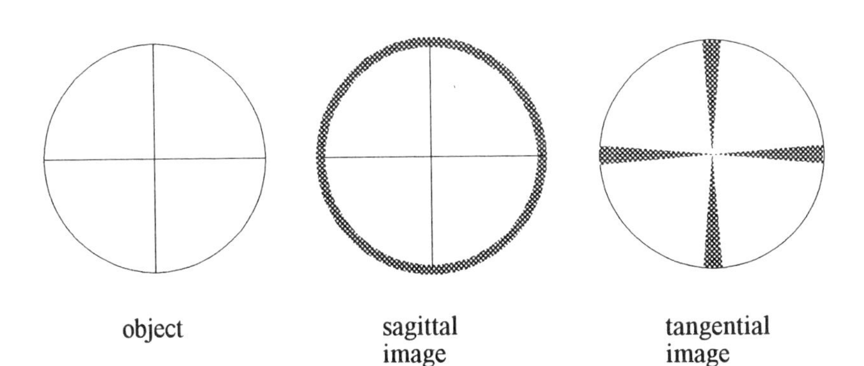

Figure 7.9 Affect of astigmatism on an axially symmetric object.

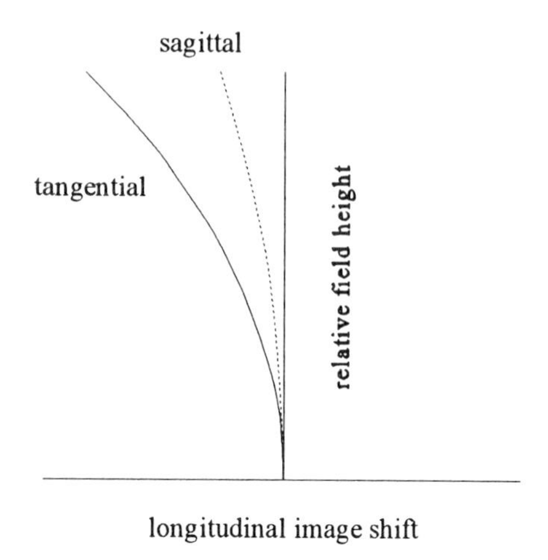

Figure 7.10 The tangential image surface lies three times further from the paraxial image plane than does the sagittal image surface.

$$A = (-0.001448) + (-0.000326) = -0.001774 ,$$

and finally,

$$\epsilon_x' = -\frac{1}{2(1.0)(-0.066667)}(-0.001774) ,$$

$$\epsilon_x' = -0.013305 ,$$

$$\epsilon_y' = -0.039915 .$$

■

Equation 7.29 shows a strong similarity between astigmatism and spherical aberration, just as coma did. However, for astigmatism we replace the dependence of spherical aberration on the incidence angle of the marginal axial ray by a dependence on the incidence angle of the full-field chief ray. Therefore, we expect that lens bending will have very little effect on astigmatism if the stop is located at the lens. The thin lens version of the astigmatism coefficient clearly shows this:

$$A_t = -\frac{(y h v)^2}{f'} + 2\frac{\bar{y}}{y} C_t^* + \left(\frac{\bar{y}}{y}\right)^2 S_t \, . \tag{7.32}$$

If the object is at infinity, then the substitution from Eq. 7.28 must be made.

Equation 7.32 shows that if the stop is at the lens, then all of the quantities in the calculation are fundamental quantities for a particular imaging system. It is difficult, therefore, to see what can be done to reduce astigmatism in a thin lens system. The only obvious approach is to use stop shifts to balance astigmatism against the other aberrations.

7.5 PETZVAL CURVATURE

Our next aberration is Petzval curvature. From the aberration polynomial, Petzval curvature has the form

$$\epsilon_x' = -\frac{1}{2 n_k u_k} P \Lambda^2 \eta^2 \rho \sin\theta \, , \tag{7.33}$$

$$\epsilon_y' = -\frac{1}{2 n_k u_k} P \Lambda^2 \eta^2 \rho \cos\theta . \tag{7.34}$$

As shown in Figure 7.11, Petzval curvature causes a blur circle similar to the spherical aberration blur circle. However, the Petzval curvature circle's radius depends only linearly on ρ. In addition, the radius depends on η^2. As with astigmatism, this dependence leads to a curved image surface—hence the name of the aberration. If Petzval curvature were the only aberration, we would have good imaging on this surface. Unfortunately, we normally want planar images (for film, etc.) not curved images. Petzval curvature is defined as a longitudinal shift in image position with object size or field angle.

Note that astigmatism and Petzval curvature have the same ρ, η and θ dependence. In practice, this generally means that astigmatism causes additional image shifts beyond that caused by Petzval curvature alone. Typically, the tangential image surface is three times further from the sagittal image surface is from the Petzval curvature surface, which is itself displaced from the paraxial image plane.

The Petzval curvature coefficient is by far the easiest to calculate:

$$P_j = \frac{n_{j-1} - n_j}{n_{j-1} n_j} c_j \, . \tag{7.35}$$

Note that no paraxial rays need to be traced to calculate this coefficient. This reflects the fundamental nature of Petzval curvature. As you know, there is a direct

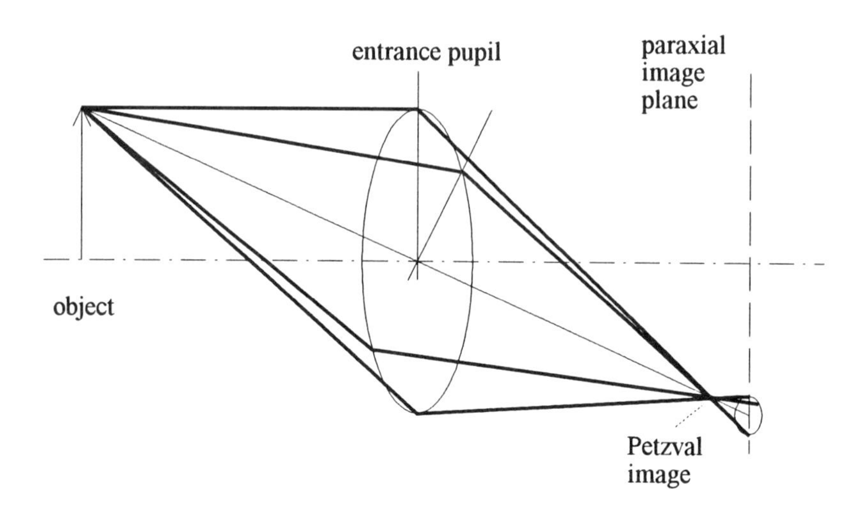

Figure 7.11 Petzval curvature produced by an idealized optical system.

relationship between the object and image distances. As a real object moves off-axis, it effectively moves away from the lens. Therefore, we must expect a corresponding shift in image position.

While it is true that no rays need to be traced to find P, the Petzval coefficient appears in Eqs. 7.33 and 7.34 with Λ. Both the marginal axial ray and the full-field chief ray need to be traced to calculate Λ.

Example 7.8

Calculate the amount of Petzval curvature present in our example lens.

The coefficient P comes directly from the lens prescription data:

$$P_1 = \frac{(1.0) - (1.50137)}{(1.0)(1.50137)}(0.199453) = -0.066606 ,$$

$$P_2 = \frac{(1.50137) - (1.0)}{(1.50137)(1.0)}(0.0) = 0.0 ,$$

$$P = (-0.066606) + (0.0) = -0.066606 .$$

Next, from the ray traces performed earlier we find

$$\Lambda = (0.133333)(0.0) - (-0.066667)(2.0) = 0.133333 .$$

Finally, we find

$$\epsilon' = -\frac{1}{2(1.0)(-0.066667)}(-0.066606)(0.133333)^2 ,$$

$$\epsilon' = -0.008881 .$$

■

Equation 7.35 indicates that a simple way to reduce Petzval curvature is to make the lens meniscus-shaped. If the two curvatures of the meniscus lens have the same sign and magnitude, then the change in sign due to the index of refraction will cause the two surface contributions to cancel. A thin meniscus lens with equal curvatures has no power, but a similar thick lens has power proportional to the lens thickness.

Furthermore, Eq. 7.34 indicates that for a compound lens with cemented surfaces the surfaces in contact with air should have smaller curvatures than surfaces between glass if Petzval curvature is to be minimized.

The thin lens version of the Petzval curvature coefficient,

$$P_t = -\frac{1}{nf'} , \tag{7.36}$$

indicates that raising the index of refraction can lower the Petzval curvature. Note that this formula does not have a stop shift contribution, which again demonstrates the fundamental character of Petzval curvature.

7.6 DISTORTION

The last aberration is called *distortion*. In everyday English a distorted image would be misshaped in some way. The third-order distortion bends the image in a very particular way. The components of the aberration are given by

$$\epsilon_x' = 0 , \tag{7.37}$$

$$\epsilon_y' = -\frac{1}{2n_k u_k} D\eta^3 . \tag{7.38}$$

These equations show that all rays from a particular object point meet at a single image point. We have perfect imaging when distortion is the only aberration. It's just that the image formed is not where the paraxial image should be. It is displaced in the paraxial image plane by a distance which depends on η^3. We can define distortion as the variation of magnification with object size or field angle.

To better appreciate the effect of distortion in the image, consider a square object centered on the optical axis. As shown in Figure 7.12, the image formed will look like a barrel or a pincushion depending on the sign of D for the imaging system. Since the magnitude of the image shift caused by distortion depends on the cube of the distance from the axis, the image shift is larger at the corners of the square than at the center of the sides. Similarly, a larger square would be even more distorted.

Even though distortion does not blur the image, it is an important aberration to try to correct in some systems. Lenses used for linear measurements or map making should be free of distortion. Another lens which should not be afflicted with distortion is a lens used in producing integrated circuits by photolithography. In some cases, however, distortion cannot be avoided. The classic example is a fish-eye camera lens. Whenever and extremely wide field of view must be forced into a narrow image space, barrel distortion will result.

Since the aberration contributed by distortion has a simple functional form, it may be considered poetic justice that the coefficient for distortion is the most complicated of all of the coefficients:

$$D \quad \sum_{j=1}^{k} \left[\bar{W}_j i_j \bar{i}_j + (\bar{u}_{j-1}^2 - \bar{u}_j^2) \Lambda \right], \tag{7.39}$$

where

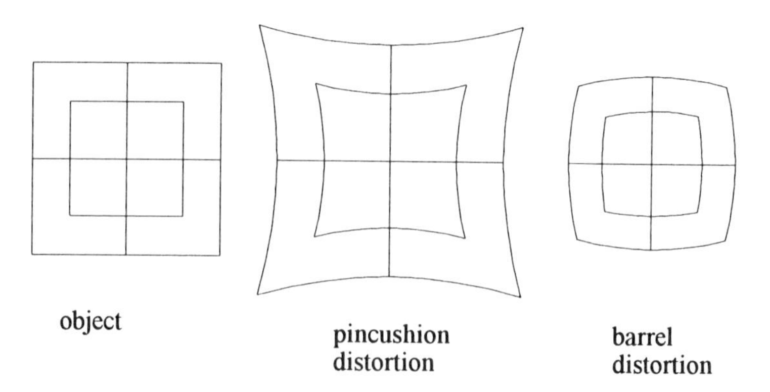

Figure 7.12 The affect of distortion on an axially symmetric object for different signs of the distortion coefficient.

$$\overline{W}_j = \overline{y}_j n_{j-1} \left[\frac{n_{j-1}}{n_j} - 1 \right] (\overline{u}_j + \overline{i}_j) . \tag{7.40}$$

Example 7.8

Calculate the distortion contribution for our example lens.

Once again, from previous examples, we have all of the necessary pieces

$$D_1 = (-0.125258) [(-0.001448) + (-0.066606) (0.133333)^2] ,$$

$$D_1 = 0.000330 ,$$

$$D_2 = (1.0) [(-0.000326) + (1.0) (0.0) (0.133333)^2] ,$$

$$D_2 = -0.000326 ,$$

$$D = (0.000330) + (-0.000326) = 0.000004 .$$

Finally, the distortion contribution is

$$\epsilon_y' = - \frac{1}{2 (1.0) (-0.066667)} (0.000004) = 0.000030 .$$

■

The thin lens version of the distortion coefficient is just as complicated:

$$D_t = \left(\frac{\overline{y}}{y} \right) (3 A_t^* + P_t \Lambda^2) + 3 \left(\frac{\overline{y}}{y} \right)^2 C_t^* + \left(\frac{\overline{y}}{y} \right)^3 S_t . \tag{7.41}$$

Note that there is no distortion contribution inherent in the thin lens itself:

$$D_t^* = 0 . \tag{7.42}$$

This can be experienced with any lens used as a simple magnifier. The pupil of the eye serves as the aperture stop, and since it is close to the lens, there will be very little distortion in the image. Instead, the distortion arises solely from the stop shift. If you take the same eyepiece or magnifier and hold it at arms length, you can still see a magnified image. But now, since the stop is far from the lens, the image will be greatly distorted.

7.7 REAL RAYS AND THE THIRD-ORDER ABERRATIONS

I have mentioned repeatedly in this chapter that the five Seidel aberrations cannot, in general, be separated from each other. But, we can trace selected real rays and combine the results of such ray traces to mimic isolating on each aberration. The equations developed in this section are only good to the third-order in small quantities. Higher-order aberrations normally exist, but are being ignored here.

To see how this might be accomplished, first consider spherical aberration. As mentioned earlier, if $\eta = 0$, then we automatically isolate on spherical aberration. Spherical aberration is an aberration of a marginal axial ray. If we trace a ray with $\rho = 1$, $\theta = 0°$, and $\eta = 0$, then the aberration that results is due to spherical aberration alone. Applying the aberrations in this form we get

$$\epsilon_y'(\rho = 1, \theta = 0°, \eta = 0) = -\frac{1}{2 n_k u_k} S . \tag{7.43}$$

A similar case is distortion. If we set $\rho = 0$, we are left only with distortion. Distortion is an aberration of the full-field chief ray. Therefore

$$\epsilon_y'(0, 0°, 1) = -\frac{1}{2 n_k u_k} D . \tag{7.44}$$

The remaining aberrations require a small amount of ingenuity. But by looking at the way the third-order aberrations behave, we can deduce what combinations of real ray aberrations will yield results. Sagittal coma is the radius of the coma blur circle. It is also the distance from the paraxial image point to the lower edge of the blur circle. To find this distance we need to trace a full-field chief ray and a full-field marginal ray at 90°:

$$\epsilon_y'(1, 90°, 1) - \epsilon_y'(0, 0°, 1) = -\frac{1}{2 n_k u_k} C . \tag{7.45}$$

Similarly, the tangential coma is the distance from the paraxial image position to the very top of the coma pattern. The top of the coma pattern is located by full-field marginal rays at 0° and 180°:

$$\frac{\epsilon_y'(1,0^\circ,1) + \epsilon_y'(1,180^\circ,1)}{2} - \epsilon_y'(0,0^\circ,1) = -\frac{3}{2\,n_k u_k} C . \tag{7.46}$$

Note that because of symmetry, both the 90° and 270° full-field marginal rays will meet at the same point, but this is not guaranteed for the 0° and 180° rays. Hence, averaging is needed in Eq. 7.46 but not in Eq. 7.45.

As we have noted, astigmatism and Petzval curvature have the same functional dependence, so we will not try to separate them here. Sagittal astigmatism is the semi-minor axis of the astigmatism ellipse. This requires calculating the x component of the aberration for a full-field marginal ray at 90° and subtracting out the position of the y axis. The y axis location is determined by the image location which, in turn, is found via a marginal axial ray. Hence, for sagittal astigmatism we have

$$\epsilon_x'(1,90^\circ,1) - \epsilon_y'(1,0^\circ,0) = -\frac{1}{2\,n_k u_k}(A + P\Lambda^2) . \tag{7.47}$$

Tangential astigmatism is given by the semi-major axis of the ellipse. To find this length, we find the total length of the ellipse by tracing full-field marginal rays at 0° and 180° and subtracting their y aberrations. Next, divide this difference by 2 and subtract our the aberration due to the marginal axial ray. Tangential astigmatism is given by

$$\frac{\epsilon_y'(1,0^\circ,1) - \epsilon_y'(1,180^\circ,1)}{2} - \epsilon_y'(1,0^\circ,0)$$
$$= -\frac{1}{2\,n_k u_k}(3A + P\Lambda^2) . \tag{7.48}$$

The equations given in this section show how a set of five real rays can be used to mimic separating the five Seidel aberrations. The five rays we need are a marginal axial ray, a full-field chief ray, and three full-field marginal rays at 0°, 90°, and 180°.

SUMMARY

Third-order aberrations are a very useful approximation to exact ray tracing. By using the third-order aberration polynomials, we can calculate approximately where any given ray will cross the paraxial image plane. More importantly, they lend insight into the causes of the aberrations and give us information on how to reduce aberrations.

The aberration polynomials determine the intersection of a third-order ray with the paraxial image plane relative to a corresponding paraxial ray:

$$\epsilon_x' = -\frac{1}{2 n_k u_k}\left[S \rho^3 \sin\theta + C \eta \rho^2 \sin 2\theta + (A + P\Lambda^2)\eta^2 \rho \sin\theta\right] \quad (7.49)$$

and

$$\epsilon_y' = -\frac{1}{2 n_k u_k}\Big[S \rho^3 \cos\theta + C \eta \rho^2 (2 + \cos 2\theta)$$
$$+ (3A + P\Lambda^2)\eta^2 \rho \cos\theta + D\eta^3\Big], \quad (7.50)$$

where η is the relative field position of the ray, ρ is the relative height of the ray in the entrance pupil, and θ is the angle between the y axis and the line connecting the center of the entrance pupil with the intersection point of the ray in the entrance pupil.

The five coefficients can be expressed as functions of the ray parameters for a full-field chief ray and a marginal axial ray. For spherical aberration we have

$$S = \sum_{j=1}^{k} W_j i_j^2 , \quad (7.51)$$

where

$$W_j = y_j n_{j-1}\left[\frac{n_{j-1}}{n_j} - 1\right](u_j + i_j) . \quad (7.52)$$

Coma is given by

$$C = \sum_{j=1}^{k} W_j i_j \bar{i}_j . \quad (7.53)$$

Astigmatism is given by

$$A = \sum_{j=1}^{k} W_j \bar{i}_j^2 . \quad (7.54)$$

The Petzval curvature coefficient does not depend directly on ray parameters

$$P_j = \frac{n_{j-1} - n_j}{n_{j-1} n_j} c_j , \tag{7.55}$$

but always appears in a term multiplied by the square of the optical invariant. Finally, distortion is found using

$$D = \sum_{j=1}^{k} \left[\bar{W}_j i_j \bar{i}_j + (\bar{u}_{j-1}^2 - \bar{u}_j^2) \Lambda \right] , \tag{7.56}$$

where

$$\bar{W}_j = \bar{y}_j n_{j-1} \left[\frac{n_{j-1}}{n_j} - 1 \right] (\bar{u}_j + \bar{i}_j) . \tag{7.57}$$

Each of these aberrations has a thin lens version. The thin lens spherical aberration coefficient is

$$S_t = -y^4 (n - 1) c \left[\frac{(n + 2)}{n} c_1^2 - (2n + 1) c c_1 - \frac{4(n + 1)}{n} v c_1 + n^2 c^2 + (3n + 1) c v + \frac{(3n + 2)}{n} v^2 \right] . \tag{7.58}$$

This formula shows that spherical aberration is strongly dependent on the shape of the lens. A minimum value for the spherical aberration coefficient can be found using

$$c_1 = \frac{(2n + 1) n c + 4(n + 1) v}{2(n + 2)} . \tag{7.59}$$

The thin lens coma coefficient is

$$C_t = C_t^* + \frac{\bar{y}}{y} S_t , \tag{7.60}$$

where

$$C_t^* = y^3 (n - 1) h v \left[n c^2 - \left(\frac{n + 1}{n} \right) c c_1 + \left(\frac{2n + 1}{n} \right) v c \right] . \tag{7.61}$$

For an object at infinity,

$$h v = u_{HFOV} , \tag{7.62}$$

Again, lens bending has an effect on coma. The lens shape which gives zero coma is given by

$$c_1 = \frac{n^2 c + (2n + 1) v}{(n + 1)} . \tag{7.63}$$

The astigmatism coefficient is given by

$$A_t = - \frac{(y h v)^2}{f'} + 2 \frac{\bar{y}}{y} C_t^* + \left(\frac{\bar{y}}{y} \right)^2 S_t . \tag{7.64}$$

Petzval curvature is given by

$$P_t = - \frac{1}{n f'} . \tag{7.65}$$

And finally, distortion is given by

$$D_t = \left(\frac{\bar{y}}{y} \right) (3 A_t^* + P_t \Lambda^2) + 3 \left(\frac{\bar{y}}{y} \right)^2 C_t^* + \left(\frac{\bar{y}}{y} \right)^3 S_t . \tag{7.66}$$

We can use a standard set of five exact rays to approximately isolate on each of the third-order aberrations. For spherical aberration, we only need to trace an exact marginal axial ray:

$$\epsilon_y' (\rho = 1, \theta = 0°, \eta = 0) = - \frac{1}{2 n_k u_k} S . \tag{7.67}$$

For sagittal coma we need a full-field chief ray and a full-field marginal ray directed at the back of the entrance pupil:

$$\epsilon_y' (1, 90°, 1) - \epsilon_y' (0, 0°, 1) = - \frac{1}{2 n_k u_k} C . \tag{7.68}$$

Tangential coma is given by

$$\frac{\epsilon_y'(1,0^\circ,1) + \epsilon_y'(1,180^\circ,1)}{2} - \epsilon_y'(0,0^\circ,1) = -\frac{3}{2\,n_k u_k} C \,, \tag{7.69}$$

where full-field rays are traced to the top and the bottom of the entrance pupil in addition to the full-field chief ray. For sagittal astigmatism we have

$$\epsilon_x'(1,90^\circ,1) - \epsilon_y'(1,0^\circ,0) = -\frac{1}{2\,n_k u_k}(A + P\Lambda^2) \,. \tag{7.70}$$

And for tangential astigmatism we have

$$\frac{\epsilon_y'(1,0^\circ,1) - \epsilon_y'(1,180^\circ,1)}{2} - \epsilon_y'(1,0^\circ,0) = -\frac{1}{2\,n_k u_k}(3A + P\Lambda^2) \,. \tag{7.71}$$

Finally, for distortion we have

$$\epsilon_y'(0,0^\circ,1) = -\frac{1}{2\,n_k u_k} D \,. \tag{7.72}$$

REFERENCES

A. E. Conrady, *Applied Optics and Optical Design* (Dover, New York, 1985).

R. Kingslake, *Lens Design Fundamentals* (Academic Press, Orlando, FL, 1978).

W. T. Welford, *Aberrations of Optical Systems* (Adam Hilger, Boston, 1986).

D. C. O'Shea, *Elements of Modern Optical Design* (Wiley, New York, 1985).

R. Barry Johnson, "Image Defects Useful in Teaching Students," *Recent Trends in Optical System Design: Computer Lens Design Workshop, Proceedings of SPIE* Vol. 766, 1987.

PROBLEMS

7.1 Consider the process of tracing one exact ray versus determining the third-order aberrations. Which do you think would take more time and effort if you were using a scientific calculator? Which do you think would take more time and effort if you were using pencil, paper, and some math tables (sine, cosine, and logarithms)? Finally, which do you think would be easier to program, exact ray tracing or third-order ray tracing?

7.2 Define each of the five third-order aberrations.

7.3 Describe the affect on a paraxial image of each of the five third-order aberrations.

7.4 Each of the aberration coefficients (for a single surface or for a thin lens) contains paraxial ray quantities such as y, i, and u in some combination to the fourth order. Why do we call this third-order optics? What happens to the fourth order?

7.5 What is the significance of the fact that all of the third-order aberrations except Petzval curvature and distortion contain the same terms related to spherical aberration?

7.6 What is the significance of the fact that $D_t = 0$ if the stop is at a thin lens and $D_t \neq 0$ in general if the stop is some distance from the lens?

7.7 Give at least two reasons why coma is usually more important to correct for than astigmatism in the design of an astronomical telescope.

7.8 Under what condition could Petzval curvature be used to correct for tangential astigmatism? Sagittal astigmatism? Consider the formulae for A_t^* and P_t. Could the condition mentioned above ever occur in a real system?

7.9 Given that $S = 0.15$ for a particular system, make an LSA plot. Assume that image space is in air and the slope angle of the marginal axial ray in image space is 0.1.

7.10 Draw a ray intercept plot for the on-axis field position given the data in Problem 7.9.

7.11 Given the following aberration coefficients, make sagittal and tangential ray intercept plots at full-field and half-field. $S = -0.000075$, $C = -0.000694$, $A = -0.000016$, $P\Lambda^2 = -0.000009$, $D = 0.0$. Assume that image space is in air and the slope angle of the marginal axial ray in image space is 0.1.

7.12 Write a spreadsheet to make the calculations for ray intercept plots on-axis and at two off-axis field positions given the aberration coefficients.

7.13 Given that $S = 0.15$, make an on-axis spot diagram. Assume that image space is in air and the slope angle of the marginal axial ray in image space is 0.1. Plot at least four rings of rays.

7.14 Given the aberration coefficients $S = -0.000007$, $C = -0.000023$, $A = -0.000586$, $P\Lambda^2 = -0.000279$, and $D = -0.000059$, make spot diagrams at full-field and half-field. Assume that image space is in air and the slope angle of the marginal axial ray in image space is 0.1. Plot at least four rings of rays.

7.15 Write a spreadsheet to make the calculations for spot diagrams on-axis and at two off-axis field positions given the aberration coefficients.

7.16 Calculate the third-order aberration coefficients for a system described with $c_1 = -0.018656$, $t_1 = 0.25$, $n_1 = 1.67003$, $c_2 = -0.055968$. The paraxial ray data are $u_0 = 0.0$, $y_1 = 1.996265$, $u_1 = 0.014942$, $y_2 = 2.0$, $u_2 = -0.050047$, $\bar{u}_0 = -0.174533$, $\bar{y}_1 = 0.026078$, $\bar{u}_1 = -0.104314$, $\bar{y}_2 = 0.0$, $\bar{u}_2 = -0.174207$.

7.17 Given the lens prescription data below, calculate the third-order aberration coefficients. Single lens made of BaF53, $c_1 = 0.018656$, $c_2 = -0.018656$, $t_0 =$ infinity, $t_1 = 0.25$, $R_1 = R_2 = 2.0$, $u_{HFOV} = 10°$ (i.e., the same lens as in problem 7.16, but with a different shape).

7.18 Calculate the third-order aberration coefficients for the lens in Problem 7.16, but with the lens turned to face the opposite direction.

7.19 Calculate the third-order aberration coefficients for the lens in Problem 7.18, but with a stop added in front of the lens. The position of the stop is 8.25 cm in front of the lens. Assume that the clear radii of the stop is 2.0 cm and that it is the aperture stop (i.e., the lens clear radii are larger than 2.0 cm). This stop position was selected to reduce astigmatism. Was I successful?

7.20 Calculate the third-order aberration coefficients for an achromatic doublet using the following values: $c_1 = 0.027268$, $c_2 = -0.026567$, $c_3 = 0.01199$, $t_1 = 0.532$, $t_2 = 0.48$, first glass is LaKN16, and the second glass is LaSFN3. Assume that the object is at infinity and the *HFOV* is 2°.

7.21 Use thin lens formulae to calculate the third-order aberration coefficients for a lens made of BaF53 with $f/10$ and $f' = 40$. The object is at infinity and first surface curvature is $c_1 = -0.018656$.

7.22 Use thin lens formulae to calculate the third-order aberration coefficients for the lens in Problem 7.21, but let $c_1 = 0.018656$.

7.23 Use thin lens formulae to calculate the third-order aberration coefficients for the lens in Problem 7.21, but with the lens turned around to face the opposite direction.

7.24 Use thin lens formulae to calculate the third-order aberration coefficients for the lens in Problem 7.21, but with a stop 8.25 in front of the lens. The stop should be the aperture stop with clear radius 2.0.

7.25 Use thin lens formulae to calculate the third-order aberration coefficients for an achromatic doublet made with LaKN16 and LaSFN3 glasses. The first lens has a focal length of 25.324. The second lens' focal length is -32.098. Let $c_1 = 0.027268$ and $c_2 = -0.026567$. Assume that the object is at infinity and the u_{HFOV} is $2°$.

7.26 For a single thin lens, find the object distance (or vergence) which gives minimum spherical aberration as a function of the curvature of the first surface c_1, the net curvature c, and the index of refraction n.

7.27 Calculate the shape of a thin lens which minimizes spherical aberration. Assume that the lens is made of FK5 glass and that the object is at infinity. The desired focal length is 51.2831 (i.e., $c = 0.04$).

7.28 Calculate the shape of a thin lens which minimizes spherical aberration. Assume that the lens is made of F2 glass and that the object 50 cm in front of the lens. The desired focal length is 25 cm.

7.29 Find the shape which gives zero coma for the system described in Problem 7.27.

7.30 Find the shape which gives zero coma for the system described in Problem 7.28.

7.31 An $f/5$ thin lens with focal length 20 cm is made of SF11 glass. The object for this system is at infinity. If the curvature of the first surface minimizes spherical aberration ($c_1 = 0.0686475$), find the value for the spherical aberration coefficient S_t and the coma coefficient C_t.

7.32 Calculate the value for the spherical aberration coefficient S_t for the same situation as in Problem 7.31, except that now the first surface is bent to give zero coma ($c_1 = 0.0728809$).

7.33 Prove algebraically that the first surface curvatures that give minimum spherical aberration and zero coma for a single thin lens get closer as the index of refraction increases.

7.34 A simple rule of thumb for single lenses is that the spherical aberration is minimized when the the ratios of the surface curvatures is -6. Prove this rule of thumb by assuming that the object is at infinity and $n = 1.5$.

7.35 Use lens doubling to reduce spherical aberration. The final system should be $f/4$ with $f' = 20$ cm and object at infinity. Use BK1 glass. Calculate the transverse spherical aberration at every step in the process.

7.36 A thick meniscus lens is made from FK5 glass. The curvature of the first surface is -0.08 and the curvature of the second surface is -0.12. The lens is 0.5 cm thick. Do a y-nu ray trace with $y_1 = 3.00$ and $u_0 = 0.0$. Calculate the incidence angle at every surface. From this data calculate the spherical aberration coefficient S. Finally, plot the third-order spherical aberration versus y_1, on the same graph with the exact data shown below.

y	TSA (exact)	TSA (third-order)
0.5	0.002	
1.0	0.016	
1.5	0.057	
2.0	0.138	
2.5	0.278	
3.0	0.502	

7.37 A thin lens is to be made from SF3 glass and must have a focal length of 10 cm. Calculate the amount of TSA and coma this lens will have if it is shaped for the minimum amount of spherical aberration. Next, calculate the amount of TSA the lens would have if it were shaped for no coma. Assume that the object is located a distance of 20 cm from the lens and that its height is 0.500 cm. Also, assume that the aperture size is 2.00 cm.

7.38 Find the two stop positions which should give zero astigmatism for a thin lens made from SK3 glass. The lens should be $f/8$ with $f' = 32$. The shape of the lens is given by $c_1 = 3\,c_2$.

7.39 Using the stop shift formulae for coma and astigmatism show that it is possible to have $C_t = A_t = 0$ simultaneously. Show that when both coma and astigmatism are corrected, there is only one appropriate stop position.

7.40 Continue the previous problem to find a formula for the lens shape which satisfies $C_t = A_t = 0$.

CHAPTER 8

FIRST-ORDER DESIGN AND y–$\bar{y}$ DIAGRAMS

Did you notice in earlier chapters that occasionally the solution to a complicated problem was incredibly simple? Did you wonder how I managed to luck into finding such problems? Well, the answer is that it was done by design.

Up until now, we have been concerned exclusively with analyzing optical systems. We are given the lens prescription consisting of thicknesses, curvatures, and indices of refraction. From this information, we are asked to determine information about the image, stops, or cardinal points.

In this chapter, we begin discussing optical system design—in particular, first-order design. We will look at how to generate an optical system with specific first-order properties.

The standard first-order design method in the past has been trial and error. The designer (you) would make an initial guess as to what lens prescription would produce the desired results. Calculations would be performed to check the initial guess. Based on the results of the calculations, changes would be made to the system to approach the desired performance. If the designer was knowledgeable and had a lot of experience, this process would quickly result in satisfactory system for most types of problems.

Trial and error is perfectly acceptable way to perform the first-order layout of a system. But if you don't have a lot of experience, it can be time-consuming and tedious. It's also not necessary since a completely analytic design method exists.

8.1 ANALYTIC SOLUTIONS FOR FIRST-ORDER DESIGN

What we are after, in the first-order layout, is the lens prescription which makes paraxial rays go where we want them to go. Some description of the desired behavior of paraxial rays will be one input to our design. We will take the indices of refraction as another input. We will be satisfied to find the separations and curvatures of the elements of the system.

At first glance, this might seem an impossible task. For one thing, how can we control where all possible paraxial rays go? That one is easy. Since we are talking about paraxial optics, we only need to specify where two paraxial rays go. Remember that we can find all first-order system properties from just two rays. The linear nature of the *y-nu* ray trace equations will guarantee that all other rays will behave themselves. A further consequence of the linear nature of paraxial optics is that the equations are easily inverted to solve for the lens prescription data.

Now any two rays could be used as long as the optical invariant calculated with those two rays is nonzero. Standard practice, however, is to use the marginal axial ray as the first ray and the full-field chief ray as the second ray. As always, chief ray data are indicated with a bar over the parameter. Therefore, for the marginal axial ray we have

$$y_j = y_{j-1} + \tau_{j-1}\,\alpha_{j-1} \tag{8.1}$$

and

$$\alpha_j = \alpha_{j-1} - \phi_j y_j \,. \tag{8.2}$$

For the full-field chief ray we have

$$\bar{y}_j = \bar{y}_{j-1} + \tau_{j-1}\,\bar{\alpha}_{j-1} \tag{8.3}$$

and

$$\bar{\alpha}_j = \bar{\alpha}_{j-1} - \phi_j \bar{y}_j \,. \tag{8.4}$$

At each surface, we have two unknowns, τ_{j-1}, and ϕ_j, but with four equations, we can solve for two additional unknowns. Therefore, we will specify as our ray parameter inputs only the heights of the two rays as they pass through our system. We will treat the reduced slope angles as unknowns to be determined.

Let's begin deriving our design procedure by combining Eqs. 8.1 and 8.3:

$$y_j \bar{y}_j = y_j(\bar{y}_{j-1} + \tau_{j-1}\,\bar{\alpha}_{j-1}) = (y_{j-1} + \tau_{j-1}\,\alpha_{j-1})\,\bar{y}_j \,. \tag{8.5}$$

By multiplying and rearranging we get

$$\tau_{j-1} = \frac{y_{j-1}\bar{y}_j - y_j \bar{y}_{j-1}}{\bar{\alpha}_{j-1} y_j - \alpha_{j-1}\bar{y}_j} \,. \tag{8.6}$$

The denominator of this equation is just another version of the optical invariant (see Problem 8.4). The result clearly shows why the two rays selected for the first-order design must have a nonzero optical invariant

$$\tau_{j-1} = \frac{\bar{y}_{j-1} y_j - y_{j-1} \bar{y}_j}{\Lambda} . \tag{8.7}$$

Finally, by using the definition of the reduced separation and the fact that we are treating the index of refraction as an input we get

$$t_{j-1} = n_{j-1} \tau_{j-1} . \tag{8.8}$$

We are now ready to find the reduced slope angles. From the two translation equations we get

$$\alpha_{j-1} = \frac{y_j - y_{j-1}}{\tau_{j-1}} \tag{8.9}$$

and

$$\bar{\alpha}_{j-1} = \frac{\bar{y}_j - \bar{y}_{j-1}}{\tau_{j-1}} . \tag{8.10}$$

Similar results can be obtained for the jth surface by simply increasing all of the subscripts by one. Remember that the ray slope angle is related to the reduced slope angle:

$$u_{j-1} = \frac{\alpha_{j-1}}{n_{j-1}} . \tag{8.11}$$

The next step is to solve for the power using Eqs. 8.2 and 8.4. The method is the same as solving for the reduced separation:

$$\phi_j = \frac{\bar{\alpha}_{j-1} \alpha_j - \alpha_{j-1} \bar{\alpha}_j}{\Lambda} . \tag{8.12}$$

Now that the power can be determined, we can find the curvature for a surface:

$$c_j = \frac{\phi_j}{n_j - n_{j-1}} . \tag{8.13}$$

Once again, the indices of refraction are considered inputs. This formula works for mirrors as well as refracting surfaces if the sign on the index changes after the mirror. For thin lenses in air, the power relates to the effective focal length of the lens:

$$f_j' = \frac{1}{\phi_j} . \tag{8.14}$$

To summarize, if we specify the optical invariant Λ, the index of refraction n_j after each surface, and the heights of the marginal axial ray and the full-field chief ray y_j and $\bar{y}_j$ at each surface, then we can uniquely solve for the separations t_j and curvatures c_j or focal lengths f_j' for each surface in the system.

Note that to make all of this work, you have to have a good idea of where you want the rays to go. You need to know how many lenses you want. You need to know the optical invariant (more on this later). And you need some idea of what the indices of refraction should be. In short, this analytic method of first-order design won't do all of the work for you. You still need to know what you are doing.

As with *y-nu* ray tracing, we can set up a table to help with our calculations. The layout of the table works best if the surface numbers run down a column instead of across a row. The second column will be the index of refraction. Next come columns for the chief ray heights and the marginal axial ray heights. Finally, the reduced separation, reduced slope angles, surface power, actual thickness, and curvature each take a column in that order. The best way to see the setup is with an example.

Example 8.1

Design (find the curvatures and thickness) a single lens. You are given an optical invariant $\Lambda = 0.5$. The index of refraction of the glass is 1.5. The full-field chief ray and marginal axial ray heights are given in Table 8.1 below. ■

It has been mentioned several times that the optical invariant is an input for our design process. But the optical invariant is calculated from the ray data, and the ray data include the reduced slope angles. As we just saw, the slope angles are determined in our calculations using the optical invariant. This appears to be a problem.

What we need to do is calculate the optical invariant by some independent means. One way to calculate Λ is by taking one of the values for τ as known. More typically, we can get Λ from information regarding the object or image. For example, if the object is at infinity, then

$$\Lambda = -\bar{\alpha}_0 y_0 . \tag{8.15}$$

Table 8.1 Calculations for Example 8.1.

	n	$\bar{y}$	y	τ	$\bar{\alpha}$	α	ϕ	t	c
0		5.0	0.0						
	1.0			20.0	-0.25	0.10		20.0	
1		0.0	2.0				0.3		0.6
	1.5			0.4	-0.25	-0.50		0.6	
2		-0.1	1.8				-0.25		0.5
	1.0			36.0	-0.275	-0.05		36.0	
3		-10	0.0						

Here, we can replace y_0 with the clear radius of the entrance pupil R_E. The reduced slope angle given above is just the half angle of the field of view, u_{HFOV}. Both of these parameters are typically known for a particular design. Therefore,

$$\Lambda = -u_{HFOV} \times R_E \tag{8.16}$$

is another way to calculate the optical invariant. If the $f/\#$ for the system is specified instead of the clear radius of the entrance pupil, then we have

$$\Lambda = -\frac{u_{HFOV} \times f'}{2 \times f/\#} . \tag{8.17}$$

One last method for finding the optical invariant is to look in image space where

$$\Lambda = h' \times NA , \tag{8.18}$$

where NA is the numerical aperture defined paraxially as $n_k u_k$ and h' is the image size.

8.2 THE y–$\bar{y}$ DIAGRAM

The equations presented in the previous section are the essence of analytic first-order design. Unfortunately, the equations do not help you to figure out what the ray heights should be. Also, you don't get an immediate understanding of what the system does or what it looks like. To overcome these limitations, we will use a diagram to help us visualize and comprehend what we are trying to design. The diagram will also help us determine what values to use for $\bar{y}$ and y. This diagram was introduced by Erwin Delano in 1963. He called it the y–$\bar{y}$ diagram. Other authors later called it a Delano diagram.

Basically, the y–$\bar{y}$ diagram is just a two-dimensional plot of the ray height pairs $\bar{y}$ and y with a solid line connecting the points. But this description of the diagram does not explain why the diagram is useful. Let's take another approach at defining the diagram.

First consider a simple optical system consisting of a single thin lens, an object, and an image. For this system we need to trace a marginal axial ray and a full-field chief ray as shown in Figure 8.1. What we are after is a two-dimensional plot of $\bar{y}$ and y, but in this figure both variables lie on the same axis. We will fix this problem by rotating the object into the negative x direction (Figure 8.2). When this is done, the image will rotate into the positive x direction since for this system the image is inverted. We do not lose any information or change any values in this process since our systems are all assumed to be axially symmetric.

The final step in this process is to think of the chief ray and the marginal ray as two components which define a single skew ray. This skew ray, by definition, starts at the top of the object, passes through the top of the aperture stop and ends at the top of the image. The y–$\bar{y}$ diagram is the projection of this skew ray into any plane perpendicular to the optical axis. We view the projection from image space beyond the image. The diagram for the simple system described above is shown in Figure 8.3. In this figure, we relabel the x axis and call it the $\bar{y}$ axis. Note that the z axis points out of the page. For this first y–$\bar{y}$ diagram I have used arrows to show the direction of propagation of the skew ray. Normally the arrows will be omitted.

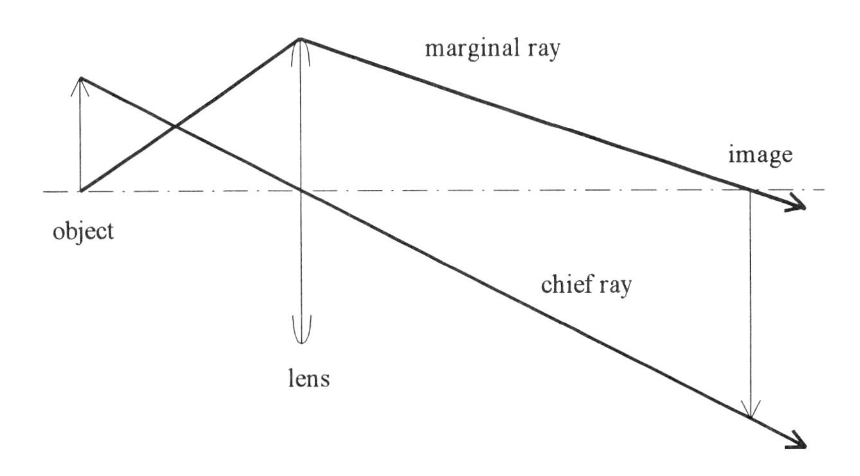

Figure 8.1 A simple optical system showing the standard marginal axial ray and full-field chief ray.

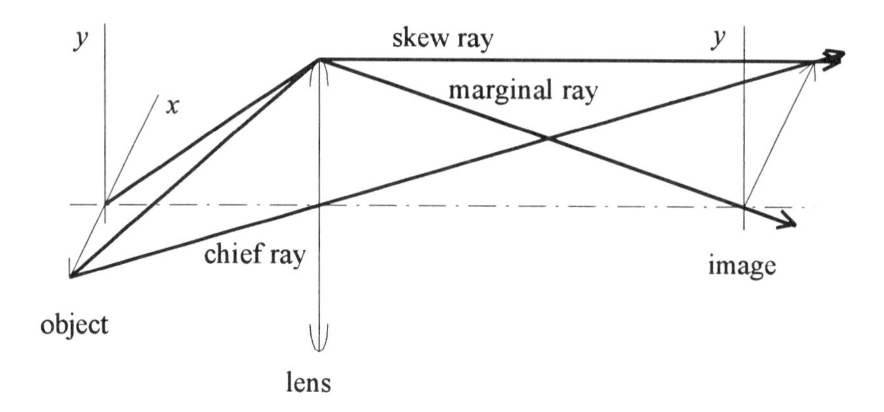

Figure 8.2 The same system as in the previous figure but now the chief ray has been rotated into the x–z plane. The resulting skew ray is also shown.

You may have asked yourself why I chose to turn the object into the negative x direction. This is done so that our y–$\bar{y}$ diagrams will look like Delano's diagrams and everyone else's. The problem is that Delano used a different set of coordinates to describe his system. Back in the early 1960's it was common to define the optical axis as the x axis. The y axis went up the page and this left the z axis to come out of the page. Delano rotated his object into the negative z direction, but viewed the skew ray from object space instead of image space. It is important for our diagrams to match everyone else's so that we do not have problems interpreting their diagrams and they do not have problems understanding our's. The only potential problem for us is that the optical invariant will now typically be a negative value.

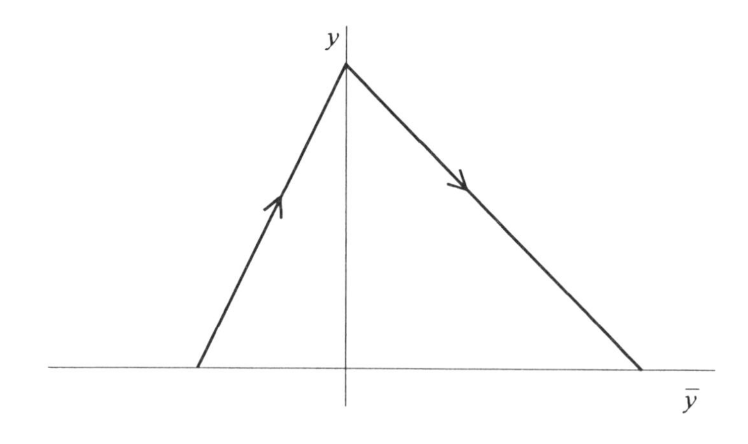

Figure 8.3 Delano diagram for a single positive thin lens.

Example 8.2

Figure 8.4 shows a thin lens triplet and the marginal and chief rays. Qualitatively, sketch a y–$\bar{y}$ diagram for this system.

The y–$\bar{y}$ diagram for this system should look something like Figure 8.5. ■

Example 8.3

A thick lens is shown in the Figure 8.6. The second surface of this lens is silvered to make a mirror. The marginal and chief rays are shown. Qualitatively, sketch the y–$\bar{y}$ diagram for this system.

The y–$\bar{y}$ diagram for the system given in Example 8.3 is identical to the diagram given for Example 8.2. The point is that the indices of refraction are a separate input into our design process. The y–$\bar{y}$ diagram by itself cannot tell us what the indices should be, or even what type of system we have. ■

Consider what the y–$\bar{y}$ diagram would look like if the object were at infinity. In this case, the marginal ray would come into the system at some fixed height y_0. Meanwhile, the chief ray height would increase without limit. The projection of the skew ray produced by this combination would be a line parallel to the $\bar{y}$ axis extending to infinity in the negative direction. By similar arguments, you can see that an image at infinity would also be represented by a horizontal line in the diagram. If the skew ray goes out in the positive direction, then the image would be inverted. An erect image would mean that the skew ray goes to infinity in the negative direction.

8.3 BASIC PROPERTIES OF THE y–$\bar{y}$ DIAGRAM

In this section, we will look deeper into interpreting and understanding y–$\bar{y}$

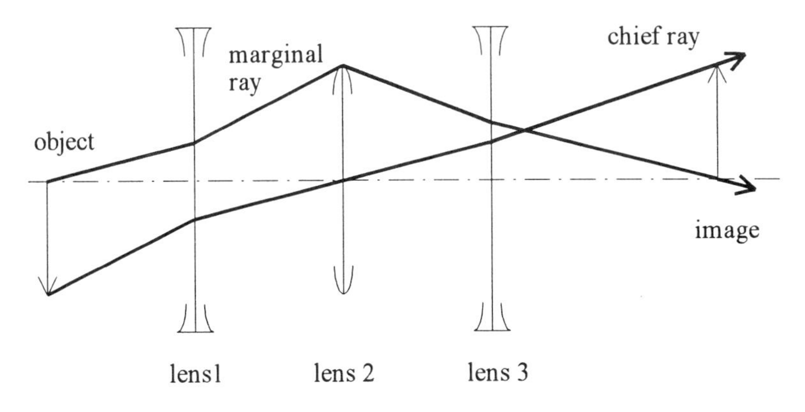

Figure 8.4 Thin lens triplet for Example 8.2.

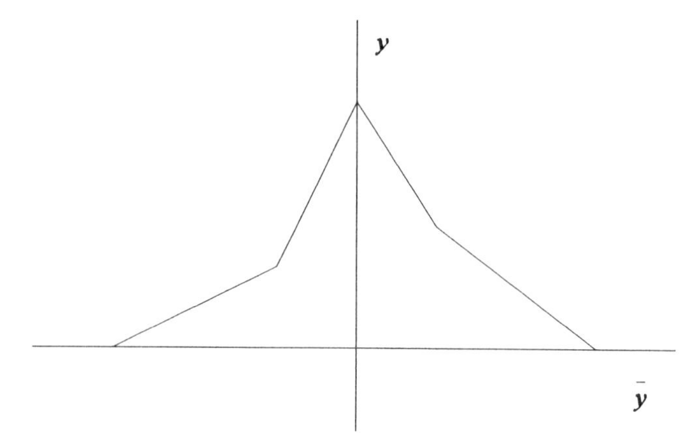

Figure 8.5 The Delano diagram for the system in Example 8.2.

diagrams. First, it is important to note that all transverse distances in an optical system are shown directly on the diagram. For example, object and image sizes are transverse distances. Object and image sizes are shown as distances along the $\bar{y}$ axis. By definition, the object and image are located where $y = 0$. Similarly, the size of the aperture stop is shown by the height of the skew ray along the y axis. Here, $\bar{y} = 0$ locates the aperture stop of the system.

On the other hand, it turns out that longitudinal distances measured along the optical axis cannot be found directly from the diagram. This makes some sense because we are looking directly back along the z axis when we view the skew ray projection. It turns out, however, that longitudinal distances are proportional to areas on the diagram.

If we consider two points on a y–$\bar{y}$ diagram, we can treat the coordinates of each point as defining a position vector. For example,

$$\vec{r}_j = \bar{y}_j \hat{x} + y_j \hat{y} . \tag{8.19}$$

Now, the two points and the origin define a triangle. The area of this triangle is given by

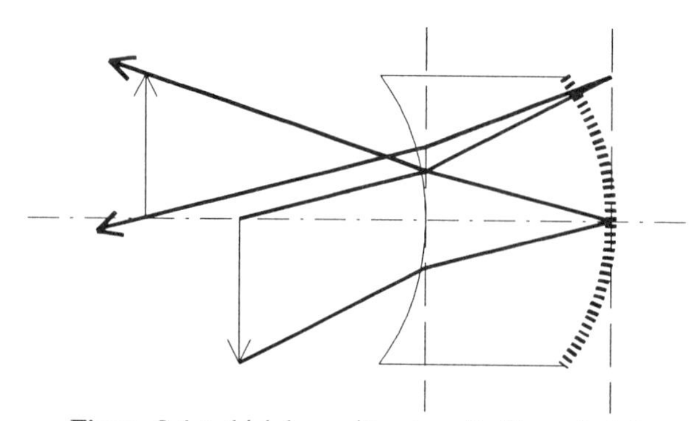

Figure 8.6 A thick lens with mirror for Example 8.3.

$$\vec{A} = \frac{1}{2}\vec{r}_{j-1} \times \vec{r}_j = \frac{1}{2}\left[\bar{y}_{j-1}y_j - y_{j-1}\bar{y}_j\right]\hat{z} . \tag{8.20}$$

Using the right-hand rule for the vector cross product we can see that the area vector actually points in the negative z direction or into the page. In other words, the z component of the area vector is

$$A = \frac{1}{2}\left[\bar{y}_{j-1}y_j - y_{j-1}\bar{y}_j\right] < 0 . \tag{8.21}$$

Recalling Eq. 8.7 and the definition of the reduced separation we see that

$$t_{j-1} = \frac{2\, n_{j-1} A}{\Lambda} . \tag{8.22}$$

In this equation it is important to note that if the skew ray is rotating clockwise, then $A < 0$. Because we must end up with $t_{j-1} > 0$ if $n_{j-1} > 0$, clockwise rotation also means that $\Lambda < 0$. As we mentioned earlier, our choice of a negative object height leads to a negative value for Λ. Equation 8.22 is confusing to work with because of the double negatives on the area and the optical invariant. The main thing to remember is that the area will be negative if it is measured in the same direction that the skew ray rotates.

Example 8.4

The y–$\bar{y}$ diagram for a thin lens system is shown in Figure 8.7. $\Lambda = -0.2$. What is the size of the object? What is the size of the image? What is the size of the aperture? What is the distance from the object to the first surface? What is the distance between the two thin lenses?

The object size is shown on the $\bar{y}$ axis where the plot crosses: $h = -5.0$. The image size is $h' = +3.0$. The aperture size is measured along the y axis: $R_{AS} = 4.67$.

The distance from the object to the first surface is proportional to the area defined by the origin, the object, and the first surface:

$$A = -\frac{1}{2}\, base \times height = -\frac{1}{2}(5.0)(6.0) = -15.0 .$$

The distance is

$$t_0 = \frac{2\, n_0 A}{\Lambda} = \frac{2(1.0)(-15.0)}{(-0.2)} = 150 .$$

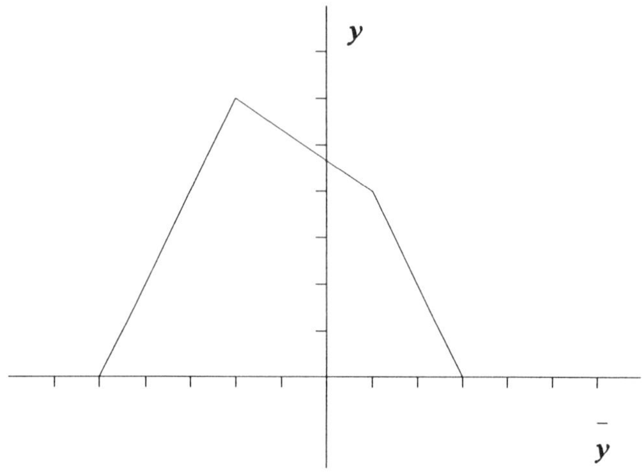

Figure 8.7 Delano diagram for Example 8.4.

The distance between the two lenses is proportional to the area defined by the origin, the first lens, and the second lens. This area consists of two parts divided by the y axis:

$$A = -\frac{1}{2}(4.67)(2.0) - \frac{1}{2}(4.67)(1.0) = -7.0 .$$

Hence, the separation is

$$t_1 + t_2 = \frac{2(1.0)(-7.0)}{(-0.2)} = 70.0 .$$

■

Since Λ is indeed invariant, we can also see that if $\Lambda < 0$, then the plot will always rotate clockwise around the origin. The skew ray can never reverse its direction of rotation. If we began with $\Lambda > 0$, then the skew ray would have to rotate counterclockwise.

It may happen, however, that the image produced by a single lens is on the same side of the origin as the object. This situation appears to violate the rule that the skew ray rotates in only one direction about the origin. When this happens, we have a virtual image and the skew ray merely appears to violate our rotation rule. Physically, the skew ray appears to come from the virtual image without ever actually passing through it.

Our skew ray can never pass through the origin. Because we have chosen the marginal axial ray and the full-field chief ray, $y = 0$ locates objects or images and $\bar{y} = 0$ locates the aperture stop or pupils. But by the definition of the aperture stop, the stop or its images cannot be located at the same place as an object or its images. The two sets of conjugates are mutually exclusive.

By the law of rectilinear propagation, we know that between optical elements rays travel in straight lines. That is why we can connect the points on the y–$\bar{y}$ diagram with straight lines. Optical elements with nonzero power bend the skew ray and its projection. It turns out that positive elements (effective focal length $f' > 0$) always bend the skew ray toward the origin of the diagram. Conversely, negative elements bend the ray away from the origin. We can now define four regions in the diagram for a particular lens (Figure 8.8). If the skew ray is bent into the first region by the element, then we know that the element is negative and the image of a real object produced by this element alone will be virtual. In the second region, the element has positive power, but not enough to produce a real image of a real object. In the third region, we have a positive lens with enough power to form a real image. If a ray entered the fourth region of the diagram, then the rotation rule would be violated. Entering the fourth region is not allowed.

8.4 FIRST-ORDER SYSTEM PROPERTIES AND THE y–$\bar{y}$ DIAGRAM

A basic tenet of geometrical optics is that for every object plane there is a single conjugate plane in image space. Here we will investigate what this means for the y–$\bar{y}$ diagram.

To begin, consider a simple optical system and some original object and its

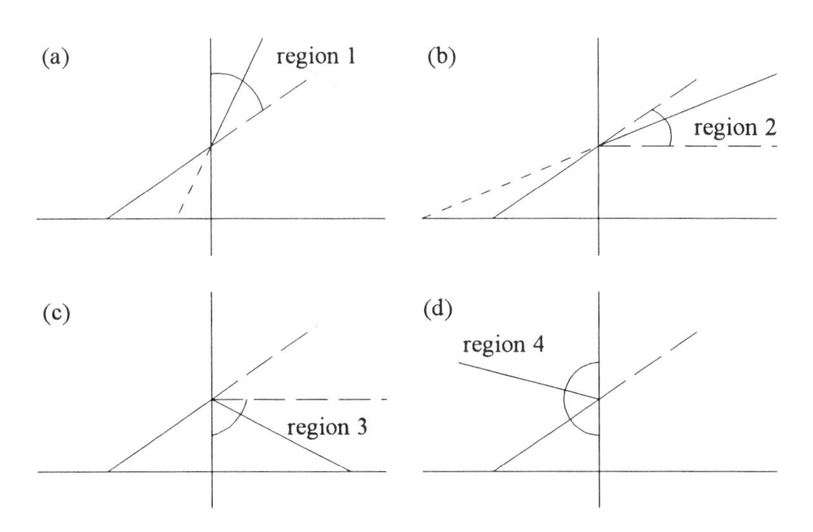

Figure 8.8 As the power of an element changes, there are four possible regions in the diagram where the skew ray could go. (a) Negative lens, virtual image. (b) Positive lens, virtual image. (c) Positive lens, real image. (d) Not allowed.

image. The y–$\bar{y}$ diagram is one we are already familiar with. Now, let's shift to a new object with a corresponding image shift. Normally, when we move the object, both the full-field chief ray and the marginal axial ray would have to change. But, for this discussion, imagine that we extend, but otherwise not alter, the chief ray or the marginal axial ray. This means that our image size must change. This is alright since what we are really after is the location of the conjugate planes and the magnification. In other words, we are leaving the y–$\bar{y}$ diagram alone. When we shift the image, the magnification will change; and as Figure 8.9 clearly shows, the new magnification can be written in terms of the original skew ray coordinates as

$$m = \frac{\bar{y}}{\bar{y}'} . \tag{8.23}$$

Equivalently the magnification can be written as

$$m = \frac{y}{y'} . \tag{8.24}$$

Combining these two equations we get

$$\frac{y}{\bar{y}} = \frac{y'}{\bar{y}'} . \tag{8.25}$$

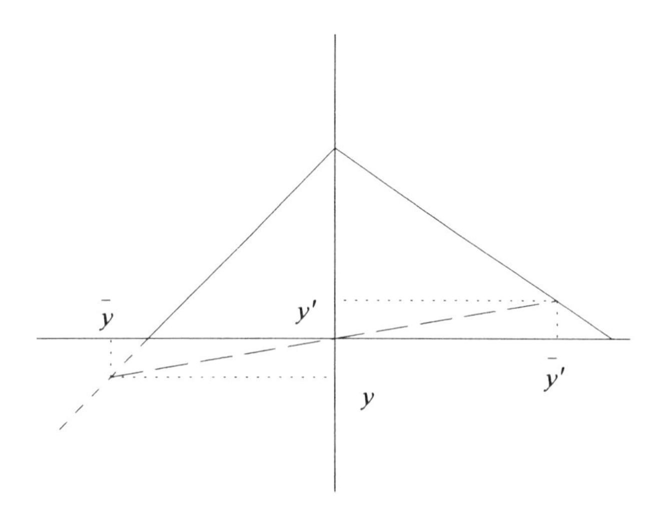

Figure 8.9 Affect of a shift of the object on the image and corresponding ray heights.

On the y–$\bar{y}$ diagram, this is an equation for a straight line through the origin. The slope of this line is given by

$$m_c = \frac{y}{\bar{y}} . \tag{8.26}$$

The line described above connects two points on the skew ray in the y–$\bar{y}$ diagram. One point is on the part of the skew ray in object space. This point locates the new object plane. The other point is on the skew ray in image space and locates the new image plane. Since this line connects two conjugate planes, it is called a *conjugate line*. Either (or both) part of the skew ray may be extended until it intersects the conjugate line.

Remember that longitudinal distances are represented as areas in the diagram. The longitudinal shift of the object is indicated by the area bounded by the original object point, the origin, and the intersection of the conjugate line with the skew ray (the new object point). In the case presented in Figure 8.9 the area vector points out of the page (positive z direction), which means that the shift is a negative longitudinal distance (index in object space is normally positive). The image location is found by looking at the area bounded by the original image, the origin, and the intersection point between the image space ray and the conjugate line. The shift shown here is positive. The y–$\bar{y}$ diagram clearly and simply shows the effects of object shifts on image location.

Now let us look at several special conjugate lines. The most obvious conjugate line is just the $\bar{y}$ axis. Here the slope of the conjugate line is 0 and the magnification of the system is the original magnification given by

$$m = \frac{\bar{y}'}{y} . \tag{8.27}$$

Another conjugate line is the y axis. This line connects the aperture stop with its images in object and image space. The aperture stop serves as the object for both the entrance and exit pupils as shown in Figure 8.10. The size of the entrance pupil is just the y value for the extension of the object space skew ray to the y axis. Similarly, the size of the exit pupil is given by the y value where the image space skew ray intersects the y axis. The locations of the pupils can also be found from the diagram. The distance from the first lens to the entrance pupil l_E is proportional to the area of the triangle defined by the origin, the first lens point, and the entrance pupil point. In calculating this area, ignore other lenses and use the index of refraction of the object space since the first surface, the entrance pupil, and the skew ray (or its extension) are all in object space. In the same way the distance from the last surface to the exit pupil $l_{E'}$ is related to the area bounded by the origin, the last surface point, and the exit pupil point. Here, everything is in image space. Remember that if the area is found by rotating in the same direction that the skew ray rotates, then the

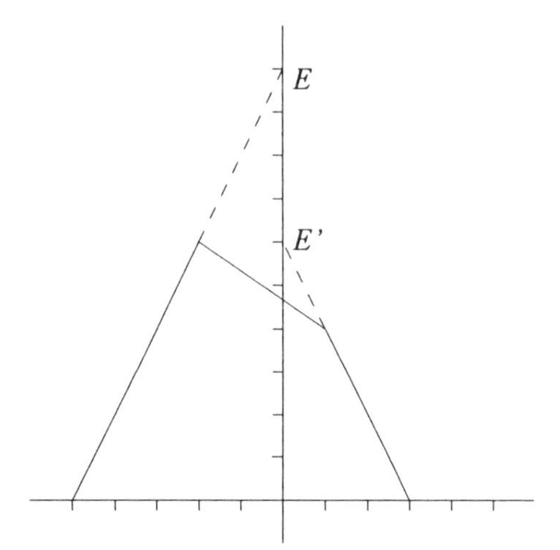

Figure 8.10 Size and position of the pupils determined from the extensions of the skew ray to the y axis for the system from Example 8.4.

value for τ will be positive; and if the area rotates in the opposite direction, then τ will be negative. In Figure 8.10, $l_E > 0$ and $l_{E'} < 0$ (assuming no mirrors).

Example 8.5

What are the sizes and positions of the entrance and exit pupils for the system given in Example 8.4?

The entrance pupil is found by extending the object space ray until it intersects the y axis. The entrance pupil size is $R_E = 10.0$. The entrance pupil location is proportional to the area bounded by the origin, the first lens, and the intersection point of the object ray and the y axis:

$$A = -\frac{1}{2}(10.0)(2.0) = -10.0 .$$

So the entrance pupil location is

$$l_E = \frac{2(1.0)(-10.0)}{(-0.2)} = 100.0 .$$

The exit pupil is found by extending the image space ray back until it intersects

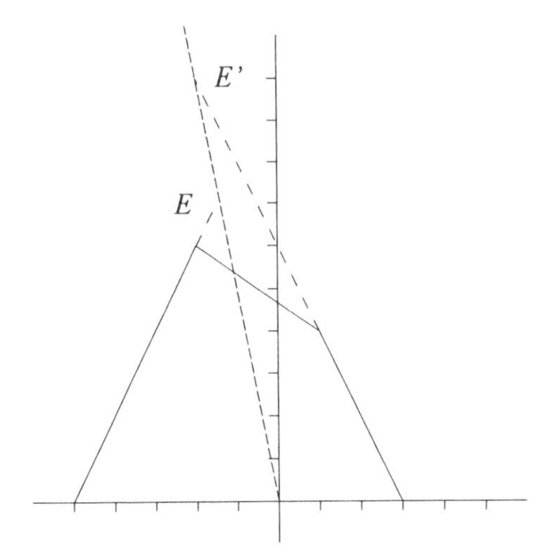

Figure 8.11 A stop shift is achieved by tilting the y axis.

the y axis. The size of the exit pupil is $R_{E'} = 6.0$. The position of the exit pupil is proportional to the area bounded by the origin, the last surface, and the exit pupil:

$$A = \frac{1}{2}(6.0)(1.0) = 3.0 \ .$$

The position is

$$l_{E'} = \frac{2(1.0)(3.0)}{(-0.2)} = -30 \ .$$

■

Now that we have seen that the y axis is a conjugate line, let's consider shifting the location of the aperture stop and pupils. We want to leave the object and image alone, so the $\bar{y}$ axis does not change. Since $\bar{y} = 0$ defines the aperture stop, tilting the y axis is equivalent to moving the aperture stop location. And, since any line through the origin of the plot connects conjugates, moving the y axis changes the size and locations of the pupils accordingly. Figure 8.11 shows the y axis tilted to move the aperture stop closer to the first lens. The new entrance and exit pupils are also shown.

We have seen that we can extend the object and image parts of the skew ray, what happens when the extensions meet? Now there is only one intersection point on the conjugate line. Clearly, the object size and image size are the same for this conjugate pair. Therefore, $m = 1$. We have already seen that the principal planes are conjugates with $m = 1$. The alternate definition of the principal planes also makes sense in this situation. When the two portions of the skew ray are extended, effectively we have only one refraction in our system. Remember that principal planes are planes of effective refraction for a system.

At first glance, Figure 8.12 appears to indicate that there is only one principal plane. But remember that this intersection point really represents two points, one in object space and the other in image space. The location of the first principal plane can be measured from the first surface (l_H) by finding the area marked by the origin, the first surface point, and the principal plane point. The second principal plane position ($l_{H'}$) is measured from the last surface by finding the area bounded by the origin, the last surface point, and the principal plane point. As always, the sign convention gives positive distances if the area is measured in the same direction as the skew ray rotation.

Next, consider shifting the object until the conjugate line is parallel to the portion of the skew ray in image space. In this situation, the new image would be at infinity because the area from the last surface to the image point is infinite. Therefore, the object must be located at the object space focal point for the system. The front focal length can now be determined by finding the area from the first surface to the front focal point. If, instead, we shift the object to infinity, then the conjugate line will be parallel to the object space portion of the skew ray. The back focal length is given the area from the last surface to the image space focal point. Finally, the effective focal length of the system is measured from the back principal plane to the back

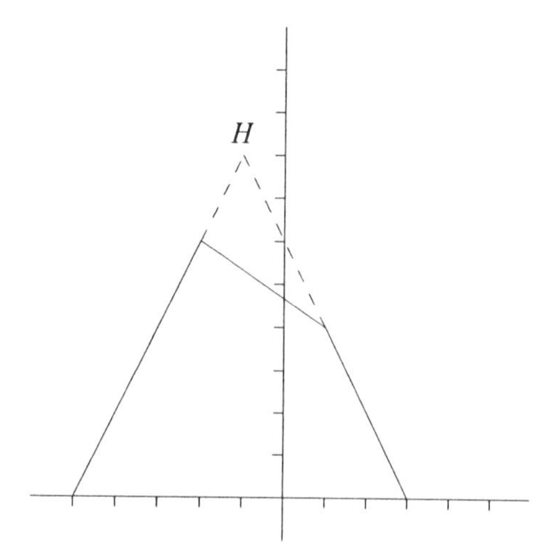

Figure 8.12 Extending the object space skew ray to intersect the extension of the image space skew ray locates the principal planes of a system.

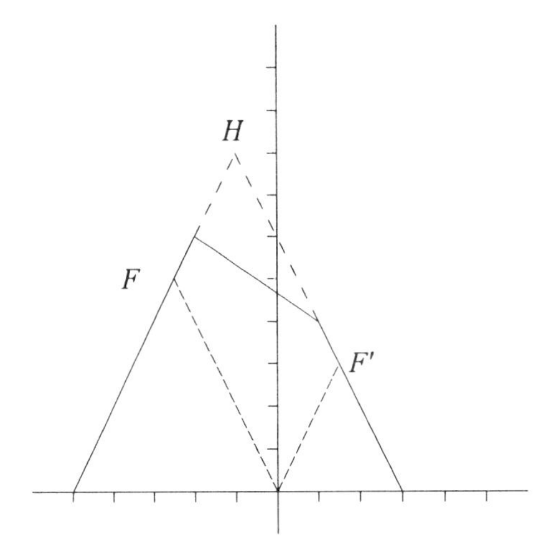

Figure 8.13 Locating the image and object space focal points.

focal point. This distance is given by the area from H' to F'. Figure 8.13 shows the front and back focal points of a system.

Example 8.6

Determine the locations of the principal planes, the focal planes, and the effective focal length of the system given in Example 8.4.

Here we need to extend the object ray and the image ray until they intersect. As shown in Figure 8.13, the intersection marks the principal planes. The distance from the first surface to the object space principal plane is proportional to the area bounded by the origin, the first surface, and the principal plane point:

$$A = -\frac{1}{2}(10.0)(2.0) + \frac{1}{2}(10.0)(1.0) = -5.0 .$$

Therefore,

$$l_H = \frac{2(1.0)(-5.0)}{(-0.2)} = 50.0 .$$

Similarly, the image space principal plane is found by

$$A = \frac{1}{2}(6.0)(1.0) + \frac{1}{2}(6.0)(1.0) = 6.0$$

and

$$l_{E'} = \frac{2(1.0)(6.0)}{(-0.2)} = -60.0 \ .$$

Next, the object space focal plane is found by drawing a line parallel to the image space ray until it intersects the object space ray. The front focal length is found by first finding the area enclosed by the first lens surface, the origin, and the front focal point

$$A = \frac{1}{2}(2.5)(10.0) - \frac{1}{2}(2.0)(10.0) = 2.5 \ .$$

The actual front focal length is

$$l_F = \frac{2(1.0)(2.5)}{(-0.2)} = -25 \ .$$

Finally, the image space focal plane is found by drawing a line through the origin which is parallel to the object space ray. The back focal length is proportional to the area defined by the origin, the last surface, and the focal point intersection:

$$A = -\frac{1}{2}(6.0)(1.5) + \frac{1}{2}(6.0)(1.0) = -1.5 \ ,$$

$$l_{F'} = \frac{2(1.0)(-1.5)}{(-0.2)} = 15.0 \ .$$

Measuring from the image space principal plane we can find the effective focal length:

$$A = -\frac{1}{2}(6.0)(1.0) - \frac{1}{2}(6.0)(1.5) = -7.5 \ ,$$

$$f' = \frac{2(1.0)(-7.5)}{(-0.2)} = 75.0 \ .$$

■

We are now in a position to ask (and answer) another question. What happens to the skew ray if we keep the object and entrance pupil fixed, but move the lens? Your first

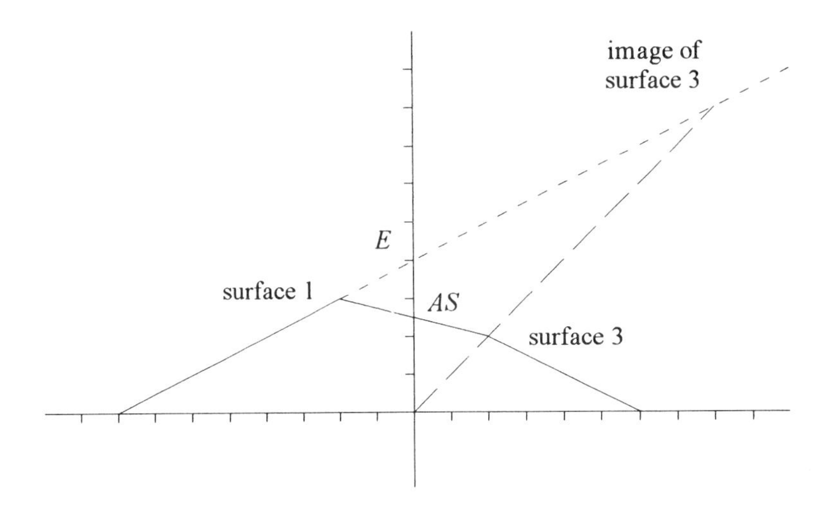

Figure 8.15 Delano diagram showing how to find the image of surface 3 in object space.

the optical axis, so the vertical axis represents lateral distances. The object height should be drawn vertically at the appropriate position long the horizontal axis. A line drawn from the top of the object through the center of each image will locate the projected center in the plane of the entrance pupil. The line just drawn will form two similar triangles. The offset of the image center from the center of the entrance pupil

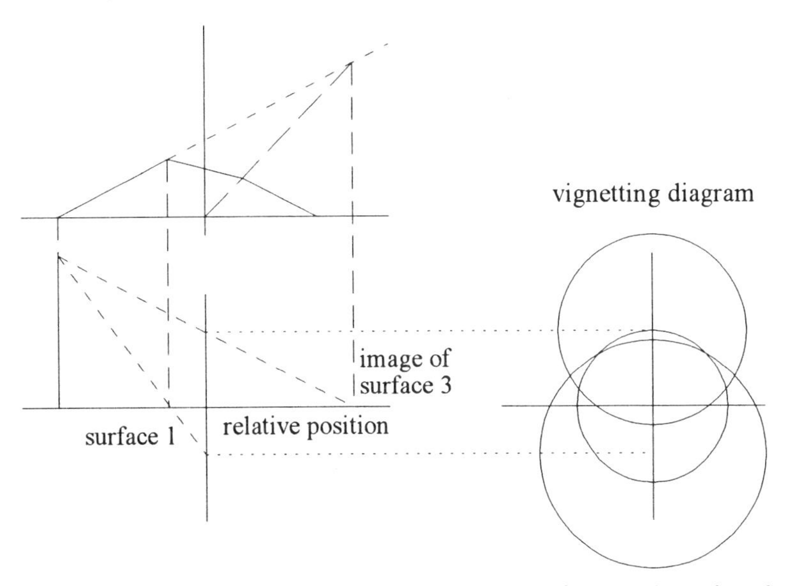

Figure 8.16 The relative positions of surfaces or images of surfaces can be projected onto an auxiliary graph. From this graph, the offsets for the various circles in the vignetting diagram are determined.

forms the horizontal side of one of the triangles. The object height forms the horizontal side of the other triangle. Therefore, the offset height is proportional to the object height as it should be. The proportionality between these two heights depends on the ratio of the positions of the object and the lens rim image relative to the entrance pupil. Because these two positions enter as a ratio, we do not need the undetermined proportionality constant for the horizontal axis. Any scaling of the horizontal axis will work. In fact, if the object size is too large or too small to be conveniently drawn on the vertical axis, this could be scaled also. We would have to remember, however, to scale the final vignetting diagram accordingly.

The final step is to draw the actual vignetting diagram. On yet another set of coordinates, draw a circle centered at the origin with radius of R_E. This distance appears directly on the y-$\bar{y}$ diagram. If the object was scaled in the previous step, then the entrance pupil radius would have to be scaled by the same amount. Next, project the centers from the auxiliary graph over to the vignetting diagram. Finally, for each lens rim image the appropriate radius must be determined. This cannot be done graphically because the original lens rim radii are not shown on the y-$\bar{y}$ diagram. In addition, we don't want to use the original lens rim radii here, but rather the radii of the lens rim images.

To understand this step, it is important to keep in mind that the lens rim images and the lens rims themselves are conjugates of each other. The marginal axial ray strikes each lens rim at a certain height y_j. The marginal axial ray or its extension must also strike each image at a certain height y_j' which is proportional to y_j:

$$\frac{y_j'}{R_j'} = \frac{y_j}{R_j} . \tag{8.30}$$

But we are really interested in the projected radius of this image R^*. Because the ray height in object space y_j' and the radius of the entrance pupil R_E lie on the same line (the skew ray), they are proportional to each other:

$$\frac{y_j'}{R_j'} = \frac{R_E}{R_j^*} . \tag{8.31}$$

Putting all of this together we get

$$R_j^* = \frac{R_j}{y_j} R_E . \tag{8.32}$$

The radii given by the previous equation are used to complete the vignetting diagram. But, remember that if we scaled the object, we will also need to scale these radii in the final vignetting diagram.

Example 8.7

Example 4.6 showed how to draw a vignetting diagram given the lens prescription. Rework that problem using only the y-$\bar{y}$ diagram information.

The y-$\bar{y}$ data are given in the Table 8.3. The diagram itself is shown in Figure 8.16.

In the figure, the object ray is extended until it intersects the conjugate line for the third surface. The first surface is already in object space so no conjugate line is needed. The conjugate line for the aperture stop is just the y axis as discussed previously.

The object, first surface, entrance pupil and image of the third surface are all projected onto the horizontal axis of the auxiliary graph. Next, the object height is marked off vertically on the auxiliary graph. Then projection lines are drawn from the top of the object to the image location for each surface. Where these projection lines cross the entrance pupil plane locates the projected centers of the lens rim images. These positions are projected into the vignetting diagram itself.

The entrance pupil can be drawn centered in the vignetting diagram with the entrance pupil radius from the y-$\bar{y}$ diagram. Next, the projected radii are calculated:

$$R_1^* = \frac{R_1}{y_1} R_E = \frac{(3.5)}{(3.0)} (4.0) = 4.667 ,$$

$$R_3^* = \frac{R_3}{y_3} R_E = \frac{(2.5)}{(2.0)} (4.0) = 5.00 .$$

Using these radii and the center offsets, the vignetting diagram is completed. ■

The final task in this section is to show how to use a vignetting diagram to achieve a certain amount of vignetting. From the previous discussion, it should be clear that y-$\bar{y}$ diagram establishes the offsets of the lens rim images in the entrance pupil. But the radii of these images require the input of the clear radii of the lens rims. We can turn this around by drawing the vignetting diagram with the desired radii and offsets. The offsets can be used to determine a possible y-$\bar{y}$ diagram, but there will

Table 8.2 Ray trace data for Example 8.7.

	0		1		2		3		4
y	0		3.0		2.5		2.0		0
α		0.125		−0.1		−0.1		−0.167	
$\bar{y}$	−8		−2		0		2		6
$\bar{\alpha}$		0.25		0.4		0.4		0.3333	

be considerable ambiguity. For example, the vignetting diagram by itself does not give any information about the effective focal length of the system Only if additional constraints are added will it be possible to determine a particular system. But after a y–$\bar{y}$ diagram is drawn, then the clear radii can be calculated from

$$R_j = \frac{R_j^*}{R_E} y_j . \tag{8.33}$$

SUMMARY

The paraxial ray trace equations are linear equations. As a result, they can easily be inverted. Originally we used the *y-nu* ray trace equations to determine ray heights and slope angles given the lens prescription. But in this chapter we have shown that given the ray heights, we can determine the lens prescription. The reduced separation is given by

$$\tau_{j-1} = \frac{\bar{y}_{j-1} y_j - y_{j-1} \bar{y}_j}{\Lambda} . \tag{8.34}$$

The actual physical separation is

$$t_{j-1} = n_{j-1} \tau_{j-1} . \tag{8.35}$$

The reduced slope angle can be found using

$$\alpha_{j-1} = \frac{y_j - y_{j-1}}{\tau_{j-1}} , \tag{8.36}$$

and the ray slope angle is given by

$$u_{j-1} = \frac{\alpha_{j-1}}{n_{j-1}} . \tag{8.37}$$

The power of an optical surface is

$$\phi_j = \frac{\bar{\alpha}_{j-1} \alpha_j - \alpha_{j-1} \bar{\alpha}_j}{\Lambda} , \tag{8.38}$$

from which we can find the curvature for a surface (reflecting or refracting)

$$c_j = \frac{\phi_j}{n_j - n_{j-1}} \tag{8.39}$$

or the focal length of a thin lens

$$f_j' = \frac{1}{\phi_j} . \tag{8.40}$$

It should be clear from the preceding equations that not only must the heights of two linearly independent rays be specified, but also the optical invariant and the indices of refraction.

The y-$\bar{y}$ diagram is a useful device for visualizing the first-order behavior and characteristics of an optical system. A y-$\bar{y}$ diagram can be interpreted merely as a plot of the height of a marginal axial ray as a function of the height of a full-field chief ray. It is more helpful, however, to look at the y-$\bar{y}$ diagram as the projection of a skew ray on to a plane perpendicular to the optical axis.

Paraxial characteristics of an optical system which are transverse distances such as the sizes of the object, image, stops, pupils, and windows will appear directly on the diagram. And the diagram can be used to determine other transverse properties such as vignetting.

Longitudinal distances turn out to be proportional to areas on the diagram

$$t_{j-1} = \frac{2\, n_{j-1} A}{\Lambda} . \tag{8.41}$$

Just looking at the diagram can tell use qualitative facts about an optical system. For example, positive lenses bend the plot toward the origin and negative lenses bend it away.

An extremely useful fact about the y-$\bar{y}$ diagram is that a line through the origin connects conjugate points on the plot. This is used to determine pupil locations and sizes, principal plane locations, and conjugate and stop shifts.

Understanding y-$\bar{y}$ diagrams means really understanding paraxial optics.

REFERENCES

E. Delano, "First-Order Design and the y, $\bar{y}$ Diagram," *Applied Optics,* Vol. 2, No. 12, pp. 1251–1256, December, 1963.

R. V. Shack, "Analytic System Design with Pencil and Ruler—The Advantages of the y-$\bar{y}$ Diagram," *Applications of Geometrical Optics, Proceedings of SPIE*, Vol. 39, pp. 127–140, August, 1973.

F. J. Lopez-Lopez, "Analytical Aspects of the y-$\bar{y}$ Diagram," *Applications of Geometrical Optics, Proceedings of SPIE*, Vol. 39, pp. 151–164, August, 1973. (This article contains an annotated bibliography of the first ten years of the y-$\bar{y}$ diagram.)

PROBLEMS

8.1 Could the stops, pupils, and cardinal points of a system be determined from a y-$\bar{y}$ diagram if two rays other than the marginal axial ray and full-field chief ray were used?

8.2 Can third order aberrations be determined from a y-$\bar{y}$ diagram if the indices of refraction are given? What if the system consisted of thin lenses?

8.3 How can you tell if a system is a thin lens system or contains mirrors by looking at the y-$\bar{y}$ diagram?

8.4 Prove that $\bar{\alpha}_{j-1}y_j - \alpha_{j-1}\bar{y}_j = \Lambda$.

8.5 What is the focal length for a single thin lens which will produce the following ray data: $\Lambda = -0.4$, $\bar{y}_0 = -5.0$, $y_0 = 0.0$, $\bar{y}_1 = 0.0$, $y_1 = 4.0$, $\bar{y}_2 = 5.0$, $y_2 = 0.0$?

8.6 What is the focal length for a single thin lens which will produce the following data: $\Lambda = -0.4$, $\bar{y}_0 = -5.0$, $y_0 = 0.0$, $\bar{y}_1 = 0.0$, $y_1 = 4.0$, $\bar{y}_2 = -10.0$, $y_2 = 0.0$?

8.7 What are the focal lengths and separation distances for a singlet with $n_1 = 1.5$ given the following ray data: $\Lambda = -0.1$, $\bar{y}_0 = -10.0$, $y_0 = 0.0$, $\bar{y}_1 = 0.0$, $y_1 = 4.0$, $\bar{y}_2 = 0.5$, $y_2 = 4.0$, $\bar{y}_3 = 4.0$, $y_3 = 0.0$?

8.8 For the same ray data given in Problem 8.7, what are the focal lengths and separations for a thin lens system?

8.9 A thin lens system produces the following ray data: $\Lambda = -0.5$, $\bar{y}_0 = -8.0$, $y_0 = 0.0$, $\bar{y}_1 = 0.0$, $y_1 = 5.0$, $\bar{y}_2 = 0.2$, $y_2 = 4.0$, $\bar{y}_3 = 10.0$, $y_3 = 0.0$. What are the focal lengths and separations for the two thin lenses?

8.10 Using the ray data from Problem 8.9, what are the curvatures and separations for a thin lens with $n = 1.5$?

8.11 A Galilean (terrestrial) telescope is made using two thin lenses. What are the focal lengths and separations for these lenses given the ray data: $\Lambda = -0.1$, $\bar{y}_1 = 0.0$, $y_1 = 8.0$, $\bar{y}_2 = 7.0$, $y_2 = 1.0$?

8.12 An astronomical telescope produces the following ray data: $\Lambda = -0.1$, $\bar{y}_1 = 0.0$, $y_1 = 10.0$, $\bar{y}_2 = 2.0$, $y_2 = -1.0$. What are the curvatures and separations for the two mirrors in this system?

8.13 What is the curvature for a single mirror which produces the ray data given in Problem 8.5?

8.14 Given the ray data: $\Lambda = -2.0$, $\bar{y}_0 = -9.0$, $y_0 = 0.0$, $\bar{y}_1 = -1.0$, $y_1 = 3.0$, $\bar{y}_2 = 0.0$, $y_2 = 4.0$, $\bar{y}_3 = 4.0$, $y_3 = 0.0$, what are the curvatures and separations for a singlet with $n = 1.6228$?

8.15 A thin lens triplet produces the following ray data: $\Lambda = -1.0$, $\bar{y}_0 = -36.0$, $y_0 = 0.0$, $\bar{y}_1 = -1.0$, $y_1 = 5.0$, $\bar{y}_2 = 0.0$, $y_2 = 3.0$, $\bar{y}_3 = 2.0$, $y_3 = 5.0$, $\bar{y}_4 = 72$, $y_4 = 0.0$. What are the focal lengths of the lenses and their separations?

8.16 A thick lens doublet produces the same ray data as given in Problem 8.15. What are the curvatures and separations for this system if $n_1 = 1.5$ and $n_2 = 1.72$?

8.17 A thick lens doublet produces the same ray data as given in Problem 8.15. What are the curvatures and separations for this system if $n_1 = 1.6$ and $n_2 = 1.5$?

8.18 A thick mirror produces the following ray data: $\Lambda = -0.5$, $\bar{y}_0 = -10.0$, $y_0 = 0.0$, $\bar{y}_1 = -1.0$, $y_1 = 4.5$, $\bar{y}_2 = 0.0$, $y_2 = 6.0$, $\bar{y}_3 = 1.0$, $y_3 = 3.5$, $\bar{y}_4 = 3.5$, $y_4 = 0.0$. What are the radii of the surfaces and their separations?

8.19 A thick mirror produces the following ray data: $\Lambda = -0.5$, $\bar{y}_0 = -18.0$, $y_0 = 0.0$, $\bar{y}_1 = -2.0$, $y_1 = 8.0$, $\bar{y}_2 = 0.0$, $y_2 = 5.0$, $\bar{y}_3 = 2.0$, $y_3 = 6.0$, $\bar{y}_4 = 6.0$, $y_4 = 0.0$. What are the radii of the surfaces and their separations?

8.20 The half angle field of view (u_{HFOV}) for a particular system is 5.0°. The clear radius of the entrance pupil (R_E) is 5.00. What is the optical invariant for this system?

8.21 A particular system is designed to have an effective focal length of 50.0, an $f/\#$ of 4.0, and a half field of view of 20°. What is the optical invariant for this system?

8.22 The first surface of a system is the aperture stop. Its clear radius is 2.0. An object 4.0 tall is located 25.0 in front of the system. What is the optical invariant for this system?

8.23 The numerical aperture (NA) for a system is 0.2 and the image size is 0.5. What is the optical invariant?

8.24 A thin lens system has the following ray data: $\Lambda = -0.1$, $\bar{y}_0 = -5.0$, $y_0 = 0.0$, $\bar{y}_1 = 5.0$, $y_1 = 5.0$, $\bar{y}_2 = 9.0$, $y_2 = 3.0$, $\bar{y}_3 = 6.0$, $y_3 = 0.0$. If the object is shifted 5.0 further from the system, what would be the new image distance as measured from the last surface? What is the new magnification?

8.25 If the original object in Problem 8.24 is shifted 5.0 closer to the system, what would be the new image distance and magnification?

8.26 A thin lens triplet produces the following ray data: $\Lambda = -0.2$, $\bar{y}_0 = -4.0$, $y_0 = 0.0$, $\bar{y}_1 = -2.0$, $y_1 = 2.0$, $\bar{y}_2 = 0.0$, $y_2 = 3.0$, $\bar{y}_3 = 3.0$, $y_3 = 4.0$, $\bar{y}_4 = 5.0$, $y_4 = 0.0$. If the image is shifted 50.0 closer to the system, what is the new distance from the object to the first surface? What is the new magnification?

8.27 A thin lens system has the following ray data: $\Lambda = -0.2$, $\bar{y}_0 = -4.0$, $y_0 = 0.0$, $\bar{y}_1 = 2.0$, $y_1 = 3.0$, $\bar{y}_2 = 0.0$, $y_2 = -2.0$, $\bar{y}_3 = -2.0$, $y_3 = 0.0$. If the object is shifted 20.0 closer to the system, what would be the new image distance as measured from the last surface? What is the new magnification?

8.28 Find the size and position of the entrance and exit pupils for the system given in Problem 8.24.

8.29 Find the size and position of the entrance and exit pupils for the system given in Problem 8.26.

8.30 Find the size and position of the entrance and exit pupils for the system given in Problem 8.27.

For each of the next five problems, ray data are given. Using these data describe the size and location of the aperture stop as well as the size and location of the entrance and exit pupils.

8.31 $\Lambda = -1.0$, $\bar{y}_0 = -4.0$, $y_0 = 0.0$, $\bar{y}_1 = 2.0$, $y_1 = 3.0$, $\bar{y}_2 = 5.0$, $y_2 = 0.0$.

8.32 $\Lambda = -0.5$, $\bar{y}_0 = -8.0$, $y_0 = 0.0$, $\bar{y}_1 = -2.0$, $y_1 = 6.0$, $\bar{y}_2 = 2.0$, $y_2 = 4.0$, $\bar{y}_3 = 4.0$, $y_3 = 0.0$.

8.33 $\Lambda = -0.2$, object at infinity, $\bar{y}_1 = 0.0$, $y_1 = 6.0$, $\bar{y}_2 = 2.0$, $y_2 = 2.0$, $\bar{y}_3 = 6.0$, $y_3 = 0.0$.

8.34 $\Lambda = -0.4$, $\bar{y}_0 = -4.0$, $y_0 = 0.0$, $\bar{y}_1 = -4.0$, $y_1 = 4.0$, $\bar{y}_2 = 0.0$, $y_2 = 2.0$, $\bar{y}_3 = 4.0$, $y_3 = 4.0$, $\bar{y}_4 = 4.0$, $y_4 = 0.0$. This type of system is called *telecentric*.

8.35 $\Lambda = -0.1$, $\bar{y}_0 = -9.0$, $y_0 = 0.0$, $\bar{y}_1 = 6.0$, $y_1 = 5.0$, $\bar{y}_2 = 0.0$, $y_2 = -7.0$, $\bar{y}_3 = -7.0$, $y_3 = 0.0$.

Ray data are given below for five problems. For each of these problems find the locations of the principal planes, the front and back focal lengths, and the effective focal length.

8.36 $\Lambda = -0.4$, $\bar{y}_0 = -10.0$, $y_0 = 0.0$, $\bar{y}_1 = -4.0$, $y_1 = 3.0$, $\bar{y}_2 = 3.0$, $y_2 = 4.0$, $\bar{y}_3 = 5.0$, $y_3 = 0.0$.

8.37 $\Lambda = -0.5$, object at infinity, $\bar{y}_1 = 0.0$, $y_1 = 5.0$, $\bar{y}_2 = 2.0$, $y_2 = 2.0$, $\bar{y}_3 = 10.0$, $y_3 = 0.0$.

8.38 $\Lambda = -1$, $\bar{y}_0 = -4.0$, $y_0 = 0.0$, $\bar{y}_1 = -2.0$, $y_1 = 2.0$, $\bar{y}_2 = 0.0$, $y_2 = 6.0$, $\bar{y}_3 = 2.0$, $y_3 = 2.0$, $\bar{y}_4 = 8.0$, $y_4 = 0.0$.

8.39 $\Lambda = -0.1$, $\bar{y}_0 = -1.0$, $y_0 = 0.0$, $\bar{y}_1 = 0.0$, $y_1 = 1.0$, $\bar{y}_2 = 2.0$, $y_2 = -1.0$, $\bar{y}_3 = 4.0$, $y_3 = 0.0$. Note that the final image is virtual.

8.40 $\Lambda = -0.2$, object at infinity, $\bar{y}_1 = -6.0$, $y_1 = 4.0$, $\bar{y}_2 = 0.0$, $y_2 = 6.0$, $\bar{y}_3 = 6.0$, $y_3 = 0.0$.

8.41 What is the minimum clear radius needed for the thin lens in Problem 8.31 to ensure that there is no vignetting?

8.42 What are the minimum clear radii needed for the two thin lenses in Problem 8.32 to ensure that there is no vignetting?

8.43 What are the minimum clear radii needed for the two thin lenses in Problem 8.33 to ensure that there is no vignetting?

8.44 Draw a vignetting diagram for the system given in Problem 8.32. Assume that $R_1 = 6.0$ and $R_2 = 4.0$.

8.45 Draw a vignetting diagram for the system given in Problem 8.33. Assume that $R_1 = 6.0$ and $R_2 = 3.0$. (*Hint*: When the object is at infinity, draw the vignetting diagram in image space.)

8.46 Draw a vignetting diagram for the system given in Problem 8.35. Assume that $R_1 = 10.0$ and $R_2 = 7.0$.

8.47 Why did I not ask you to draw a vignetting diagram for Problem 8.34?

The design problems described below may have more than one solution. Just find one.

8.48 Design a system consisting of two positive thin lenses which will give an overall magnification of -1.0 and produce a vignetting diagram as follows: $R_E = R_1^* = 5.0$, $R_2^* = 6.0$, $\Delta = +3.0$. Assume the object height $h = 6.0$.

8.49 Design a system consisting of two thin lenses (neither of which is the aperture stop) which gives a magnification of $-2/3$ and produces a vignetting diagram as follows: the entrance pupil radius is 6.0, one lens image has a radius of 8.0 and an offset of -4.0, the other lens image has a radius of 12.0 and an offset of +8.0. Take the object to be 4.0 high.

8.50 Design a telephoto lens consisting of two thin lenses using the following inputs: object at infinity, image height is 2.0, overall focal length is +20.0, $f/\#$ is 2.0, distance from first lens to image is 8.0, distance from second lens to image is 4.0.

8.51 Design an optical system with two thin lenses described as follows: magnification of +1.0, object size is -8.0, the second lens is the aperture stop with clear radius 4.0, entrance pupil has radius 2.0 and is located 2.0 in front of the first lens, the optical invariant is -1.

CHAPTER 9

OPTIMIZATION

So far you have learned how to thoroughly analyze a given optical system. You have also been introduced to some of the initial steps in designing optical systems. The majority of the work involved in optical design consists of fine tuning the initial design to improve image quality. Optimization is the process of modifying optical systems to reduce aberrations. The present chapter will describe automatic optimization where a special computer program will take an initial system and modify it to improve the overall image quality.

The amount of calculation involved in optimization can be staggering. The image quality is checked by tracing exact rays, a process we already know to be computationally intensive. But, the exact rays must be traced more than once as the system is modified. An important consideration in automatic optimization is how to reduce this computational load.

Limitations of the automatic optimization process will also be discussed. While we normally think of computers as being very precise, the results of automatic optimization are only as good as the initial inputs. As you will see, there is no guarantee that software will find the best overall solution to a given problem.

It is extremely important for the novice lens designer to realize that the ability of the software to do vast amounts of calculation does not eliminate the need for the designer to think. The designer needs to fully understand the automatic optimization process to utilize it to full advantage and sometimes to even get it to work at all. Automatic optimization is an extremely powerful tool which the designer must be trained to use well. It is not an automatic solution to any given problem.

Therefore, the automatic design process will be discussed in great detail in this chapter. First, ways of measuring image quality will be discussed. Then, the general idea behind optimization will be illustrated with a simple, one-parameter function. Because optical systems generally have many free parameters available, the optimization discussion will be expanded by first showing a two-parameter optimization and then the general, multiparameter case. Finally, dealing with constraints will be discussed.

9.1 MEASURING PERFORMANCE

The first requirement in improving optical system performance is to define a measure of image quality. There is no one correct way to measure image quality. Several methods will be discussed here, and then a standard method will be described.

One way to measure image quality is to calculate the third-order aberration coefficients. It should be clear that optical system performance generally improves as these coefficients approach zero. Unfortunately, third-order aberrations do not tell the whole story. A system may be well corrected for third-order aberrations but still have poor image quality due to excessive higher-order aberrations. Equations have been developed for fifth- and higher-order aberrations. In theory, we could approach exact ray tracing to any level of accuracy by merely including more terms.

There is a major problem to this approach. The functions for higher-order aberrations become very complex. First, the formulae for the individual coefficients become increasingly complicated as the order increases. Second, the number of coefficients increases with the order. The number of coefficients is given by

$$\sum_{i=1}^{(n+1)/2} (i+1) , \tag{9.1}$$

where n is the order. There are 5 third-order aberrations, 9 fifth-order, 14 seventh-order, and so on. And finally, the functional dependence on ρ, θ, and η gets more complicated. At some point, it just becomes easier to trace exact rays rather than deal with the expansions.

One very good measure of image quality is the modulation transfer function (MTF). As discussed in Chapter 6, MTF is a way to measure and quantify the ability of a system to image details of an object as a function of the spatial frequency of the image. The biggest problem with using MTF as a measure of image quality is that MTF does not give one single number as the measure of quality for a particular system. (The fact that MTF is not just a single number is its greatest strength as a measure of image quality. There is obviously more information contained in the MTF, than in a simpler measure of image quality.) It is not entirely clear to me how to optimize a function. In addition, calculating MTF is not a simple task. A large number of exact rays must be traced to get good results. One of our goals is to reduce the number of calculations that are needed. Despite these drawbacks, there are some commercial software packages which have, as an option, optimization with the MTF. Usually, this technique is used as a way to fine tune a well-corrected design and not as an initial optimization step.

Back in Chapter 2, I told you that any object, no matter how complicated, could be thought of a collection of point sources. The same was true of images. Therefore, a good measure of image quality should be how well a point object is imaged. The standard procedure is to calculate an aberration function by tracing a certain number of exact rays N and calculating the aberration for each ray:

$$\vec{\epsilon}_i = (X_i - x_i)\hat{i} + (Y_j - y_j)\hat{j} \,. \tag{9.2}$$

Here the capital letters indicate the coordinates for the exact ray in the paraxial image plane and the lowercase letters are the coordinates for the paraxial image. Next, square each component so that exact rays which are symmetrically placed around the paraxial image do not cancel. This step has the added benefit of ensuring that the aberration function is always positive. The optimization process will allow us to find the minimum of this function rather than trying to find where the unsquared function might be zero. In addition, as we will see later, squaring the aberrations will permit a great deal of simplification and reduce the amount of calculations required. Finally, some rays may be more important for image quality than others. We will want the more important rays to contribute more to our calculation so we multiply by a weighting factor to get a weighted aberration function:

$$\psi = \sum_{i=1}^{N} \left[w_{x,i}\,\epsilon_{x,i}^2 + w_{y,i}\,\epsilon_{y,i}^2 \right] . \tag{9.3}$$

Since we will be using the aberration function a lot, we can simplify our notation by introducing a_i to be a single weighted aberration component. Therefore,

$$\psi = \sum_{i=1}^{N} a_i^2 \,, \tag{9.4}$$

where N is no longer the number of rays that are traced, but rather the number of components of those rays which are actually used.

In its simplest form, optimization consists of finding the minimum value of ψ. The software will do this by changing certain system parameters. Therefore, in a very real sense, ψ is a function of the design parameters of the system:

$$\psi = \psi(c_j, t_j, n_j, V_j) \,. \tag{9.5}$$

Our notation can again be simplified by introducing a general system parameter p_j which could represent a curvature, a thickness, or any other system parameter which we wish to allow to vary in order to optimize the system. In the general case, there will be M free parameters.

9.2 ONE-PARAMETER CASE

To get a clear idea of how the aberration function will be minimized, we will begin with an artificial, one-dimensional case. Assume that the function shown in Figure 9.1 is an aberration curve which depends on only one lens parameter p. This particular function does not correspond to a real optical situation. It was chosen because it looked nice and will clearly illustrate the important points of this discussion. The function has a global minimum near $p = 1.0$ and a local minimum at $p = 5.0$. I found these values by the standard method of differentiating the function and solving for the zeros of the derivative function.

Unfortunately, finding the minima for real aberration functions is not that easy or straightforward. To begin with, we don't know what the aberration function looks like. We do not know its analytic form, so we can't take the derivative and find the minima. But notice that near the minima, the function looks like a parabola. This is true for any function if you get close enough to the minimum and only look at the function in a small region. The plan is to find a parabola which approximates our aberration function near a minimum. Once we find this parabola, we can find the derivative of the parabola. The derivative will be a linear function of p. Finally, we solve the linear function to find its zero. This point will approximate the minimum of the function. We can repeat this procedure to find a better approximating parabola and a better approximation for the minimum until we are satisfied that we are close enough to the actual minimum.

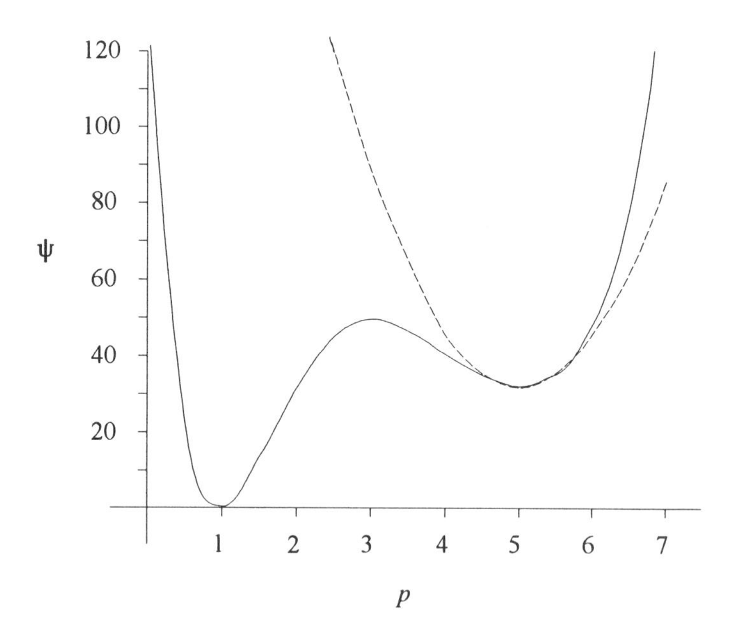

Figure 9.1 A one-dimensional aberration function is shown as a solid curve, while an approximating parabolic function is shown as a dashed curve.

How do we find the approximating parabola? We use something called a *Taylor's series approximation*. Any well-behaved function can be approximated at a point $x = a$ by

$$f(x) \approx f_{TS}(x) = f(a) + (x-a)\left.\frac{df}{dx}\right|_{x=a} + \frac{(x-a)^2}{2!}\left.\frac{d^2 f}{dx^2}\right|_{x=a} + \cdots, \quad (9.6)$$

or, more succinctly,

$$f_{TS}(x) = \sum_{n=0}^{\infty} \frac{(x-a)^n}{n!} f^{(n)}(a), \quad (9.7)$$

where $f^{(n)}(a)$ is the nth derivative with respect to x evaluated at $x = a$. We get our parabola by truncating this series at the quadratic term.

Example 9.1

Find the minimum for the function given in Figure 9.1 starting at $p = 5.2$. The aberration function is given by

$$\psi(p) = \left(\frac{p^3}{3} - 4p^2 + 15p - 11\right)^2 .$$

First, we evaluate the function for the initial system $p = 5.2$ (reasonably close to one of the minima) which gives

$$\psi(5.2) = 32.5965 .$$

Next, we evaluate the first derivative of the aberration function with respect to the parameter:

$$\left.\frac{d\psi}{dp}\right|_{p=5.2} = 5.02421 ;$$

and then we evaluate the second derivative:

$$\left.\frac{d^2\psi}{dp^2}\right|_{p=5.2} = 27.792 .$$

The approximating parabola is given by

$$\psi_{parabola} = \psi(5.2) + (p - 5.2)\frac{d\psi}{dp}\bigg|_{p=5.2} + \frac{1}{2}(p - 5.2)^2 \frac{d\psi^2}{dp^2}\bigg|_{p=5.2} .$$

Plugging in the numbers and simplifying gives

$$\psi_{parabola} = 13.896\,p^2 - 139.494\,p + 382.218 .$$

We find our first approximation to the minimum by differentiating this function to get

$$\psi'_{parabola} = 27.792\,p - 139.494 = 0 ,$$

from which we find that the minimum is at approximately $p = 5.01922$. Not bad for the first estimate. Using this value, we could reevaluate the aberration function and its derivatives to get a better approximation to the actual minimum. Without going into details, the second iteration gives $p = 5.00018$. ■

At this point, if you have really been paying attention, you should be crying "foul!" After all, I used the first and second derivatives of the function as part of the procedure for finding the approximating parabola. But we don't know the analytic form of the function, so how can we find the derivatives? The answer is that we actually make another approximation. We estimate the derivatives by a numerical technique called *finite differences*. We already have the value for the function ψ at one particular parameter value $p = \rho$. Next, we make a small change in p and re-evaluate ψ at this new value. Using these two calculations we can estimate the first derivative by

$$\frac{d\psi}{dp}\bigg|_{p=\rho} \approx \frac{\delta\psi}{\delta p} = \frac{\psi(\rho+\delta\rho) - \psi(\rho)}{\delta\rho} , \tag{9.8}$$

where $\delta\rho$ is the small change in the parameter p. For the second derivative we could use

$$\frac{d\psi^2}{d^2p}\bigg|_{p=\rho} \approx \frac{\delta\psi^2}{\delta^2 p} = \frac{\psi(\rho+2\,\delta\rho) - 2\,\psi(\rho+\delta\rho) + \psi(\rho)}{\delta\rho^2} . \tag{9.9}$$

A word of caution is in order here. We are getting into an area of mathematics called *numerical analysis*. This is a large field in its own right and well beyond the scope of this book. There are a number of subtleties involved. For instance, there are actually a large number of different ways to calculate the first and second derivatives. There is a trade off between the time required by the different methods for the calculation and the accuracy of the calculation. Another consideration is selecting the size of $\delta\rho$. Generally, the smaller this value, the more accurate the calculation. But if $\delta\rho$ is too small, then accuracy of the calculation will suffer due to round-off errors. The results of our calculations are not extremely sensitive to the choice of $\delta\rho$, but it is a consideration. You should consult a book on numerical analysis for more information on these topics.

Example 9.2

Use finite differences to calculate the first and second derivatives of the aberration function given in Example 9.1. Use finite step sizes of 5.0×10^{-6}, 5.0×10^{-4}, and 5.0×10^{-8}.

For 5.0×10^{-6} we get

$$\frac{\delta\psi}{\delta p} = \frac{(32.5965122325) - (32.5964871111)}{5.0 \times 10^{-6}}$$

$$= 5.02428$$

and

$$\frac{\delta^2\psi}{\delta p^2} = [(32.59653735463394) - 2(32.59651223252514)$$

$$+ (32.59648711111095)] / (5.0 \times 10^{-6})^2$$

$$= 27.7848 .$$

Similar calculations with 5.0×10^{-4} give

$$\frac{\delta\psi}{\delta p} = 5.03116$$

and

$$\frac{\delta^2\psi}{\delta p^2} = 27.8066 .$$

And finally, with 5.0×10^{-8} we get

$$\frac{\delta\psi}{\delta p} = 5.02422$$

and

$$\frac{\delta^2\psi}{\delta p^2} = -65.3699 \ .$$

Comparing these results with the analytical derivatives

$$\left.\frac{d\psi}{dp}\right|_{p=5.2} = 5.02421$$

and

$$\left.\frac{d^2\psi}{dp^2}\right|_{p=5.2} = 27.792 \ ,$$

we see that first derivative does not come out as well with the larger value of $\delta\rho$. The second derivative is completely wrong when calculated with the smallest value. Thus, the middle value is probably the best choice. ■

Example 9.1 also clearly indicates another very important point. Because of our initial choice for the value of the parameter p, our numerical process for finding the minimum found only the nearby local minimum. In fact, the method outlined above does not even guarantee that any minimum will be found. For example, if we start with $p = 3.2$, we quickly converge on the local *maximum* at $p = 3.0$. This is a general feature of optimization. You must start the optimization procedure with a good initial system if you want to find a really good final design. Selecting a good initial system is at the heart of optical system design.

9.3 TWO-PARAMETER CASE

Now that you have the general idea behind optimization, we will move on to more general cases. First, we will look at two variable parameters in this section and then the general, multiparameter case in the next section.

As before, we will replace the actual aberration function with an approximation based on the Taylor series expansion. In two dimensions, the Taylor series looks like

$$f_{TS}(x,y) = f(a,b) + (x - a)\frac{\partial f}{\partial x} + (y - b)\frac{\partial f}{\partial y}$$

$$+ \frac{1}{2!}\left[(x - a)^2 \frac{\partial^2 f}{\partial x^2} + 2(x - a)(y - b)\frac{\partial^2 f}{\partial x\, \partial y} + (y - b)^2 \frac{\partial^2 f}{\partial y^2}\right]$$

$$+ \frac{1}{3!}\left[(x - a)^3 \frac{\partial^3 f}{\partial x^3} + 3(x - a)^2 (y - b)\frac{\partial^3 f}{\partial x^2\, \partial y}\right.$$

$$\left. + 3(x - a)(x - b)^2 \frac{\partial^3 f}{\partial x\, \partial y^2} + (y - b)^3 \frac{\partial^3 f}{\partial y^3}\right] + \cdots, \tag{9.10}$$

where all of the partial derivatives are evaluated at the point $x = a$, $y = b$. When we apply the Taylor series to the aberration function we make two changes. First, we drop any terms which are higher than second order in the two parameters. Second, we replace the partial derivatives with finite differences. Thus, we have

$$\psi_{parabola}(p_1, p_2) = \psi(\rho_1, \rho_2) + (p_1 - \rho_1)\frac{\delta\psi}{\delta p_1}$$

$$+ (p_2 - \rho_2)\frac{\delta\psi}{\delta p_2} + \frac{1}{2}\left[(p_1 - \rho_1)^2 \frac{\delta^2\psi}{\delta p_1^2}\right.$$

$$\left. + 2(p_1 - \rho_1)(p_2 - \rho_2)\frac{\delta^2\psi}{\delta p_1\, \delta p_2} + (p_2 - \rho_2)\frac{\delta^2\psi}{\delta p_2^2}\right]. \tag{9.11}$$

Adding the second free parameter leads to a finite difference where both parameters are varied. One possible way to calculate this finite difference is to use

$$\frac{\delta^2\psi}{\delta p_1\, \delta p_2} =$$

$$\frac{\psi(\rho_1 + \delta\rho_1, \rho_2 + \delta\rho_2) - \psi(\rho_1 + \delta\rho_1, \rho_2 - \delta\rho_2)}{4\delta\rho_1\, \delta\rho_2}$$

$$- \frac{\psi(\rho_1 - \delta\rho_1, \rho_2 + \delta\rho_2) - \psi(\rho_1 - \delta\rho_1, \rho_2 - \delta\rho_2)}{4\delta\rho_1 \delta\rho_2}. \tag{9.12}$$

It should be clear that this step requires a lot of calculation.

Our task now, is to minimize the aberration function in both dimensions. The minimum is found by differentiating Eq. 9.11 and solving the resulting two equations:

$$\frac{\partial \psi_{parabola}}{\partial p_1} \approx \frac{\delta \psi}{\delta p_1} + (p_1 - \rho_1)\frac{\delta^2 \psi}{\delta p_1^2} + (p_2 - \rho_2)\frac{\delta^2 \psi}{\delta p_1\, \delta p_2} = 0 \,, \tag{9.13}$$

$$\frac{\partial \psi_{parabola}}{\partial p_2} \approx \frac{\delta \psi}{\delta p_2} + (p_1 - \rho_1)\frac{\delta^2 \psi}{\delta p_1\, \delta p_2} + (p_2 - \rho_2)\frac{\delta^2 \psi}{\delta p_2^2} = 0 \,. \tag{9.14}$$

We now have two equations in the two unknowns p_1 and p_2. Solving for these two unknowns will give us our first approximation to the nearest minimum. We can simplify the equations somewhat by introducing as our unknown variables the change required to get from our starting values to the minimum approximation values:

$$\Delta p_i = p_i - \rho_i \,. \tag{9.15}$$

Optimization which uses all of the second derivative information is called *Newton's method.* It will normally find a minimum if the initial set of parameters is sufficiently close to a local minimum. It does not necessarily find the global minimum, as we have seen. Unfortunately, it also requires the calculation of the second derivatives of the aberration function. We have already seen (Eq. 9.9) that the second derivative with respect to one variable requires an additional calculation of the aberration function. The second derivative with respect to two variables (Eq. 9.12) requires several additional calculations of the aberration function because two parameters must be changed at a time. As you can see, the number of calculations goes up rapidly as the number of variables increases. And remember that each calculation, itself, is quite complicated. We cannot simply ignore the second derivatives, because then our approximating function would no longer even remotely approximate the aberration function. However, several methods have been developed to save on the amount of calculation required to optimize a system.

The most helpful thing that we can do to reduce our calculations is to take advantage of the special form of the aberration function. Remember that the aberration function is the sum of the squares of the aberrations. For two aberrations the function would look like

$$\psi(a_1, a_2) = a_1^2 + a_2^2 \,. \tag{9.16}$$

For the first partial derivatives, we have

$$\frac{\partial \psi}{\partial p_1} = 2\, a_1 \frac{\partial a_1}{\partial p_1} + 2\, a_2 \frac{\partial a_2}{\partial p_1} \tag{9.17}$$

and

$$\frac{\partial \psi}{\partial p_2} = 2\, a_1 \frac{\partial a_1}{\partial p_2} + 2\, a_2 \frac{\partial a_2}{\partial p_2} \; . \tag{9.18}$$

Notice, that in these equations we calculate the change in each aberration with respect to each variable parameter separately. The aberration function itself is not determined.

For second derivatives we have equations like

$$\frac{\partial^2 \psi}{\partial^2 p_1^2} = 2\left[\left(\frac{\partial a_1}{\partial p_1}\right)^2 + a_1 \frac{\partial^2 a_1}{\partial p_1^2} + a_2 \frac{\partial^2 a_2}{\partial p_1^2} + \left(\frac{\partial a_2}{\partial p_1}\right)^2\right], \tag{9.19}$$

and

$$\frac{\partial^2 \psi}{\partial p_1\, \partial p_2} = 2\left[a_1 \frac{\partial^2 a_1}{\partial p_1\, \partial p_2} + \frac{\partial a_1\, \partial a_1}{\partial p_1\, \partial p_2} + \frac{\partial a_2\, \partial a_2}{\partial p_1\, \partial p_2} + a_2 \frac{\partial^2 a_2}{\partial p_1\, \partial p_2}\right]. \tag{9.20}$$

Substituting all of the first and second derivatives into Eqs. 9.13 and 9.14 gives

$$\left[a_1 \frac{\delta a_1}{\delta p_1} + a_2 \frac{\delta a_2}{\delta p_1}\right] + \Delta p_1 \left[\left(\frac{\delta a_1}{\delta p_1}\right)^2 + a_1 \frac{\delta^2 a_1}{\delta p_1^2} + a_2 \frac{\delta^2 a_2}{\delta p_1^2} + \left(\frac{\delta a_2}{\delta p_1}\right)^2\right]$$
$$\Delta p_2 \left[a_1 \frac{\delta^2 a_1}{\delta p_1\, \delta p_2} + \frac{\delta a_1\, \delta a_1}{\delta p_1\, \delta p_2} + \frac{\delta a_2\, \delta a_2}{\delta p_1\, \delta p_2} + a_2 \frac{\delta^2 a_2}{\delta p_1\, \delta p_2}\right] = 0 \tag{9.21}$$

and

$$\left[a_1 \frac{\delta a_1}{\delta p_2} + a_2 \frac{\delta a_2}{\delta p_2}\right] + \Delta p_2 \left[\left(\frac{\delta a_1}{\delta p_2}\right)^2 + a_1 \frac{\delta^2 a_1}{\delta p_2^2} + a_2 \frac{\delta^2 a_2}{\delta p_2^2} + \left(\frac{\delta a_2}{\delta p_2}\right)^2\right]$$
$$\Delta p_1 \left[a_1 \frac{\delta^2 a_1}{\delta p_1 \delta p_2} + \frac{\delta a_1\, \delta a_1}{\delta p_1\, \delta p_2} + \frac{\delta a_2\, \delta a_2}{\delta p_1\, \delta p_2} + a_2 \frac{\delta^2 a_2}{\delta p_1\, \delta p_2}\right] = 0 \; . \tag{9.22}$$

The previous two equations look a whole lot more complicated than our original equations, but remember that each individual derivative calculation is simpler here. Also, this form will eventually allow for further simplification.

Example 9.3

Find a minimum for the aberration function given by

$$a_1(p_1, p_2) = \left(\frac{1}{3}p_1^3 - 4p_1^2 + 15p_1 - 11\right)(p_2 - 4)$$

and

$$a_2(p_1, p_2) = \left(\frac{1}{3}p_2^3 - 3p_2^2 + 8p_2 - 3\right)(p_1 - 3),$$

starting at $p_1 = 3.2$, $p_2 = 3.8$.

The first step is to calculate the aberrations

$$a_1 = \left(\frac{1}{3}(3.2)^3 - 4(3.2)^2 + 15(3.2) - 11\right)((3.8) - 4),$$

$$a_1 = -1.39253,$$

$$a_2 = \left(\frac{1}{3}(3.8)^2 - 3(3.8)^2 + 8(3.8) - 3\right)((3.2) - 3),$$

$$a_2 = 0.474133.$$

Next, we find the first derivatives

$$\frac{\partial a_1}{\partial p_1} = \left(p_1^2 - 8p_1 + 15\right)(p_2 - 4)$$

$$= \left((3.2)^2 - 8(3.2) + 15\right)((3.8) - 4) = 0.072,$$

$$\frac{\partial a_1}{\partial p_2} = \left(\frac{1}{3} p_1^3 - 4 p_1^2 + 15 p_1 - 11 \right)$$

$$= \left(\frac{1}{3} (3.2)^3 - 4(3.2)^2 + 15(3.2) - 11 \right) = 6.96267 ,$$

$$\frac{\partial a_2}{\partial p_1} = \left(\frac{1}{3} p_2^3 - 3 p_2^2 + 8 p_2 - 3 \right)$$

$$= \left(\frac{1}{3} (3.8)^3 - 3(3.8)^2 + 8(3.8) - 3 \right) = 2.37067 ,$$

$$\frac{\partial a_2}{\partial p_2} = \left(p_2^2 - 6 p_2 + 8 \right)\left(p_1 - 3 \right)$$

$$= \left((3.8)^2 - 6(3.8) + 8 \right)\left((3.2) - 3 \right) = -0.072 .$$

We also need to calculate the second derivatives for the first aberration

$$\frac{\partial^2 a_1}{\partial p_2^2} = \left(2 p_1 - 8 \right)\left(p_2 - 4 \right)$$

$$= \left(2(3.2) - 8 \right)\left((3.8) - 4 \right) = 0.32 ,$$

$$\frac{\partial^2 a_1}{\partial p_1 \, \partial p_2} = p_1^2 - 8 p_1 + 15$$

$$= (3.2)^2 - 8(3.2) + 15 = -0.36 ,$$

$$\frac{\partial^2 a_1}{\partial p_2^2} = 0 ,$$

and the second derivatives for the second aberration

$$\frac{\partial^2 a_2}{\partial p_1^2} = 0 ,$$

$$\frac{\partial^2 a_2}{\partial p_1 \, \partial p_2} = p_2^2 - 6\,p_2 + 8$$

$$= (3.8)^2 - 6(3.8) + 8 = -0.36 ,$$

$$\frac{\partial^2 a_2}{\partial p_2^2} = \left(p_1 - 3\right)\left(2\,p_2 - 6\right)$$

$$= \left((3.2) - 3\right)\left(2(3.8) - 6\right) = 0.32 .$$

Note that it is merely coincidence that the values for several of the second derivatives turn out to be the same.

The next step is to substitute these values into Eqs. 9.21 and 9.22. Because these equations are so complicated, let's do the calculations one step at a time. The constant terms are

$$a_1 \frac{\partial a_1}{\partial p_1} + a_2 \frac{\partial a_2}{\partial p_1}$$

$$= (-1.39253)(0.072) + (0.474133)(2.37067) = 1.02375 ,$$

$$a_1 \frac{\partial a_1}{\partial p_2} + a_2 \frac{\partial a_2}{\partial p_2}$$

$$= (-1.39253)(6.96267) + (0.474133)(-0.072) = -9.72988 .$$

The coefficients for the Δp_1 terms are

$$\left(\frac{\partial a_1}{\partial p_1}\right)^2 + a_1 \frac{\partial^2 a_1}{\partial p_1^2} + a_2 \frac{\partial^2 a_2}{\partial p_1^2} + \left(\frac{\partial a_2}{\partial p_1}\right)^2 =$$

$$(0.072)^2 + (-1.39253)(0.32) + (0.474133)(0) + (2.37067)^2$$

$$= 5.17963 ,$$

$$a_1 \frac{\partial^2 a_1}{\partial p_1 \partial p_2} + \frac{\partial a_1}{\partial p_1}\frac{\partial a_1}{\partial p_2} + \frac{\partial a_2}{\partial p_1}\frac{\partial a_2}{\partial p_2} + a_2 \frac{\partial^2 a_2}{\partial p_1 \partial p_2} =$$

$$(-1.39253)(-0.36) + (0.072)(6.96267)$$

$$+ (2.37067)(-0.072) + (0.474133)(-0.36)$$

$$= 0.661248 \; .$$

One of the coefficients for the Δp_2 terms is identical to the previous value and the other coefficient is given by

$$\left(\frac{\partial a_1}{\partial p_2} \right)^2 + a_1 \frac{\partial^2 a_1}{\partial p_2^2} + a_2 \frac{\partial^2 a_2}{\partial p_2^2} + \left(\frac{\partial a_2}{\partial p_2} \right)^2 =$$

$$(6.96267)^2 + (-1.39253)(0) + (0.474133)(0.32) + (-0.072)^2$$

$$= 48.6356 \; .$$

Putting all of this together gives two equations:

$$(1.02375) + (5.17963)\,\Delta p_1 + (0.661248)\,\Delta p_2 = 0 \; ,$$

and

$$(-9.72988) + (0.661248)\,\Delta p_1 + (48.6356)\,\Delta p_2 = 0 \; .$$

Solving these equations gives $\Delta p_1 = -0.223577$ and $\Delta p_2 = 0.203096$. Therefore, for the next iteration we will start at $p_1 = 2.97642$ and $p_2 = 4.0031$. After one more iteration, we will converge to the final values: $p_1 = 3.0$ and $p_2 = 4.0$. ■

Newton's method is obviously complicated. We will now discuss how to simplify these calculations. The most obvious thing to do is to simply ignore the second derivatives and try to solve

$$\left[a_1 \frac{\delta a_1}{\delta p_1} + a_2 \frac{\delta a_2}{\delta p_1} \right] + \Delta p_1 \left[\left(\frac{\delta a_1}{\delta p_1} \right)^2 + \left(\frac{\delta a_2}{\delta p_1} \right)^2 \right]$$

$$+ \Delta p_2 \left[\frac{\delta a_1}{\delta p_1}\frac{\delta a_1}{\delta p_2} + \frac{\delta a_2}{\delta p_1}\frac{\delta a_2}{\delta p_2} \right] = 0 \qquad (9.23)$$

and

$$\left[a_1 \frac{\delta a_1}{\delta p_2} + a_2 \frac{\delta a_2}{\delta p_2}\right] + \Delta p_2 \left[\left(\frac{\delta a_1}{\delta p_2}\right)^2 + \left(\frac{\delta a_2}{\delta p_2}\right)^2\right]$$

$$+ \Delta p_1 \left[\frac{\delta a_1}{\delta p_1}\frac{\delta a_1}{\delta p_2} + \frac{\delta a_2}{\delta p_1}\frac{\delta a_2}{\delta p_2}\right] = 0 \ . \tag{9.24}$$

Note that back in Eqs. 9.13 and 9.14 we could not ignore the second derivatives because there would not have been anything left to solve. But here we still have combinations of the first derivatives multiplied together. Dropping the second derivatives is justified because these terms are generally much smaller than the terms containing first derivatives. Ignoring the second derivatives is called the *Gauss–Newton algorithm*.

Example 9.4

Work Example 9.3 again using the Gauss–Newton method.

The constants do not change. We only need to recalculate the coefficients. The coefficients for the Δp_1 terms are

$$\left(\frac{\partial a_1}{\partial p_1}\right)^2 + \left(\frac{\partial a_2}{\partial p_1}\right)^2 = (0.072)^2 + (2.37067)^2 = 5.62524 \ ,$$

$$\frac{\partial a_1}{\partial p_1}\frac{\partial a_1}{\partial p_2} + \frac{\partial a_2}{\partial p_1}\frac{\partial a_2}{\partial p_2} =$$

$$(0.072)(6.96267) + (2.37067)(-0.072) = 0.330624 \ .$$

The remaining coefficient for the Δp_2 term is given by

$$\left(\frac{\partial a_1}{\partial p_2}\right)^2 + \left(\frac{\partial a_2}{\partial p_2}\right)^2 = (6.96267)^2 + (-0.072)^2 = 48.4839 \ .$$

Putting all of this together gives two equations:

$$(1.02375) + (5.62524)\Delta p_1 + (0.330624)\Delta p_2 = 0 \ ,$$

and

$$(-9.72988) + (0.330624)\,\Delta p_1 + (48.4839)\,\Delta p_2 = 0\ .$$

Solving these equations gives $\Delta p_1 = -0.193865$ and $\Delta p_2 = 0.202005$. Therefore, for the next iteration we will start at $p_1 = 3.00613$ and $p_2 = 4.002$. After one more iteration, we will converge to the final values: $p_1 = 3.0$ and $p_2 = 4.0$. ■

Unfortunately, optical systems are so highly nonlinear that sometimes the Gauss–Newton method will not work satisfactorily. Solutions to Eqs. 9.23 and 9.24 often result in large parameter changes which cause the solution to diverge rather than converge.

The Levenberg–Marquardt algorithm (or damped least squares method) reduces the size of the changes by replacing the second order derivatives with a damping factor α. The equations then become

$$\left[a_1 \frac{\delta a_1}{\delta p_1} + a_2 \frac{\delta a_2}{\delta p_1}\right] + \Delta p_1 \left[\left(\frac{\delta a_1}{\delta p_1}\right)^2 + \left(\frac{\delta a_2}{\delta p_1}\right)^2 + \alpha\right]$$
$$\Delta p_2 \left[\frac{\delta a_1}{\delta p_1}\frac{\delta a_1}{\delta p_2} + \frac{\delta a_2}{\delta p_1}\frac{\delta a_2}{\delta p_2}\right] = 0 \tag{9.25}$$

and

$$\left[a_1 \frac{\delta a_1}{\delta p_2} + a_2 \frac{\delta a_2}{\delta p_2}\right] + \Delta p_2 \left[\left(\frac{\delta a_1}{\delta p_2}\right)^2 + \left(\frac{\delta a_2}{\delta p_2}\right)^2 + \alpha\right]$$
$$\Delta p_1 \left[\frac{\delta a_1}{\delta p_1}\frac{\delta a_1}{\delta p_2} + \frac{\delta a_2}{\delta p_1}\frac{\delta a_2}{\delta p_2}\right] = 0\ , \tag{9.26}$$

where the damping factor α is a positive constant.

Clearly, the choice of the damping factor α should be critical to the success of this method. If the damping factor is too large, the parameter changes will be small and convergence will be slow. If the damping factor is too small, the parameter changes may be too large and result in divergence. There are many different methods for choosing the damping factor.

One approach is to look at choosing the damping factor as another optimization problem. Choose an initial value for α which is small ($\alpha = 0$) and calculate the parameter changes and the aberration function ψ. Next, make a small change in the damping factor $\delta\alpha$ and repeat. We now have enough information to make an approximation to the first derivative:

$$\frac{\partial \psi}{\partial \alpha} \approx \frac{\psi(\alpha + \delta\alpha) - \psi(\alpha)}{\delta\alpha} . \tag{9.27}$$

We can approximate the second derivative by changing the damping factor once more and calculating the aberration function which results from changing the parameters. Thus,

$$\frac{\partial^2 \psi}{\partial \alpha^2} \approx \frac{\psi(\alpha + 2\,\delta\alpha) - 2\,\psi(\alpha + \delta\alpha) + \psi(\alpha)}{\delta\alpha^2} . \tag{9.28}$$

We can now generate an estimate for the best damping factor:

$$\alpha_{new} = \alpha_{old} - \frac{\left(\dfrac{\partial \psi}{\partial \alpha}\right)}{\left(\dfrac{\partial^2 \psi}{\partial \alpha^2}\right)} . \tag{9.29}$$

Using the new value for α should result in a smaller aberration function than we would otherwise expect. The cost is that the amount of calculation has increased again. The increase in calculations is not large, however, because we are not calculating derivatives by changing each parameter separately. In fact, if there are more than two free parameters, the number of exact ray traces required is less for the damped least squares method than for Newton's method.

Example 9.5

Determine a good value for the damping factor and perform the first iteration of the damped least squares method for the problem given in Example 9.3.

The solution picks up on the work done in Example 9.4. There we had

$$(1.02375) + (5.62524)\,\Delta p_1 + (0.330624)\,\Delta p_2 = 0 ,$$

and

$$(-9.72988) + (0.330624)\,\Delta p_1 + (48.4839)\,\Delta p_2 = 0 .$$

Solutions to these two equations gave $p_1 = 3.006135$ and $p_2 = 4.002005$. With these values we can evaluate the aberration function generated without damping: $\psi(\alpha{=}0) = a_1^{\,2} + a_2^{\,2} = 0.00039385$.

To find a good damping factor, we need to make small changes in the damping factor. If our initial damping factor is $\alpha = 0$, we can make a small change, $\delta\alpha = 0.01$, to the previous equations to get

$$(1.02375) + (5.63524)\,\Delta p_1 + (0.330624)\,\Delta p_2 = 0$$

and

$$(-9.72988) + (0.330624)\,\Delta p_1 + (48.4939)\,\Delta p_2 = 0\,,$$

a subtle, yet significant, difference. Solving these equations leads to $p_1 = 3.006482$ and $p_2 = 4.001961$. Again, we calculate the aberration function, but this time with damping: $\psi(\alpha{=}0.01) = 0.00037675$.

Making another small change in the damping factor ($\alpha = 0.02$) leads to two slightly different equations with solutions: $p_1 = 3.006827$ and $p_2 = 4.001917$ and $\psi(\alpha{=}0.02) = 0.00036003$.

We are now in a position to calculate a good choice for the damping factor.

$$\alpha = -\frac{\left(\dfrac{\delta\psi}{\delta\alpha}\right)}{\left(\dfrac{\delta^2\psi}{\delta\alpha^2}\right)} = -\frac{\left(\dfrac{\psi(\delta\alpha) - \psi(0)}{\delta\alpha}\right)}{\left(\dfrac{\psi(2\,\delta\alpha) - 2\,\psi(\delta\alpha) + \psi(0)}{\delta\alpha^2}\right)}\,,$$

$$\alpha = \frac{\left(\psi(0) - \psi(\delta\alpha)\right)\delta\alpha}{\left(\psi(2\,\delta\alpha) - 2\,\psi(\delta\alpha) + \psi(0)\right)}\,,$$

$$\alpha = \frac{\left((0.00039385) - (0.00037675)\right)(0.01)}{\left((0.00036003) - 2\,(0.00037675) + (0.00039385)\right)}\,,$$

$$\alpha = 0.4387\,.$$

Now, we use this damping factor to calculate new parameters. The equations we solve look like

$$(1.02375) + (6.06394)\,\Delta p_1 + (0.330624)\,\Delta p_2 = 0$$

and

$$(-9.72988) + (0.330624)\,\Delta p_1 + (48.9226)\,\Delta p_2 = 0\,.$$

The solutions are $\Delta p_1 = -0.179736$ and $\Delta p_2 = 0.200098$. The parameters become

$p_1 = 3.020264$ and $p_2 = 4.000098$. With these parameters, the aberration function is $\psi = 0.94 \times 10^{-6}$. This represents more than two orders of magnitude improvement. Not bad for a little extra calculation. ■

The number of iterations required for the damped least squares method to converge on a solution may be more than that for the Newton method, but each individual iteration requires considerably less calculation if the number of free parameters is large. Because of the relative slow convergence of the damped least squares method, several alternative methods have been used such as multiplicative damping and pseudo-second-derivative methods. Each of these methods attempt to avoid calculating second derivatives of the aberrations by different means (see the article by Hayford for more information on other methods).

9.4 THE GENERAL CASE

The next step is to generalize the two-parameter case to any number of free parameters. To show how this is done, we first write the two initial equations from the two-parameter case in matrix form:

$$\begin{bmatrix} \dfrac{\delta\psi}{\delta p_1} \\[2ex] \dfrac{\delta\psi}{\delta p_2} \end{bmatrix} + \begin{bmatrix} \dfrac{\delta^2\psi}{\delta p_1^2} & \dfrac{\delta^2\psi}{\delta p_1\,\delta p_2} \\[2ex] \dfrac{\delta^2\psi}{\delta p_1\,\delta p_2} & \dfrac{\delta^2\psi}{\delta p_2^2} \end{bmatrix} \times \begin{bmatrix} \Delta p_1 \\[2ex] \Delta p_2 \end{bmatrix} = 0 \,. \tag{9.30}$$

The matrix form shows us how to expand from the two-parameter case to the multi-parameter case. In matrix notation, the previous equation has the form

$$\vec{g} + \overleftrightarrow{G}^T \Delta\vec{p} = 0 \,. \tag{9.31}$$

In the general equation $\vec{g}$ and $\Delta\vec{p}$ are vectors of dimension M, where M is the number of free parameters in the system.

$$\vec{g} = \begin{bmatrix} \frac{\delta\psi}{\delta p_1} \\ \frac{\delta\psi}{\delta p_2} \\ \vdots \\ \frac{\delta\psi}{\delta p_M} \end{bmatrix} \tag{9.32}$$

is called the *gradient vector of the aberration function*. The parameter change vector is

$$\Delta\vec{p} = \begin{bmatrix} \Delta p_1 \\ \Delta p_2 \\ \vdots \\ \Delta p_M \end{bmatrix} . \tag{9.33}$$

Finally, we have what is called the *Hessian matrix of the aberration function*:

$$\vec{G} = \begin{bmatrix} \frac{\delta^2\psi}{\delta p_1^2} & \frac{\delta^2\psi}{\delta p_1\,\delta p_2} & \cdots & \frac{\delta^2\psi}{\delta p_1\,\delta p_M} \\ \frac{\delta^2\psi}{\delta p_1\,\delta p_2} & \frac{\delta^2\psi}{\delta p_2^2} & \cdots & \frac{\delta^2\psi}{\delta p_1\,\delta p_M} \\ \vdots & \vdots & \cdots & \vdots \\ \frac{\delta^2\psi}{\delta p_1\,\delta p_M} & \frac{\delta^2\psi}{\delta p_2\,\delta p_M} & \cdots & \frac{\delta^2\psi}{\delta p_M^2} \end{bmatrix} . \tag{9.34}$$

In Eq. 9.31 we used the transpose of this matrix. A transpose matrix has its rows and columns interchanged.

Recall that the general aberration function was given in Eq. 9.4. We can rewrite this equation using vector notation as

$$\psi\,(a_1, a_2, \cdots, a_N) = \vec{a}^{\,T}\vec{a} , \tag{9.35}$$

where

$$\vec{a} = \begin{bmatrix} a_1 \\ a_2 \\ \vdots \\ a_N \end{bmatrix} \tag{9.36}$$

is the aberration vector and there are N aberrations. Again, in Eq. 9.35 we use the transpose of a vector to form a single row matrix from the vector for multiplication.

If we now make the change to calculating the derivatives of the aberrations, we can write

$$\vec{g} = \begin{bmatrix} \frac{\delta\psi}{\delta p_1} \\ \frac{\delta\psi}{\delta p_2} \end{bmatrix} = 2 \begin{bmatrix} \frac{\delta a_1}{\delta p_1} & \frac{\delta a_2}{\delta p_1} \\ \frac{\delta a_1}{\delta p_2} & \frac{\delta a_2}{\delta p_2} \end{bmatrix} \times \begin{bmatrix} a_1 \\ a_2 \end{bmatrix} . \tag{9.37}$$

Or in matrix notation,

$$\vec{g} = 2 \overleftrightarrow{J}^T \vec{a} , \tag{9.38}$$

for the general case. Here, we have introduced the Jacobian matrix given by

$$\overleftrightarrow{J} = \begin{bmatrix} \frac{\delta a_1}{\delta p_1} & \frac{\delta a_1}{\delta p_2} & \cdots & \frac{\delta a_1}{\delta p_M} \\ \frac{\delta a_2}{\delta p_1} & \frac{\delta a_2}{\delta p_2} & \cdots & \frac{\delta a_2}{\delta p_M} \\ \vdots & \vdots & \vdots & \vdots \\ \frac{\delta a_N}{\delta p_1} & \frac{\delta a_N}{\delta p_2} & \cdots & \frac{\delta a_N}{\delta p_M} \end{bmatrix} . \tag{9.39}$$

The 2×2 Hessian matrix of the aberration function now can be written as

$$\begin{bmatrix} \dfrac{\delta^2 \psi}{\delta p_1^2} & \dfrac{\delta^2 \psi}{\delta p_1 \, \delta p_2} \\ \dfrac{\delta^2 \psi}{\delta p_1 \, \delta p_2} & \dfrac{\delta^2 \psi}{\delta p_2^2} \end{bmatrix} = 2 \left[\begin{bmatrix} \dfrac{\delta a_1}{\delta p_1} & \dfrac{\delta a_2}{\delta p_1} \\ \dfrac{\delta a_1}{\delta p_2} & \dfrac{\delta a_2}{\delta p_2} \end{bmatrix} \times \begin{bmatrix} \dfrac{\delta a_1}{\delta p_1} & \dfrac{\delta a_1}{\delta p_2} \\ \dfrac{\delta a_2}{\delta p_1} & \dfrac{\delta a_2}{\delta p_2} \end{bmatrix} \right.$$

$$\left. + \, a_1 \begin{bmatrix} \dfrac{\delta^2 a_1}{\delta p_1^2} & \dfrac{\delta^2 a_1}{\delta p_1 \, \delta p_2} \\ \dfrac{\delta^2 a_1}{\delta p_1 \, \delta p_2} & \dfrac{\delta^2 a_1}{\delta p_2^2} \end{bmatrix} + a_2 \begin{bmatrix} \dfrac{\delta^2 a_2}{\delta p_1^2} & \dfrac{\delta^2 a_2}{\delta p_1 \, \delta p_2} \\ \dfrac{\delta^2 a_2}{\delta p_1 \, \delta p_2} & \dfrac{\delta^2 a_2}{\delta p_2^2} \end{bmatrix} \right] . \tag{9.40}$$

In general, we have

$$\overleftrightarrow{G} = \overleftrightarrow{J}^T \overleftrightarrow{J} + \sum_{i=1}^{N} a_i \overleftrightarrow{H}_i \, . \tag{9.41}$$

We can now write Eq. 9.22 in the form

$$\overleftrightarrow{J}^T \vec{a} + \left(\overleftrightarrow{J}^T \overleftrightarrow{J} + \sum_{i=1}^{N} a_i \overleftrightarrow{H}_i \right) \Delta\vec{p} = 0 \, . \tag{9.42}$$

This equation looks a lot more complicated than Eq. 9.22, but it is exactly the same. As far as I know, some variation of Eq. 9.42 is used in all commercial lens design programs. (Note, I have to say "as far as I know" because exactly how commercial lens design programs work is usually proprietary information.) As with Eq. 9.22 we can approximate the second derivatives with the damping factor. Equation 9.42 becomes

$$\overleftrightarrow{J}^T \vec{a} + \left(\overleftrightarrow{J}^T \overleftrightarrow{J} + \alpha \overleftrightarrow{I} \right) \Delta\vec{p} = 0 \, , \tag{9.43}$$

where $\overleftrightarrow{I}$ is the identity matrix.

9.5 CONSTRAINTS

Besides dealing with aberrations of the exact rays, optimization must also deal with additional restrictions on the system parameters. These additional restrictions are called *constraints*. An example of a constraint would be that the difference between the two parameters in Example 9.3 remained constant. We would express this mathematically as

$$p_1 - p_2 - \Delta = 0 , \tag{9.44}$$

where Δ is a specified constant. This constraint dramatically changes the problem.

When optimizing an optical system there are two basic types of constraints: physical constraints and optical constraints. Typically we want the focal length of the system being optimized to end up with a certain value:

$$f' - f'_{target} = 0 . \tag{9.45}$$

This is an example of an optical constraint. As we change the system parameters, other physical parameters may be affected. For example, if we change c_1 and c_2 as part of our optimization, then the edge thickness between surfaces 1 and 2 will also change even if the center thickness t_1 is not a variable parameter. It is possible for the two curvatures to change in such a way that the edge thickness becomes zero or even negative for a given clear radius. Clearly, this cannot be allowed to happen if the system is to be realized physically. This is an example of a physical constraint.

As we have just seen, constraints are functions of the variable system parameters and can be written as equalities of the form

$$c_i(p) = 0 , \tag{9.46}$$

as with the focal length constraint. Or, constraints may be inequalities of the form

$$c_i(p) \geq 0 , \tag{9.47}$$

as with the edge thickness constraint. Constraints which are equalities or inequalities on their boundaries are called *active constraints*. Inequality-type constraints do not need to be included in optimization until they become active by trying to exceed their boundaries.

There are two basic approaches for handling active constraints. The first is to include the constraint as part of the aberration function. Variations of this approach are called *penalty function methods*. The second approach is to solve a set of equations based upon the constraints in addition to minimizing the aberration function. These are called *quadratic programming methods*. Quadratic programming methods are beyond the scope of this text.

In the penalty function method we include the active constraints in the aberration function as additional weighted aberrations:

$$\psi^*(p) = \psi(p) + \sum_i (w_i c_i(p))^2 . \tag{9.48}$$

Optimization proceeds normally with the added aberrations. There is no need for additional programming which is a major advantage. Another advantage is that this method can converge from starting points which are very different from the final system. Unfortunately, this method also has some drawbacks. First, penalty functions can sometimes miss solutions. Second, the solutions are dependent on the weights used for the constraints. Finally, it is difficult to implement inequality constraints via a penalty function. Again, a detailed discussion of the drawbacks of including the constraints as part of the aberration function is beyond the scope of this text.

Example 9.6

Design a single lens using K10 glass which will have a focal length of 10.0 cm, $f/\# = 5.0$, and a minimum amount of spherical aberration for an object at infinity.

Typically, we begin the design procedure by using paraxial optics and third order aberration theory to get a good starting point for the design. This procedure is called *predesign*. For this problem we can determine the value for c from

$$c = \frac{1}{(n-1)f'} = \frac{1}{(0.50137)(10.0)} = 0.199453 \; .$$

A first guess at the optimum shape for the lens is found by using third-order aberration theory. From Eq. 7.19 we have

$$c_1 = \frac{(2n+1)nc + 4(n+1)v}{2(n+2)} \; .$$

For this problem $v = 0$, so

$$c_1 = \frac{(2(1.50137)+1)(1.50137)(0.199453)}{2(3.50137)} = 0.171166 \; .$$

Solving for the second curvature we get

$$c_2 = c_1 - c = (0.171166) - (0.199453) = -0.028287 \; .$$

The final solution to this problem should lie close to these values.

But to demonstrate the difference between the Gauss–Newton method and the damped least squares method, let's begin with a system which is far from these values. Let $c_1 = -0.2\ \mathrm{cm}^{-1}$ and $c_2 = -0.4\ \mathrm{cm}^{-1}$. Since $f/\# = 5.0$, the clear radius of the lens should be 1.0 cm. This implies a minimum thickness of about 0.3 cm. If we increase the lens thickness a little, we will not have to worry about the edge thickness. Therefore, let $t_1 = 0.5$ cm. This lens has an effective focal length of about 9.35 cm and transverse spherical aberration of about -0.20 cm.

The aberration function consists of two parts: $a_1 = f' - f'_{target}$ and $a_2 = \epsilon'_y(\rho=1, \theta=0°, \eta=0)$. Initially,

$$\vec{a} = \begin{bmatrix} -0.651684 \\ -0.204812 \end{bmatrix},$$

and $\psi = 0.4666$.

If we make small changes in the two curvatures, we can calculate the partial derivatives:

$$\frac{\partial(f' - f'_{target})}{\partial c_1} = -40.0880 ,$$

$$\frac{\partial(f' - f'_{target})}{\partial c_2} = 45.2894 ,$$

$$\frac{\partial \epsilon'_y}{\partial c_1} = -0.017644 ,$$

$$\frac{\partial \epsilon'_y}{\partial c_2} = 1.234870 .$$

For the Gauss–Newton method the optimization equations look like

$$\vec{J}^T \vec{a} + \vec{J}^T \vec{J} = 0 ,$$

$$\begin{bmatrix} 26.6444 \\ -29.7673 \end{bmatrix} + \begin{bmatrix} 1{,}671.17 & -1{,}851.45 \\ -1{,}851.45 & 2{,}052.65 \end{bmatrix} \times \begin{bmatrix} \Delta c_1 \\ \Delta c_2 \end{bmatrix} = 0 .$$

Solving these equations gives $\Delta c_1 = 0.170504$ and $\Delta c_1 = 0.168293$. The new values for the curvatures are $c_1 = -0.029495$ and $c_2 = -0.231707$. Comparing these results to our expectations from paraxial optics and the third-order aberrations, we see that we are heading in the right direction, but we have a long way to go. The results of further iterations are given in the table below.

For the damped least squares method, the aberrations and derivatives are calculated as shown above. A good damping factor is found to be 0.1224 for this first iteration. With this damping factor, new parameter values of $c_1 = -0.057184$ and $c_2 = -0.256689$ are found. Because of the damping, we have not progressed as far as we did in the first iteration of the Gauss–Newton method. The results of subsequent iterations are shown in the table below.

The table shows that on iteration number 3, the Gauss–Newton method slightly overshoots the solution and then drastically overcompensates on the next iteration. This is the major drawback of the Gauss–Newton method. On the other hand, the damped least squares method slowly but continuously converges to a solution. Clearly, many iterations could have been saved by starting out with the curvatures given by the predesign. ■

SUMMARY

Optimization is the process of modifying an existing optical system to improve image quality. The very first step in this process is to define image quality. Here, we use the aberration function which consists of the sums of selected weighted aberrations:

$$\psi = \sum_{i=1}^{N} \left[w_{x,i}\, \epsilon_{x,i}^2 + w_{y,i}\, \epsilon_{y,i}^2 \right] = \sum_{i=1}^{N} a_i^2 \,. \tag{9.49}$$

Constraints on the optical system such as target focal length can be included in the aberration function.

The next step in the optimization process is to select an initial optical system. The starting system must have enough free parameters p_j and the available parameters must be appropriate for reducing the aberration function.

Several mathematical techniques exist for minimizing the aberration function. Newton's method uses first and second derivatives to approximate the slope and curvature of the function at some point. Unfortunately, calculating the second derivatives can become time-consuming and cumbersome. The Gauss–Newton method drops the second derivatives entirely. This method can lead to a solution in some cases, but in general the resulting parameter changes oscillate or diverge from the solution. A compromise is the damped least squares method. Like the Gauss–Newton method, the second derivatives are not calculated. But the parameter changes are reduced by means of a damping factor.

The damped least squares method consists of solving the set of equations

$$\vec{J}^{\,T} \vec{a} + \left(\vec{J}^{\,T} \vec{J} + \alpha \vec{I} \right) \Delta \vec{p} = 0 \,, \tag{9.50}$$

where

$$\vec{a} = \begin{bmatrix} a_1 \\ a_2 \\ \vdots \\ a_N \end{bmatrix} \tag{9.51}$$

and

$$\vec{J} = \begin{bmatrix} \frac{\delta a_1}{\delta p_1} & \frac{\delta a_1}{\delta p_2} & \cdots & \frac{\delta a_1}{\delta p_M} \\ \frac{\delta a_2}{\delta p_1} & \frac{\delta a_2}{\delta p_2} & \cdots & \frac{\delta a_2}{\delta p_M} \\ \vdots & \vdots & \vdots & \vdots \\ \frac{\delta a_N}{\delta p_1} & \frac{\delta a_N}{\delta p_2} & \cdots & \frac{\delta a_N}{\delta p_M} \end{bmatrix} . \tag{9.52}$$

The elements in the matrix in Eq. 9.52 are finite differences which may be determined using

$$\frac{\delta\psi}{\delta p} = \frac{\psi(\rho+\delta\rho) - \psi(\rho)}{\delta\rho} . \tag{9.53}$$

It remains to determine the best damping factor α. The method used here is to treat finding the best damping factor as a one-dimensional optimization problem. Using Newton's method we have

$$\alpha_{new} = \alpha_{old} - \frac{\left(\frac{\partial\psi}{\partial\alpha}\right)}{\left(\frac{\partial^2\psi}{\partial\alpha^2}\right)} . \tag{9.54}$$

REFERENCES

S. D. Conte and C. de Boor, *Elementary Numerical Analysis: An Algorithmic Approach* (McGraw-Hill, New York, 1972).

R. Fletcher, *Practical Methods of Optimization, Volume 1: Unconstrained*

Optimization; and Volume 2: Constrained Optimization (John Wiley & Sons, New York, 1980 and 1981.

P. E. Gill, W. Murray, and M. H. Wright, *Practical Optimization*, (Academic Press, Orlando Fl, 1981).

M. J. Hayford, "Optimization Methodology," *Geometrical Optics, SPIE*, Vol. 531, pp. 68–81, 1985.

PROBLEMS

9.1 If the number of variable parameters in a system is large, which of the optimization methods—Gauss, Gauss–Newton, or damped least squares—converges fastest? Which of these requires the least calculation?

9.2 How many ninth-order aberrations are there? How many eleventh-order?

Answer the following questions for an aberration function given by

$$\psi = \left(p^2 - 6p + 10\right)^2 .$$

9.3 Use the standard methods of calculus to determine the parameter value for which this aberration function is a minimum. What is the value of the aberration function at minimum?

9.4 Calculate the first and second derivatives of the aberration function analytically. Evaluate these derivatives at $p = 3.3$ and determine what change is needed to minimize the aberration function. Repeat these calculations until the solution converges. What are the final values for p and ψ?

9.5 Use Newton's method with analytic first and second derivatives of just the weighted aberration to calculate the minimum of the aberration function. Start with $p = 3.3$. Repeat until the solution converges. What are the final values for p and ψ?

9.6 Try using the Gauss–Newton method with analytic first derivatives to calculate the minimum of the aberration function. Start with $p = 3.3$. Show that this method oscillates or diverges by repeating a couple of iterations.

9.7 Minimize the aberration function using the method of damped least squares. Use a damping factor of 2.0 for each iteration.

9.8 Starting with a damping factor of 2.0, determine a better damping factor for the first iteration of Problem 9.7.

Answer the following questions for an aberration function given by

$$\psi = \left(p^3 - 9p^2 + 15p + 1\right)^2 .$$

9.9 Use the standard methods of calculus to determine the parameter value for which this aberration function is a minimum. What is the value of the aberration function at minimum?

9.10 Calculate the first and second derivatives of the aberration function analytically. Evaluate these derivatives at $p = 3.3$ and determine what change is needed to minimize the aberration function. Repeat these calculations until the solution converges. What are the final values for p and ψ?

9.11 Use Newton's method with analytic first and second derivatives of just the weighted aberration to calculate the minimum of the aberration function. Start with $p = 3.3$. Repeat until the solution converges. What are the final values for p and ψ?

9.12 Try using the Gauss–Newton method with analytic first derivatives to calculate the minimum of the aberration function. Start with $p = 3.3$. Show that this method oscillates or diverges by repeating a couple of iterations.

9.13 Minimize the aberration function using the method of damped least squares. Use a damping factor of 2.0 for each iteration.

9.14 Starting with a damping factor of 2.0, determine a better damping factor for the first iteration of Problem 9.13.

Answer the following questions for an aberration function given by

$$\psi = \left(p_1^2 - 4p_1 + 5\right)^2 + \left(p_2^2 + 6p_2 + 7\right)^2 .$$

9.15 Minimize this aberration function using Newton's method. Start with $p_1 = 1.9$ and $p_2 = -2.8$.

9.16 Attempt to minimize the aberration function using the Gauss–Newton method. Start with $p_1 = 1.9$ and $p_2 = -2.8$.

9.17 Minimize the aberration function using damped least squares. Determine the optimum damping factor at each iteration. Start with $p_1 = 1.9$ and $p_2 = -2.8$.

Answer the following questions for an aberration function given by

$$\psi = \left(p_1^2 - 2p_1 + 3\right)^2 + \left(4p_2^2 - 12p_2 + 10\right)^2 + (p_3 - 5)^2 .$$

9.18 Minimize this aberration function using Newton's method. Start with $p_1 = 1.2$, $p_2 = 1.8$, and $p_3 = 4.5$.

9.19 Attempt to minimize the aberration function using the Gauss–Newton method. Start with $p_1 = 1.2$, $p_2 = 1.8$, and $p_3 = 4.5$.

9.20 Minimize the aberration function using damped least squares. Determine the optimum damping factor at each iteration. Start with $p_1 = 1.2$, $p_2 = 1.8$, and $p_3 = 4.5$.

9.21 Use thin lens equations for the focal length and spherical aberration to calculate analytically the first derivative of each aberration for each variable for the initial system in Example 9.6. Your calculations should roughly approximate the numerical derivatives found in the example.

9.22 For the initial system given in Example 9.6, determine which value for δc_1 is best for calculating the first derivatives of the two aberrations. Try 10^{-3}, 10^{-6}, and 10^{-9}. Since we do not have an analytic calculation to compare our results with, try other values for δc_1 (10^{-4}, 10^{-7}, and 10^{-10}) to see if the result is sensitive to the choice. A good choice for δc_1 will not be sensitive.

9.23 Find the lens shape which minimizes marginal spherical aberration for an object at infinity if the lens is made of BaLF3 glass. The lens should have an effective focal length of 20.0 cm and *f* / 4. Begin the optimization with the shape for minimum third order spherical aberration. Be sure to include a generous lens thickness so that there will not be any edge violations.

9.24 Find the lens shape which minimizes marginal spherical aberration for an object 40.0 cm from a lens, if the lens is made of SK2 glass. The lens should have an effective focal length of 20.0 cm and *f* / 4. Begin the optimization with the shape for minimum third order spherical aberration.

9.25 The gamma function $\Gamma(x)$ is a type of mathematical function which is easily calculated using a computer algebra program. Unfortunately, even computer algebra programs cannot differentiate the function. Therefore, to find a minimum, we must use numerical techniques as outlined in this chapter. Use Newton's method to find the minimum of the gamma function for $x > 0$. *Hint*: In Maple, the gamma function is evaluated by entering GAMMA(x).

9.26 The gamma function described in Problem 9.25 has several minima and maxima for $x < 0$. Use the damped least squares method to find the minimum with the smallest absolute value for $-x$.

CHAPTER 10

INTRODUCTION TO LENS DESIGN

It has been said many times that lens design is as much of an art as it is a science. It is a science, in that we use mathematics and the laws of science (geometrical optics) to measure and quantify our designs. But, it is also an art, in that valid results are often a matter of personal choice. If you give a lens design problem to a dozen different designers, you will probably get a dozen different final designs. The explanation is that the designers' background, experience, and even personality enter into the choices he or she makes in the design process. For this reason, a knowledgeable lens designer might not like this chapter. He or she might not like my choice of examples. Too bad, but this chapter is written merely to introduce the novice to some of the considerations involved in lens design. I don't expect the reader to become a competent lens designer just based on reading this chapter.

In the previous two chapters, you learned about an analytical method for preparing the first-order layout of an optical system and the method used by lens design software to minimize aberrations in an optical system. If you really know and understand that material, you might assume that you are fully prepared for the task of lens design. That would be a poor assumption. There is a lot more to lens design than merely entering a thick lens version of your first-order layout into a computer program with automatic optimization.

First, and most importantly, the initial system that you enter into the software has a dramatic impact on the quality of the final system that the software puts out. Remember that the calculations described in Chapter 9 can only find the nearest local minimum. If you put in a poor system to begin with, you will get out a poor system. The key to good lens design is to understand geometrical optics well enough so that you can choose a good starting system.

Second, the designer must determine which of the optical system parameters should be variables. Some parameters do not affect some aberrations, while other parameters may have very strong affects. Again, the designer must be very knowledgeable to make intelligent and effective choices.

Finally, the designer must choose the structure of the error function. If the designer includes aberrations which are unimportant or which cannot be corrected, the final design will be compromised. Clearly, the designer must have a thorough knowledge of aberration theory to design a really good system.

The purpose of this chapter is to give you a brief introduction to each of these topics and to prepare you for further study with more advanced texts.

10.1 DESIGN OF A TELESCOPE OBJECTIVE

The basic layout of an astronomical telescope is a long-focal-length objective lens and a short-focal-length eyepiece. The overall angular magnification of the system when focused at infinity is given by Eq. 2.67:

$$M = -\frac{f_o'}{f_e'} . \tag{10.1}$$

Normally, the objective lens is the aperture stop and the eyepiece (or some part of the eyepiece) is the field stop. Here, we are concerned only with the design of the objective lens.

From Eq. 10.1 we might conclude that the larger the focal length of the objective, the better. But this ignores the field of view which will be limited by the field stop. Since the size of the field stop is fixed by other considerations such as the size of the eyepiece being used or the size of the detector, the focal length must be adjusted to give the desired field of view. If the objective lens focal length is too large, only a small portion of the desired field of view may be observed. And if the focal length is too small, fine detail will not be sufficiently magnified. Therefore, in general, we want the longest focal length which gives the desired field of view for a telescope objective.

Example 10.1

We want to design an objective lens to be used for astrophotography. What is the appropriate focal length for the objective lens so that the field of view of the system can cover a large nebula such as M8, the Lagoon Nebula in Sagittarius? The camera is a standard 35 mm attached directly to the telescope without eyepiece or camera lens.

First, the angular size of the Lagoon Nebula is approximately 60 arc-minutes by 35 arc-minutes. Therefore, our HFOV must be somewhat larger than 30 arc-minutes. The focal length is determined by the image size and the field of view:

$$f_o' = \frac{h'}{\tan(\bar{u}_0)} .$$

Directly mounting the camera body to the telescope means that the film stop will serve as the field stop of the system. The frame size for 35-mm film is 24 mm by 37 mm. (That's why they call it 35-mm film, right? The 35-mm dimension is the total width of the film including the sprocket holes.) If we associate the 37-mm dimension with the 60 arc-minute angular size, we find

$$f_o' = \frac{18.5 \text{ mm}}{\tan(30')} = 2120 \text{ mm} .$$

In the other direction we have

$$f_o' = \frac{12 \text{ mm}}{\tan(17.5')} = 2360 \text{ mm} .$$

Therefore, we can safely round this value down to a convenient $f_o' = 2000$ mm.■

Next, let's discuss which aberrations will be important in the design of a telescope objective. To see which aberrations are important, let's assume that at some particular field of view all of the third-order aberrations produce the same magnitude of aberration (not a realistic assumption, but useful). Figure 10.1 shows how the various aberrations vary with the field of view. A telescope objective will have a long focal length and small field of view. Since spherical aberration is independent of the field of view, it will obviously be important. The other, off-axis aberrations will vary in importance from coma through field curvature (astigmatism and Petzval curvature) and finally distortion. This sequence holds at small field of view angles. At large fields of view the situation reverses. Although not shown in Figure 10.1, axial color (independent of field of view) will be more important than lateral color (varies linearly with field of view).

In Chapter 4, we saw that axial color could be corrected with a cemented doublet. So that we don't confuse surfaces with lenses, we will use the subscript a for the first

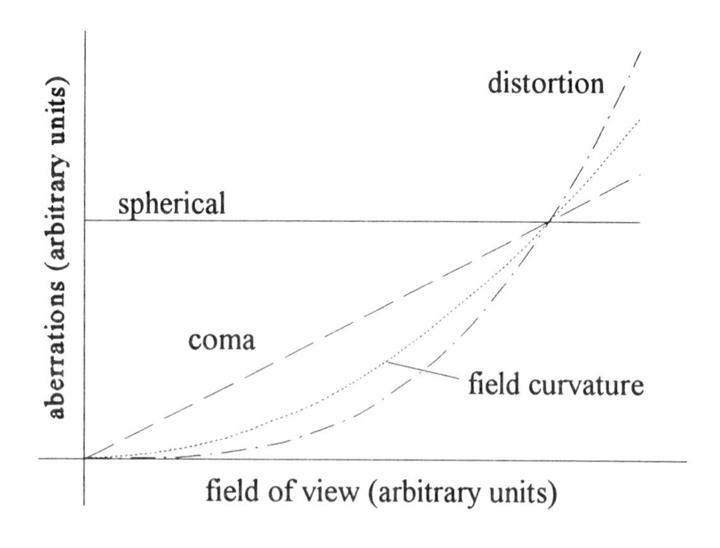

Figure 10.1 The dependence of the third-order aberrations on field of view.

lens and b for the second lens, then the focal lengths for the two thin lenses of an achromatic doublet with the target focal length are

$$f_a' = f_T' \left(\frac{V_a - V_b}{V_a} \right) \tag{10.2}$$

and

$$f_b' = f_T' \left(\frac{V_b - V_a}{V_b} \right) . \tag{10.3}$$

Also, as you should recall, the shape of the cemented doublet is still a degree of freedom. Now, in Chapter 7 we saw that spherical aberration varies dramatically with the shape of the lens. But even with the dependence of spherical aberration on lens shape, a single thin lens could not achieve zero third-order spherical. Can we bend an achromatic cemented doublet to get zero third-order spherical aberration? Remembering that

$$c = \frac{1}{(n - 1) f'} , \tag{10.4}$$

note that the thin lens equation for spherical aberration from Eq. 7.15, namely

$$S_t = -y^4 (n - 1) c \left[\frac{(n + 2)}{n} c_1^2 - (2n + 1) c c_1 - \frac{4(n + 1)}{n} v c_1 + n^2 c^2 + (3n + 1) c v + \frac{(3n + 2)}{n} v^2 \right] , \tag{10.5}$$

suggests that the spherical aberration coefficient may change sign when the focal length changes sign. The sign change should allow us to combine a positive lens with negative or undercorrected spherical aberration with a negative lens with positive or overcorrected spherical aberration to effectively cancel out the overall spherical aberration. A similar effect is what allows the cemented doublet to be corrected for axial color.

To account for spherical aberration in our cemented doublet, we need to apply Eq. 10.5 to each lens by adding the appropriate subscript to each parameter. Because the two lenses are thin and in contact, we obtain

$$y_a = y_b = y \tag{10.6}$$

and

$$v_b = v_a + \frac{1}{f_a'} . \tag{10.7}$$

Equation 10.7 is just an application of the Gaussian form of the thin lens equation where the image produced by the first lens is used as the object for the second lens. Also, because the two lenses are cemented together, the second curvature of the first lens must be the same as the first curvature of the second lens. Therefore,

$$c_{1b} = c_{1a} - \frac{1}{(n_a - 1) f_a'} . \tag{10.8}$$

Combining all of this gives a very complicated equation for the spherical aberration coefficient. We will simplify the equation by writing out its general form:

$$S_t = -y^4 \left(a c_{1a}^2 - b c_{1a} + c \right) , \tag{10.9}$$

where

$$a = \left[\frac{n_a + 2}{n_a f_a'} + \frac{n_b + 2}{n_b f_b'} \right] , \tag{10.10}$$

$$b = \left[\frac{4 (n_a + 1) v_a}{n_a f_a'} + \frac{2 n_a + 1}{(n_a - 1) f_a'^2} + \frac{4 (n_b + 1) \left(v_a + \frac{1}{f_a'} \right)}{n_b f_b'} + \frac{2 n_b + 1}{(n_b - 1) f_b'^2} + \frac{2 (n_b + 2)}{n_b (n_a - 1) f_a' f_b'} \right] , \tag{10.11}$$

and

$$c = \left[\frac{(3n_a + 2)v_a^2}{n_a f_a'} + \frac{n_a^2}{(n_a - 1)^2 f_a'^3} + \frac{(3n_a + 1)v_a}{(n_a - 1) f_a'^2} \right.$$

$$+ \frac{n_b + 2}{n_b (n_a - 1)^2 f_a'^2 f_b'} + \frac{n_b^2}{(n_b - 1)^2 f_b'^3}$$

$$+ \frac{(3n_b + 1)\left(v_a + \frac{1}{f_a'} \right)}{(n_b - 1) f_b'^2} + \frac{4(n_b + 1)\left(v_a + \frac{1}{f_a'} \right)}{n_b (n_a - 1) f_a' f_b'}$$

$$\left. + \frac{2n_b + 1}{(n_a - 1)(n_b - 1) f_a' f_b'^2} + \frac{(3n_b + 2)\left(v_a + \frac{1}{f_a'} \right)^2}{n_b f_b'} \right]. \tag{10.12}$$

Please do not confuse the coefficient c with a curvature.

As you can see, the spherical aberration coefficient is still quadratic in the first curvature of the lens c_{1a}. But here, the coefficients are much more complicated. To see how this function behaves, let's choose particular values for the index ($n_a = 1.5$) and Abbe value ($V_a = 70$) for the first lens. Let us also choose a target focal length of $f_T' = 10$ and an aperture size of $y = 1$. Next, choose $V_b = 60$. And finally, for various values of n_b I plot the spherical aberration coefficient as a function of lens shape (c_{1a}) in Figure 10.2

This figure shows several interesting features. It is quadratic in c_{1a} as expected. In fact, there are two first surface curvatures for which the third-order spherical aberration coefficient is zero. Note that as the index difference ($n_b - n_a$) increases, the curve moves upward and the two solutions move further apart. At some point, the curvatures of the lenses will become so large that the lens cannot actually be made with a given clear radius. Before that happens, the positive lens may become so thick that our thin lens assumptions break down. Therefore, there are generally upper and lower limits on the index difference which will give systems with zero third-order spherical aberration.

In a similar manner, we can fix the first lens ($n_a = 1.5$, $V_a = 70$) and the index of refraction of the second lens ($n_b = 1.6$) while allowing the Abbe value of the second lens to vary. These plots are shown in Figure 10.3.

Here we see that as the Abbe difference ($V_a - V_b$) decreases, the curve raises. Again, if the curve gets too high, the curvatures which yield zero spherical aberration can get large. Hence, there is a lower limit on practical Abbe difference values. On the other hand, as the Abbe difference increases, the curve moves downward until there are no solutions.

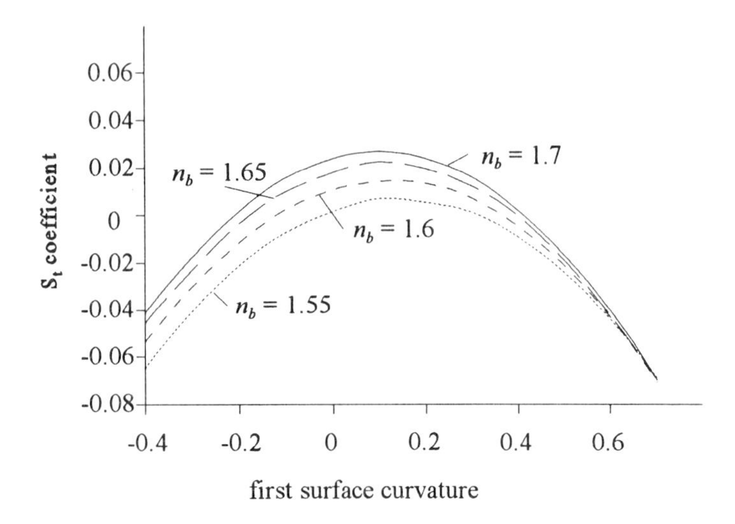

Figure 10.2 The thin lens coefficient for spherical aberration is plotted as a function of the first surface curvature for a cemented doublet.

Example 10.2

Begin the design of a spherically corrected achromat with $f'_T = 200$ cm and $f/10$, and let the first glass be BaK1 ($n_a = 1.5725$, $V_a = 57.55$).

The first step is to choose the second glass for this lens. From the S_t plots shown above, we see that an index difference of about 0.1 and an Abbe value difference of about 20 seem to work. Therefore our target glass should have $n_b = 1.67$ and $V_b = 38$. From the glass map we see that BaSF2 comes close with $n_b = 1.66446$ and $V_b = 35.83$. Because of the wide range of values over which solutions exist, we do not need to come any closer to matching values. In fact, any of a dozen or so glasses could be used with similar success. Besides, the graphs shown were based on a first glass with index 1.5 and Abbe value 70. Changing the base values should change the ranges somewhat.

Now we are ready to calculate the two thin lens focal lengths, which will make the lens achromatic:

$$f'_a = f'_T \frac{V_a - V_b}{V_a} = (200) \frac{(57.55) - (35.83)}{(57.55)} = 75.48219 ,$$

$$f'_b = f'_T \frac{V_b - V_a}{V_b} = (200) \frac{(35.83) - (57.55)}{(35.83)} = -121.2392 .$$

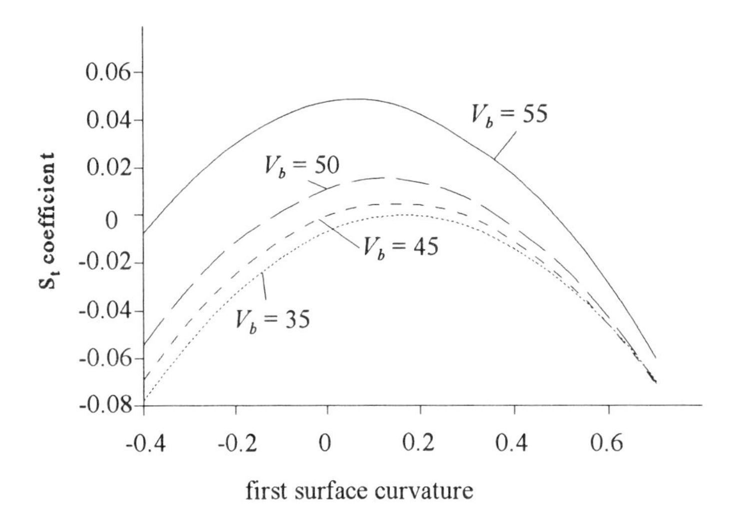

Figure 10.3 The spherical aberration coefficient versus the first surface curvature for various values of the Abbe value for the second lens.

Next, we calculate the a coefficient for the S_t equation:

$$a = \frac{n_a + 2}{n_a f_a'} + \frac{n_b + 2}{n_b f_b'}$$

$$= \frac{(3.5725)}{(1.5725)(75.48219)} + \frac{(3.66446)}{(1.66446)(-121.2392)}$$

$$= 0.01193889 .$$

The other coefficients will take up too much space to show completely. But, you can show that

$$b = 0.000173847$$

and

$$c = 0.40965841 \times 10^{-6} .$$

The solutions for Eq. 10.9 which give zero spherical aberration are

$$c_1 = 0.002957 \text{ cm}^{-1}$$

and

$$c_1 = 0.011604 \text{ cm}^{-1} .$$

The first solution, the one with the smaller curvature, can be called the "left" solution because it occurs on the left side of the S_I graph. The solution with the larger value for the first surface curvature is the "right" solution (but not necessarily "correct").

Next, the curvatures for the other two surfaces need to be calculated:

$$c_2 = c_1 - \frac{1}{(n_a - 1) f_a'}$$

$$= (0.002957) - \frac{1}{(0.5725)(75.482189)}$$

$$= -0.020184 \text{ cm}^{-1} .$$

In a similar fashion,

$$c_3 = c_2 - \frac{1}{(n_b - 1) f_b'}$$

$$= (-0.020184) - \frac{1}{(0.66446)(-121.239185)}$$

$$= -0.007771 \text{ cm}^{-1} .$$

For the "right" solution we get $c_2 = -0.011536 \text{ cm}^{-1}$ and $c_3 = 0.000877 \text{ cm}^{-1}$.

It is possible that neither of these forms may exist if the curvatures are too large for the desired clear radius.

The final step in our predesign is to insert appropriate thicknesses. From Chapter 4, we can calculate sags and determine the thicknesses to be $t_1 = 1.9$ and $t_2 = 1.3$. In general, the thickness of the first lens will not be the same for the two different solutions. We make the thicknesses the same to reduce the number of variables in the problem. The lenses are shown in Figure 10.4. ■

You should recall from Chapter 4 that there is no particular reason why the crown glass (low index, high Abbe value) must come first in the doublet. Indeed, if we

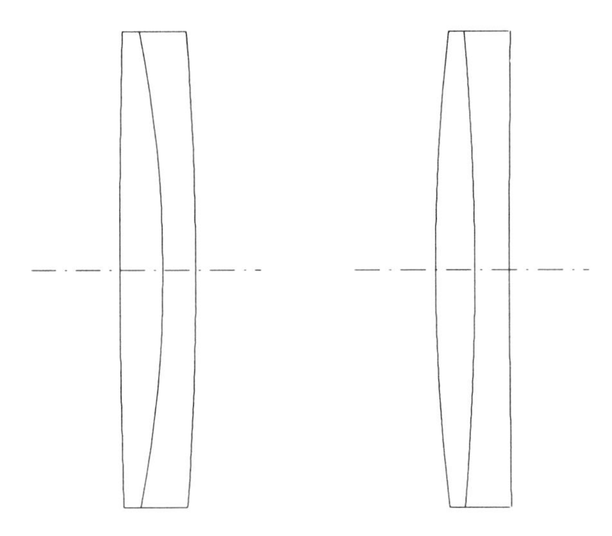

Figure 10.4 Lens drawing showing the crown first, "left" solution on the left; and the crown first, "right" solution on the right.

make the flint glass (high index, low Abbe value) the glass for the first lens, we get a new equation for S_t with new solutions.

Example 10.3

Find the flint first solutions for the problem posed in Example 10.2.

In this case, the two focal lengths are merely reversed: $f_a = -121.239185$ cm and $f_b = 75.482189$ cm. Indeed, the first coefficient $a = 0.01193889$ is the same, but the other two coefficients are different. Here, $b = -0.00025169$ and $c = 0.1153897 \times 10^{-5}$. The solutions to this quadratic are shown in the Table 10.1 below. Note that for the flint first solutions, appropriate thicknesses are $t_1 = 1.3$ cm and $t_2 = 1.9$ cm. Figure 10.5 shows lens drawings of these two solutions. ■

Longitudinal spherical aberration plots are shown in Figure 10.6 for the four achromats we just laid out. You can see that there is still some axial color and spherical aberration in each system. This is because the solutions are for thin lenses and we have inserted small but finite thicknesses. Besides, the axial color and spherical aberration equations are only approximations. When real rays are traced for the LSA plot, we should expect some difference.

In addition, each plot shows some higher-order aberrations. In particular, spherochromatism is present in each design. Spherochromatism can be defined as the chromatic variation of spherical aberration, or equivalently as the variation of

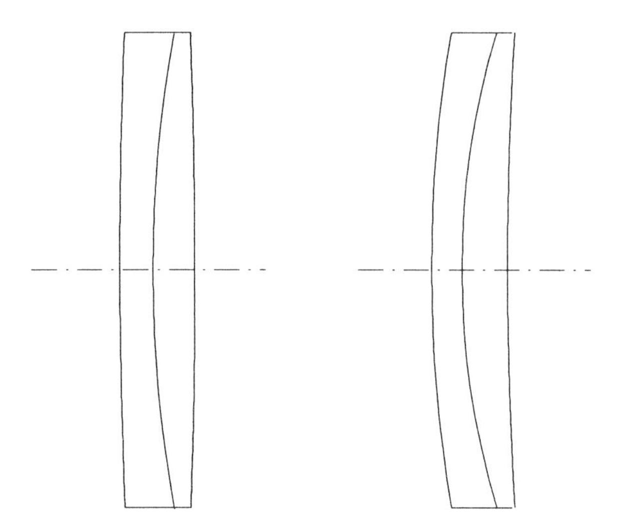

Figure 10.5 Lens drawing of the flint first, "left" solution on the left; and the flint first, "right" solution on the right.

chromatic aberration with aperture. We should not expect to correct for these higher order aberrations with such a simple system.

To complete the design of our telescope objective, we need to optimize each system using automatic optimization and then choose the best overall system from the four alternatives.

Example 10.4

Optimize the four preliminary designs using the software supplied by the publisher (Rose.exe) and choose the best overall design for the telescope objective.

The first question we need to answer is, What system parameters will be variables for the optimization? We have already been varying the three curvatures with great

Table 10.1 Summary of the preliminary solutions for spherically corrected achromat.

first glass / solution	c_1 (cm^{-1})	c_2 (cm^{-1})	c_3 (cm^{-1})
crown / "left"	0.002957	−0.020184	−0.007771
crown / "right"	0.011604	−0.011536	0.000877
flint / "left"	0.006738	0.019152	−0.003989
flint / "right"	0.014343	0.026757	0.003616

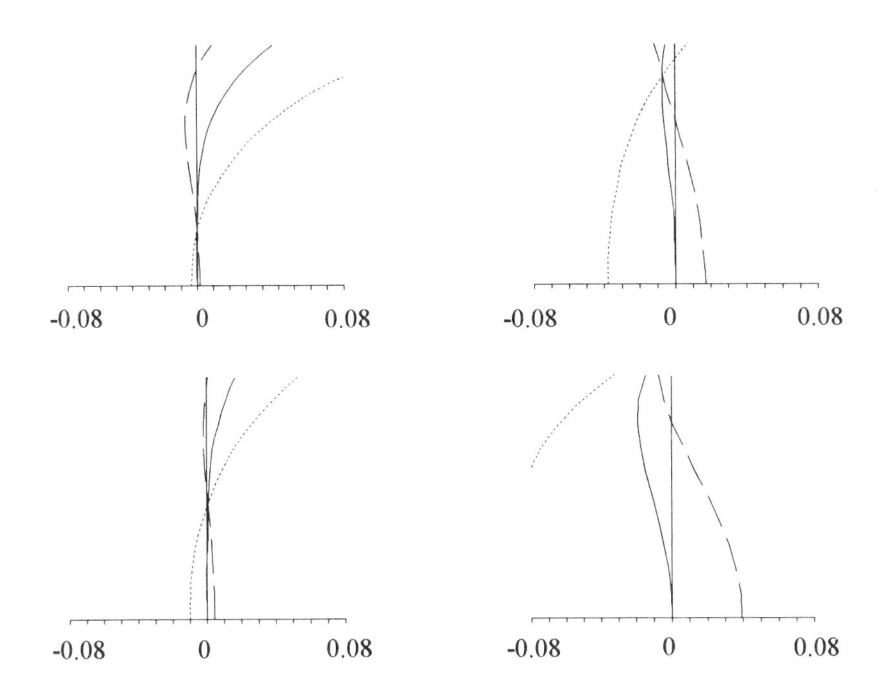

Figure 10.6 Spherical aberration plots for the four predesigns. The top plots are the crown first systems and the bottom plots are the flint first systems. The left two plots are for the lenses we have been calling the "left" solutions. The solid lines are for *d* light. The dashed curve is for *C* light, and *F* light data are shown with the dotted line.

success, but what about the other parameters? Varying the glass types is of course possible, but this choice amounts to abandoning our predesign and starting over since the predesign is so strongly dependent on the initial glass choice. We could also vary the lens thickness. Neither spherical aberration nor chromatic aberration are strongly dependent on glass thickness. It takes really thick pieces of glass to make any appreciable change in the aberrations. So we are left with the three surface curvatures.

The next question is, What should the weighted aberration function look like? Clearly we want to set the target focal length to be 200 cm. We also want to correct the spherical aberration, but at what zone? Since the majority of the light will be coming in from the edge of the lens, we want to correct the marginal spherical aberration. We also want to correct the chromatic aberration. Our predesign was intended to correct the first-order axial color, but for the final optimization it is better to try to correct the marginal chromatic aberration because more light comes through near the edge of the lens than near the center. The choice of relative weights for these three aberrations is not critical, since we can get all three to be nearly equal to zero at the same time. I used 1.0 for all three weights.

The resulting curvatures for all four systems after optimization are shown in Table 10.2. Note that the curvatures have not changed very much from the predesigns given in Table 10.1. There isn't enough of a change to show up on a lens drawing. Our predesigns were pretty good, but there was enough change in each lens to correct the focal length and to make the marginal spherical aberration and the marginal chromatic both very small.

We are now in a position to decide which of these four lenses is "best." Figure 10.7 shows LSA plots for the optimized systems. From these figures we see that the crown first, right solution and the flint first, left solution have less axial color and zonal spherical aberration than the other two solutions. Indeed, the crown first, right solution is slightly better than the flint first, left solution in terms of the zonal spherical aberration but worse in terms of axial color. In addition, the crown first, right solution has a nearly equiconvex shape for the crown element. If this element is made exactly equiconvex, there is only a small change in the aberrations, but the lens becomes much easier to make and mount. All of the lenses except for the crown first, left solution have at least one surface with a curvature larger than 0.02 cm^{-1}. This is important, because the larger the curvature, the fewer the lenses that can be mounted at one time on a polishing tool and the lower the amount of material that needs to be removed. Typically, smaller curvatures are less expensive to manufacture than larger curvatures. For these reasons, I would go with the crown first, right solution. ■

It is very important to note that there are actually four unique systems which correct for axial color and spherical aberration. If we had merely entered an arbitrary system into our optimization code, we would have found one of these solutions. But, because

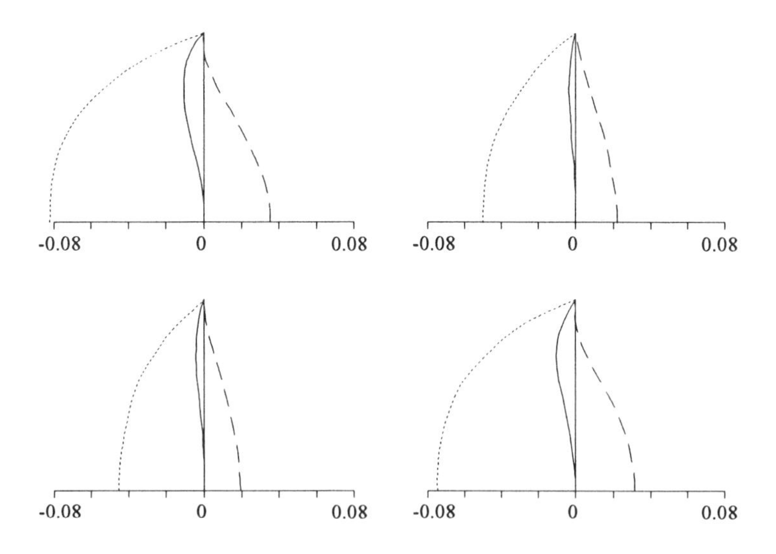

Figure 10.7 Spherical aberration plots for the four optimized systems. Again, the top plots are the crown first systems and the bottom plots are the flint first systems. The left two plots are for the lenses we have been calling the "left" solutions. The solid lines are for *d* light. The dashed curve is for *C* light, and *F* light data are shown with the dotted line.

Table 10.2 Summary of curvatures for optimized systems.

first glass / solution	c_1 (cm^{-1})	c_2 (cm^{-1})	c_3 (cm^{-1})
crown / "left"	0.003174	-0.019523	-0.007503
crown / "right"	0.011488	-0.011503	0.000804
flint / "left"	0.006781	0.019042	-0.003938
flint / "right"	0.014231	0.026610	0.003598

the optimization routine only finds the nearest local minimum, we would not have found the other solutions. For example, if we had started out with the first surface being planar and the crown glass coming first, we would find the crown first, left solution. As we have seen, this solution is inferior to the crown first, right solution. The existence of other solutions was discovered only because we applied a more fundamental understanding of geometrical optics.

10.2 THE LANDSCAPE LENS

Our next problem is to design a lens with a much larger field of view than a telescope objective. Let's assume we want a very simple photographic lens with HFOV = 25°. This lens is intended for taking pictures of landscapes, so the speed of the lens is not important (we can have a large $f/\#$). As shown in Figure 10.1, as the field angle increases, first coma and then field curvature become increasingly important to correct. At 25° we can continue to ignore distortion. Since we want to keep our lens simple and we can stop the lens down, we will not even try to correct spherical aberration.

To see where to begin the design, let's look at the thin lens aberration coefficients. Equation 10.5 is the coefficient for spherical aberration. Remember that for a single thin lens the spherical aberration coefficient varies quadratically with the shape of the lens and that it is always negative.

Next, consider coma. From Eq. 7.25 we have

$$C_t = C_t^* + \left(\frac{\bar{y}}{y} \right) S_t \, , \tag{10.13}$$

where $\bar{y}$ is the height of the full-field chief ray at the lens, y is the height of the marginal axial ray at the lens, and also from Chapter 7 we have

$$C_t^* = y^3(n-1)u_{HFOV}\left[nc^2 - \left(\frac{n+1}{n}\right)c_1 + \left(\frac{2n+1}{n}\right)v\right]. \tag{10.14}$$

In this equation, we introduce u_{HFOV} as the half field of view angle.

Next, for astigmatism the equation is

$$A_t = A_t^* + 2\left(\frac{\bar{y}}{y}\right)C_t^* + \left(\frac{\bar{y}}{y}\right)^2 S_t\,, \tag{10.15}$$

where

$$A_t^* = -\frac{(y\,u_{HFOV})^2}{f'} = -y\,u_{HFOV}(n-1)c\,. \tag{10.16}$$

This equation shows that the astigmatism for a thin lens acting as the aperture stop is independent of the shape of the lens.

Finally, for Petzval curvature we have

$$P_t = -\frac{1}{n f'}\,. \tag{10.17}$$

Again, we note that Petzval curvature is independent of the shape of the lens. But we also see that the Petzval curvature is independent of the location of the stop, unlike coma and astigmatism.

From these equations, we see that it may be possible to correct coma and astigmatism by placing a stop in a position away from the lens itself. The key will be to ensure that the spherical aberration is not too small. Therefore, our system will consist of single thin lens with a separate stop. The stop may come before or after the lens. For the moment, we will assume that the stop is in front of the lens. Since the lens is in the rear, this is called a *rear landscape lens*. The shape of the lens c_1 and the location of the stop t_1 will be our free parameters.

The next step is to look carefully at exactly what we want to correct. Remember that astigmatism and Petzval curvature have the same functional dependence on ray parameters. If we arrange our system so that $A_t = 0$, then we are still left with an aberration of the order of $P_t\Lambda^2$. On the other hand, we can try to correct the sagittal field curvature so that

$$A_t + P_t\Lambda^2 = 0\,. \tag{10.18}$$

If this is the case, then the tangential field curvature is not well corrected and we have

$$3A_t + P_t\Lambda^2 = -2P_t\Lambda^2 . \tag{10.19}$$

Remember that a single thin lens has an inward curving Petzval surface. Therefore, if we try to correct the sagittal field curvature, we end up with a strong outward curving tangential field. On the other hand, if we try to correct the tangential field so that

$$3A_t + P_t\Lambda^2 = 0 , \tag{10.20}$$

then we end up with a weak inward curving sagittal field given by

$$A_t + P_t\Lambda^2 = \frac{2}{3}P_t\Lambda^2 . \tag{10.21}$$

Therefore, the best choice to minimize field curvature overall is to correct the tangential field curvature.

The next step is to find a starting lens prescription based on our third-order analysis. We want to set coma (Eq. 10.13) to zero. We also want the tangential field curvature (Eq. 10.20) to be zero. This gives us two equations, but neither equation contains the unknown stop position t_1. The stop position is given by

$$t_1 = \bar{y}\, u_{HFOV} . \tag{10.22}$$

Equation 10.16 can be solved for t_1:

$$t_1 = \frac{nc - \dfrac{(n+1)}{n}c_1}{n^2c^2 - (2n+1)cc_1 + \dfrac{(n+2)}{n}c_1^2} . \tag{10.23}$$

This can be substituted into Eq. 10.20 and after some simplification we get a quadratic in c_1 as expected:

$$(n-1)c_1^2 - (n-1)ncc_1 + n^3c^2 = 0 . \tag{10.24}$$

The only problem with this equation is that it doesn't have any real solutions. As you can see after some manipulation, the discriminant is given by

$$-(3n+1)(n-1)n^2c^2\,, \tag{10.25}$$

which is negative for any real glass. A negative discriminant means that there are no real solutions. It turns out that the problem lies in using the thin lens versions of the third-order aberration coefficients.

What we need to do is look at the system in terms of individual surfaces rather than as a thin lens and a stop. We continue to assume that the first surface is a stop. The focal length of the lens is given by

$$\frac{1}{f'} = (n-1)\left[c_2 - c_3 + \frac{(n-1)}{n}c_2c_3t_2\right]. \tag{10.26}$$

We want coma to be zero:

$$C = \sum_{j=1}^{3} W_j i_j \bar{i}_j = 0\,, \tag{10.27}$$

where

$$W_j = y_j n_{j-1}\left(\frac{n_{j-1}}{n_j} - 1\right)(u_j + i_j)\,. \tag{10.28}$$

And the tangential field curvature should be zero:

$$\sum_{j=1}^{3}\left[3W_j\bar{i}_j^2 + P_j\Lambda^2\right] = 0\,. \tag{10.29}$$

Each individual Petzval term is given by

$$P_j = \frac{n_{j-1} - n_j}{n_{j-1}n_j}c_j\,. \tag{10.30}$$

To make use of these equations, we need trace a marginal axial ray and a full-field chief ray. The marginal axial ray is defined by $u_0 = 0$ and $y_1 = R_E$. Using the standard paraxial ray tracing equations allows us to determine the values for the ray parameters at each surface. Thus, $u_1 = 0$, $y_2 = R_E$ and

$$u_2 = -\frac{(n-1)}{n}R_E c_2\,. \tag{10.31}$$

Furthermore,

$$i_2 = R_E c_2 \tag{10.32}$$

and

$$y_3 = R_E \left[1 - \frac{(n - 1)}{n} c_2 t_2 \right] . \tag{10.33}$$

Finally,

$$u_3 = -(n - 1) R_E \left[c_2 - c_3 + \frac{(n - 1)}{n} c_2 c_3 t_2 \right] \tag{10.34}$$

and

$$i_3 = R_E \left[-\frac{(n - 1)}{n} c_2 + c_3 - \frac{(n - 1)}{n} c_2 c_3 t_2 \right] . \tag{10.35}$$

With the aperture in front of the lens the full-field chief ray is given by $\bar{u}_0 = u_{HFOV}$ and $\bar{y}_1 = 0$. Ray tracing gives $\bar{u}_1 = u_{HFOV}$ and

$$\bar{y}_2 = t_1 u_{HFOV} . \tag{10.36}$$

Continuing on we have

$$\bar{u}_2 = \frac{u_{HFOV}}{n} \left[1 - (n - 1) c_2 t_1 \right] \tag{10.37}$$

and

$$\bar{i}_2 = u_{HFOV} (1 + c_2 t_1) . \tag{10.38}$$

At the third surface we have

$$\bar{y}_3 = u_{HFOV} \left[t_1 + \frac{t_2}{n} \left(1 - (n - 1) c_2 t_1 \right) \right] , \tag{10.39}$$

$$\bar{u}_3 = u_{HFOV}\left[1 - (n - 1)t_1\left(c_2 - c_3\right) + \frac{c_3 t_2}{n}\left(1 - (n - 1)c_2 t_1\right)\right], \quad (10.40)$$

and

$$\bar{i}_3 = u_{HFOV}\left[\frac{1}{n} - \frac{(n - 1)}{n}c_2 t_1 + c_3 t_1 + \frac{c_3 t_2}{n}\left(1 - (n - 1)c_2 t_1\right)\right]. \quad (10.41)$$

We can now substitute all of our ray trace results into Eq. 10.27 to get

$$\begin{aligned} & n c_2^2(1 + c_2 t_1) - \left[n - (n - 1)c_2 t_2\right] \\ & \times \left[(n^2 - 1)c_2 - n^2 c_3 + n(n - 1)c_2 c_3 t_2\right] \\ & \times \left[(n - 1)c_2 - c_3(n - (n - 1)c_2 t_2)\right] \\ & \times \left[1 - (n - 1)c_2 t_1 + c_3\left(n t_1 + t_2(1 - (n - 1)c_2 t_1)\right)\right] = 0 . \quad (10.42) \end{aligned}$$

Making the same substitutions into Eq. 10.29 gives

$$\begin{aligned} & 3 n c_2(1 + c_2 t_1)^2 + 3\left(n - (n - 1)c_2 t_2\right) \\ & \times \left[(n^2 - 1)c_2 - n^2 c_3 + n(n - 1)c_2 c_3 t_2\right] \\ & \times \left[1 - (n - 1)c_2 t_1 + c_3\left(n t_1 + t_2(1 - (n - 1)c_2 t_1)\right)\right]^2 \\ & + n^2(c_2 - c_3) = 0 . \quad (10.43) \end{aligned}$$

The solution to Eqs. 10.26, 10.42 and 10.43 for c_2, c_3, and t_2 will be the predesign for our front landscape lens.

Example 10.5

Design an $f/$ 15, 50-mm rear landscape lens.

A 50-mm lens used for photography with 35-mm film gives a half field of view of just under 24°. To be conservative we will use an HFOV angle of 25°. The focal length and $f/\#$ determine R_E = 1.6667 mm.

For our glass, we arbitrarily choose BK10.

Next, we need to determine an appropriate lens thickness. The lens thickness will depend on the clear radius of the lens which in turn depends on how far the stop is from the lens (assuming no vignetting is allowed). Since we do not yet know the stop position, we can only guess at the appropriate lens thickness. Let's assume that the clear radius of the lens is about three times larger than the stop (i.e., R_{lens} = 5.00 mm). Furthermore, assume a meniscus shape for the lens such that c_1 = -0.01 mm^{-1}. The thin lens formula gives $c_2 = -0.05$ mm^{-1}. Our lens thickness, then, is about 1.50 mm. Substituting values for n, f', and t_2 into Eqs. 10.26, 10.42 and 10.43 leaves us with three fairly complicated equations for the three unknowns c_2, c_3, and t_2.

The only practical way to solve these equations is with a computer algebra system such as Maple. There are actually a total of nine solutions. Four of the solutions are complex. Of the remaining five solutions, three have extremely large curvatures. The remaining solutions are: t_1 = 7.272291 cm, c_2 = -0.0466189 cm^{-1}, c_3 = -0.0848226 cm^{-1}, and t_1 = -6.550634 cm, c_2 = 0.1352323 cm^{-1}, c_3 = 0.1019291 cm^{-1}. Clearly, the first of these two solutions is the one we want, but the second solution indicates that it may be possible to design a landscape lens with the aperture behind the lens.

To optimize, we make the separation between the stop and the lens and the two lens curvatures variables. Optimize on tangential coma and tangential field curvature with weights of 10.0 each. Set the target focal length to 50 and set its weight to 1.0. You can experiment with different weights. But if the aberration weights are much smaller relative to the focal length weight, convergence becomes very slow. After three iterations we obtain t_1 = 7.909561 cm, c_2 = -0.038862 cm^{-1}, and c_3 = -0.077535 cm^{-1}.

The next step is to go back and check the clear radius and thickness of the lens. If we want to avoid vignetting, we would need to make the clear radius a little larger and make the lens a little thicker. The change in thickness will throw off our optimization slightly. After optimizing again we get a final system t_1 = 7.920738 cm, t_2 = 1.75, $R_2 = R_3$ = 6.0 cm, c_2 = -0.037075 cm^{-1}, and c_3 = -0.075620 cm^{-1}. After having done this once, I could have made it all look easier by simply starting with t_2 = 1.75 and $R_2 = R_3$ = 6.0 cm, but I wanted to demonstrate that you can stop in the middle of an optimization and change the system if needed.

Figure 10.8 shows sagittal and tangential ray intercept plots for the final optimized system on axis and at full-field. As these plots clearly show, there is some undercorrected spherical aberration in the system and some inward sagittal field curvature, but no hint of tangential field curvature.

As mentioned earlier, one solution to our equations indicates there should be another solution to our problem with the stop behind the lens. This is called a *front landscape lens*. We should derive a new set of equations for the front landscape lens, but the negative solution for the rear landscape lens will be close enough to begin the optimization.

The first step is to enter a lens system with t_1 = 1.75 cm, t_2 = 6.550634, R_1 = R_2 = 6.0 cm, c_1 = 0.135232 cm^{-1}, and c_2 = 0.101929 cm^{-1}. We could do a calculation to determine the proper radius for the aperture stop, but it is easier to first guess and then use the software to find R_{AS} = 1.336 cm. To optimize this system, we set the first and second curvatures as variables as well as the separation between the lens and the stop. The distance between the stop and the image will change as the optimization proceeds, but we do not need to mark this separation as a variable. It is automatically

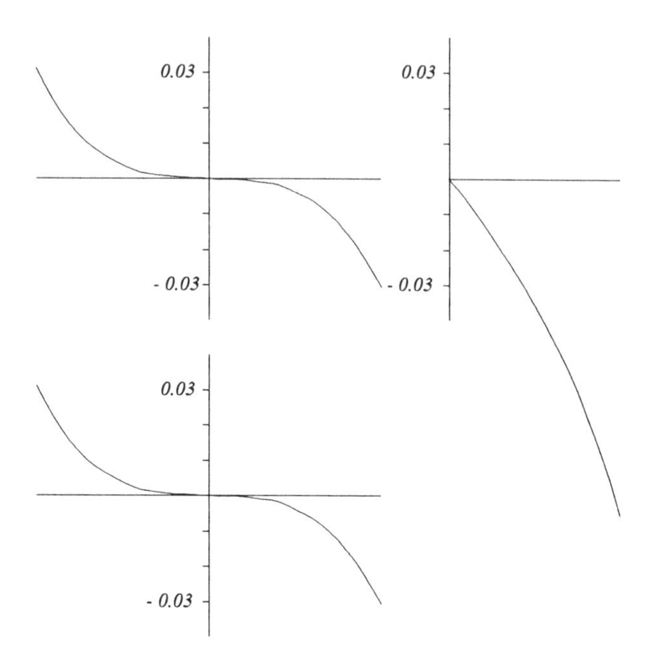

Figure 10.8 Ray intercept plots for the rear landscape lens. The top two plots are off axis.

adjusted by the software. Using the same weights as with the rear landscape lens (coma weight = 10.0, tangential field weight = 10.0, target focal length weight = 1.0), we get a solution after five iterations. We have t_2 = 4.541339, $c_1 = 0.123407$ cm^{-1}, and $c_2 = 0.089668$ cm^{-1}. Because the aperture stop has moved, it must be resized. We now have R_{AS} = 1.595 cm. Changing the size of the aperture stop causes only a minor change in the aberrations, and it really is not necessary to optimize again.

The ray intercept plot for the front landscape lens is shown in Figure 10.9.

Finally, to compare the two systems, let's look first at the aberrations. From the ray intercept plots, it is clear that the rear landscape lens is better than the front landscape lens. Also, the front landscape lens has more distortion than the rear landscape. The deeper surface curvatures on the front landscape is responsible for the larger aberrations. The deeper curvatures on the front landscape lens also means it will be more expensive to manufacture. Both lenses are shown in Figure 10.10.

But the front landscape does have its own advantages. If the stop is to be a mechanical iris rather than a fixed stop, then locating the stop behind the lens would protect the iris. Also, the front landscape lens is physically shorter than the rear landscape lens. ■

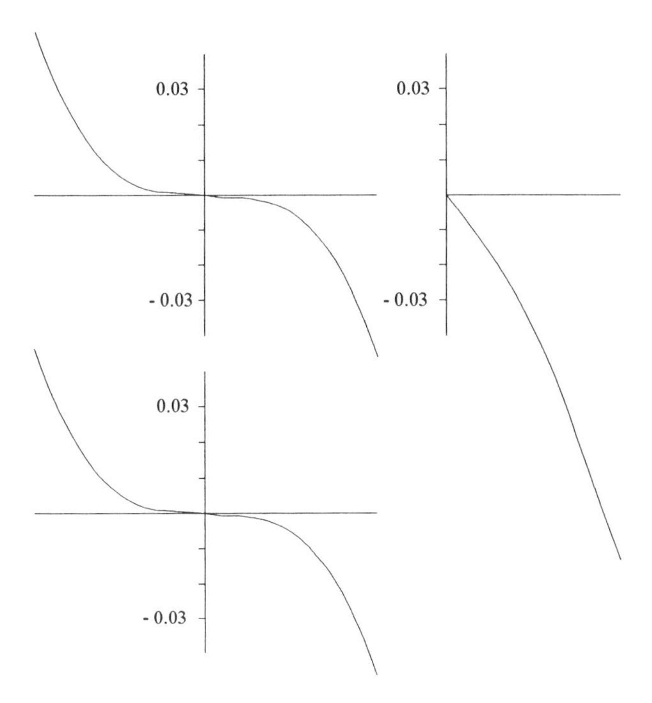

Figure 10.9 Ray intercept plots for the front landscape lens.

10.3 THE COOKE TRIPLET

For our final lens, let's design a lens with a longer focal length and smaller $f/\#$ than the landscape lens. Because this lens will have both a fairly large aperture and large field of view, it must be well corrected for all of the first- and third-order aberrations. Since there are two first-order chromatic aberrations and five third-order aberrations and the focal length to be considered, we will need a system with a total of at least eight free parameters. The Cooke triplet is one way of gaining these eight free parameters. This system consists of three separate lenses. Thus we have three lens powers, three lens shapes (or six curvatures), and two separations as our free parameters. We will not use glass choice as a free parameter.

We begin our analysis by writing out the thin lens equation for the overall power (reciprocal of the focal length) of our system

$$y_a \phi_a + y_b \phi_b + y_c \phi_c = y_a \phi_T , \tag{10.44}$$

where the subscripts a, b, and c refer to the first, second, and third lenses, respectively. We don't number the lenses, because we will eventually number the surfaces and this notation avoids confusion. The subscript T refers to the total power of the system. The y values refer to a marginal axial ray. This equation can be

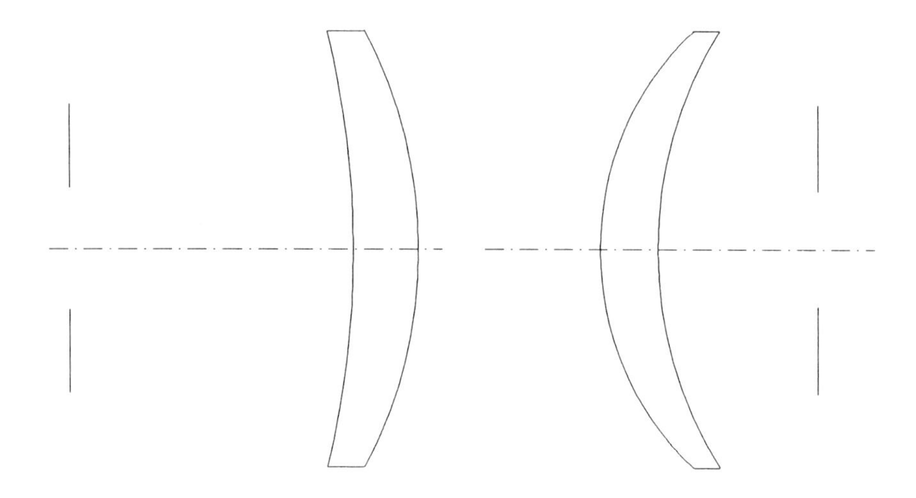

Figure 10.10 The rear landscape lens is shown on the left and the front landscape lens is shown on the right.

derived most easily by setting up a *y-nu* ray trace to determine the focal length of the system. Note that because the lenses are separated we have $y_a \neq y_b \neq y_c$.

Next, we can write out the equation for the coefficient of axial color for a system consisting of three thin lenses. From Appendix D we have

$$AC_t = -\frac{y_a^2 \Phi_a}{V_a} - \frac{y_b^2 \Phi_b}{V_b} - \frac{y_c^2 \Phi_c}{V_c} . \tag{10.45}$$

For lateral color the equation is

$$LC_t = -\frac{y_a \bar{y}_a \Phi_a}{V_a} - \frac{y_b \bar{y}_b \Phi_b}{V_b} - \frac{y_c \bar{y}_c \Phi_c}{V_c} . \tag{10.46}$$

In this equation the $\bar{y}$ values refer to the full-field chief ray.

Next, come the third-order aberrations. These equations become quite lengthy, so I will break them down into smaller parts. The coefficient for spherical aberration for one lens can be written

$$S_a = -y_a^4 \left(\frac{n_a^2 \phi_a^2}{(n_a - 1)^2} - \frac{(2n_a + 1) c_1 \phi_a}{n_a - 1} + \frac{(3n_a + 1) v_a \phi_a}{n_a - 1} + \frac{(n_a + 2) c_1^2}{n_a} - \frac{4(n_a + 1) v_a c_1}{n_a} + \frac{(3n_a + 2) v_a^2}{n_a} \right), \tag{10.47}$$

where c_1 is the curvature of the first surface of the first lens. The reciprocal of the object distance is v_a. The spherical aberration coefficient for the entire system consists of three terms, one for each lens

$$S = S_a(y_a, n_a, \phi_a, c_1, v_a) + S_b(y_b, n_b, \phi_b, c_3, v_b) + S_c(y_c, n_c, \phi_c, c_5, v_c) . \tag{10.48}$$

In this formula, I have shown all of the variables that appear in each of the terms.

The coma coefficient for one lens is given by

$$C_a = y_a^3 \left(\bar{u}_0 + v_a \bar{y}_a \right) \left(\frac{n_a \phi_a}{n_a - 1} - \frac{(n_a + 1) c_1}{n_a} + \frac{(2n_a + 1) v_a}{n_a} \right) \phi_a . \tag{10.49}$$

Even though this formula is slightly simpler than the corresponding formula for spherical aberration, the overall system equation is much more complicated because of the stop shift. The system coma coefficient is given by

$$C = C_a(y_a, \bar{y}_a, n_a, \phi_a, c_1, v_a, \bar{u}_0) + \left(\frac{\bar{y}_a}{y_a} \right) S_a + C_b(y_b, \bar{y}_b, n_b, \phi_b, c_3, v_b, \bar{u}_a) + \left(\frac{\bar{y}_b}{y_b} \right) S_b + C_c(y_c, \bar{y}_c, n_c, \phi_c, c_5, v_c, \bar{u}_b) + \left(\frac{\bar{y}_c}{y_c} \right) S_c . \tag{10.50}$$

As you can imagine, writing this equation out in its full form would take up a lot of space. Don't worry, things just get worse.

For astigmatism, the coefficient for a thin lens looks like

$$A_a = -y_a^2 (\bar{u}_0 + v_a \bar{y}_a)^2 \phi_a , \tag{10.51}$$

which is relatively simple, but for the entire system we get

$$A = A_a(y_a, \bar{y}_a, \phi_a, \bar{u}_0) + 2\left(\frac{\bar{y}_a}{y_a}\right) C_a + \left(\frac{\bar{y}_a}{y_a}\right)^2 S_a$$

$$+ A_b(y_a, \bar{y}_a, \phi_a, \bar{u}_0) + 2\left(\frac{\bar{y}_b}{y_b}\right) C_b + \left(\frac{\bar{y}_b}{y_b}\right)^2 S_b$$

$$+ A_c(y_a, \bar{y}_a, \phi_a, \bar{u}_0) + 2\left(\frac{\bar{y}_c}{y_c}\right) C_c + \left(\frac{\bar{y}_c}{y_c}\right)^2 S_c \,. \tag{10.52}$$

Next, for Petzval curvature, we have

$$P = \frac{\phi_a}{n_a} + \frac{\phi_b}{n_b} + \frac{\phi_c}{n_c} \,. \tag{10.53}$$

for the entire system.

And finally, for distortion we have

$$D = \left(\frac{\bar{y}_a}{y_a}\right)\left(3A_a + \phi_a \Lambda^2\right) + 3\left(\frac{\bar{y}_a}{y_a}\right)^2 C_a + \left(\frac{\bar{y}_a}{y_a}\right)^3 S_a$$

$$+ \left(\frac{\bar{y}_b}{y_b}\right)\left(3A_b + \phi_b \Lambda^2\right) + 3\left(\frac{\bar{y}_b}{y_b}\right)^2 C_b + \left(\frac{\bar{y}_b}{y_b}\right)^3 S_b$$

$$+ \left(\frac{\bar{y}_c}{y_c}\right)\left(3A_c + \phi_c \Lambda^2\right) + 3\left(\frac{\bar{y}_c}{y_c}\right)^2 C_c + \left(\frac{\bar{y}_c}{y_c}\right)^3 S_c \,. \tag{10.54}$$

We now have our complete set of eight equations, but there is a lot more work which needs to be done. We must establish which of the numerous parameters in these equations are known and which are unknown. First, we can give all of the total, system aberration coefficients zero values. As with the telescope objective, our final lens will not have zero first- and third-order aberrations, rather we will use these aberrations to counteract higher order aberrations. Setting these coefficients to zero merely gives us a place to start. Another parameter that we know is the overall power of the system ϕ_T.

Next, we can arbitrarily select glasses for our lenses since we have enough other free parameters. So we will assume that the indices and Abbe values for all three lenses are known. Also, we will take the original object at infinity. This leads to

$v_a = 0$ and $u_0 = 0$. Setting the $f/\#$ of the system allows us to determine the value for y_a.

With the initial values for the marginal axial ray determined, we can use *y-nu* ray tracing to determine the ray parameters at the other surfaces in terms of system parameters. After the first lens we have

$$u_a = -y_a \phi_a \tag{10.55}$$

and

$$y_b = y_a - t_a y_a \phi_a . \tag{10.56}$$

After the second lens we obtain

$$u_b = -y_a \phi_a - \left(y_a - t_a y_a \phi_a\right)\phi_b \tag{10.57}$$

and

$$y_c = y_a - t_a y_a \phi_a - t_b (y_a \phi_a + (y_a - t_a y_a \phi_a)\phi_b) . \tag{10.58}$$

Substituting Eqs. 10.56 and 10.58 into our set of equations allows us to express these equations in terms of the system parameters t_a and t_b.

We can also express the vergences for the second two lenses in the following ways

$$v_b = -\frac{u_a}{u_b} = \frac{y_a \phi_a}{y_a - t_a y_a \phi_a} \tag{10.59}$$

and

$$v_c = -\frac{u_b}{y_c} = \frac{y_a \phi_a + (y_a - t_a y_a \phi_a)\phi_b}{y_a - t_a y_a \phi_a - t_b (y_a \phi_a + (y_a - t_a y_a \phi_a)\phi_b)} . \tag{10.60}$$

Finally, we look at the full-field chief ray. Our set of equations can be greatly simplified by requiring one of the lenses to be the aperture stop. If we choose the second lens as the aperture stop, then $\bar{y}_b = 0$ and several terms in our equations can simply be dropped. We choose the second lens as the stop, because we know distortion and lateral color will be reduced the more symmetrical the system is about the aperture stop. Also, if the full-field angle $\bar{u}_0$ is specified, then the other chief ray parameters can be determined. After the first lens we have

$$\bar{y}_a = \frac{t_a \bar{u}_0}{t_a \phi_a - 1} \tag{10.61}$$

and

$$\bar{u}_a = \bar{u}_0 - \frac{t_a \bar{u}_0 \phi_a}{t_a \phi_a - 1} . \tag{10.62}$$

After the second lens we have

$$\bar{u}_b = \bar{u}_a \tag{10.63}$$

and

$$\bar{y}_c = t_b \left(\bar{u}_0 - \frac{t_a \bar{u}_0 \phi_a}{t_a \phi_a - 1} \right) . \tag{10.64}$$

After all of these substitutions and some simplification, we are left with eight equations in the eight unknowns ϕ_a, ϕ_b, ϕ_c, c_1, c_3, c_5, t_a, and t_b. This system of equations is much too complicated to try to list here, let alone try to solve by hand. As a matter of fact, it was too complicated for my computer algebra program to solve as it stands. The program ran out of memory. But there is hope. The equations for the overall power, axial and lateral color, and Petzval curvature do not contain the shape parameters c_1, c_3, or c_5. This means we have a set of four equations in five unknowns. We need another equation! Remember that it is desirable for correction of distortion and lateral color to have a symmetrical system, therefore let

$$\frac{t_a}{t_b} = 1.0 . \tag{10.65}$$

Then, we can solve the remaining four equations in four unknowns. This is simple enough that the computer algebra program can handle the calculations.

Since our equations are not linear in the unknowns, there exists the possibility that there will be more than one solution. We will pick the solution which gives the shortest overall length (smallest value for t_a). This is not an arbitrary choice. The other solutions are ridiculously long. A very long system will prevent us from achieving our goals for the field of view without a lot of vignetting.

The next step is to substitute the thin lens powers and separations into the remaining equations. We now have four equations in the three remaining unknowns.

If we use only the equations for spherical aberration, coma, and astigmatism, we can have the computer algebra system solve for the three lens shapes.

Finally, we evaluate the remaining aberration, distortion. We don't expect distortion to be zero, but we can change the ratio of the lens separations t_a and t_b and solve for a new set of lens powers and shapes. By repeating this process, we can find a ratio of the lens separations which gives a small (if not exactly zero) value for the distortion.

Example 10.6

Design a Cooke triplet with $f' = 10.0$ cm and $f/4.5$. Use SK4 glass for the first and third lens and F7 for the middle lens. The middle lens should be the aperture stop.

To begin the design process, we first calculate the overall power and the height of the marginal axial ray at the first lens:

$$\phi_T = \frac{1}{f'} = \frac{1}{10.0} = 0.1 ,$$

$$y_a = \frac{f'}{2 f/\#} = \frac{10.0}{2 \times 4.5} = 1.111111 .$$

Next, we write out the first five equations that need to be solved. The power equation becomes

$$\phi_a + \phi_b + \phi_c - t_c \phi_a (\phi_b + \phi_c) - t_b \phi_c (\phi_a + \phi_b - t_a \phi_a \phi_b) = 0.1 .$$

The equation for axial color is

$$2084.8828\, \phi_a + 3437.4769\, \phi_b (1 - t_a \phi_a)^2$$
$$+ 2084.8828\, \phi_c (1 - t_a \phi_a - t_b (\phi_a + \phi_b - t_a \phi_a \phi_b))^2 = 0 .$$

The equation for lateral color simplifies to

$$58.62\, t_a \phi_a - 58.63\, t_b \phi_c (1 - t_a \phi_a - t_b (\phi_a + \phi_b - t_a \phi_a \phi_b)) = 0 .$$

The Petzval equation is relatively simple:

$$0.620070\, \phi_a + 0.615248\, \phi_b + 0.620070\, \phi_c = 0 .$$

The last of this set of equations is the fudging equation:

$$t_a = t_b \ .$$

The Maple symbolic math program gives two solutions to these equations. One solution has t_a = 19.9987 cm, which is much too large. The other solution is $t_a = t_b$ = 1.175258 cm, ϕ_a = 0.230803 cm^{-1}, ϕ_b = −0.496203 cm^{-1}, and ϕ_a = 0.261541 cm^{-1}.

Next, we substitute the separations and powers we just found into three of the remaining equations. The spherical aberration equation is

$$-0.047295 + 0.559916\, c_1 - 0.788038\, c_1^2 + 0.282952\, c_3$$

$$+ 0.475770\, c_3^2 + 0.203827\, c_5 - 0.541568\, c_5^2 = 0 \ .$$

The coma equation is

$$0.192956 - 0.462723\, c_1 + 0.399258\, c_1^2 + 0.203813\, c_3$$

$$- 0.040978\, c_5 - 0.310926\, c_5^2 = 0$$

and the astigmatism equation is

$$-0.082925 + 0.325149\, c_1 - 0.202283\, c_1^2 - 0.114238\, c_5$$

$$- 0.178510\, c_5^2 = 0 \ .$$

Again, there are two solutions. One solution has a very large curvature on the first surface. The first lens is a meniscus shape. This shape will not be good for higher-order aberrations, so we will keep only the solution with smaller curvatures. The second solution has c_1 = 0.278378 cm^{-1}, c_3 = −0.472800 cm^{-1}, and c_5 = −0.081051 cm^{-1}. With these curvatures and the individual lens powers, we can calculate the remaining curvatures:

$$c_2 = c_1 - \frac{\phi_a}{n_a - 1} = 0.278378 - \frac{0.230803}{1.61271 - 1} = -0.098308 \text{ cm}^{-1} \ .$$

Similarly, c_4 = 0.320668 cm^{-1} and c_6 = −0.507903 cm^{-1}. The last lens here is a meniscus lens, but the curvature of the last surface is not as large as the first surface curvature of the other solution.

The next step is to evaluate the distortion coefficient. With the separations, powers and curvatures given above the distortion coefficient has the value 0.028539. This is not zero, so we try another value for the separation ratio. Repeating all of the calculations described above with $t_a = 1.5\ t_b$ we find that distortion is reduced to 0.015722. A few more trials along these lines shows that for $t_a = 1.9\ t_b$ the distortion coefficient is reduced to 0.000895 which is close enough for our preliminary design.

At this point we have $c_1 = 0.244616\ \text{cm}^{-1}$, $c_2 = -0.014276\ \text{cm}^{-1}$, $c_3 = -0.445654\ \text{cm}^{-1}$, $c_4 = 0.343634\ \text{cm}^{-1}$, $c_5 = 0.063224\ \text{cm}^{-1}$, $c_6 = -0.477191\ \text{cm}^{-1}$, $t_a = 1.705916$ cm, and $t_b = 0.897850$ cm. This completes the first step in our design process. ■

We now have a thin lens system with very small third-order aberrations. The system can be easily tested by entering the curvatures and separations we just found into a lens analysis program. We also need to enter zero thickness after the first, third and fifth surfaces since the lenses are thin. With the system properly entered, we can see that, indeed, the chromatic and third-order aberrations are zero with the exception of distortion.

But real systems need finite thicknesses. Of course the lens thickness depends on the clear radius of the lens. We will purposely introduce some vignetting into our system to keep the lenses relatively thin and also to eliminate some rays with large aberrations. Introducing finite lens thicknesses will throw off all of our aberrations (except for Petzval curvature) and the focal length. But that is all right because the final step is to optimize the system with software. For optimization, all of the curvatures and the two lens separations will be variables. Our aberration function consists of all of the first- and third-order aberrations and the target focal length.

Example 10.7

We are now ready to finish the design of the lens we started in Example 10.6.

The curvatures of the predesign system are given in Example 10.6. So are the separations between the three lenses. Here we need to add reasonable clear radii and lens thicknesses. For the first lens, $R_1 = R_2 = 1.20$ cm, $t_1 = 0.40$ cm. For the second and third lenses, $R_3 = R_4 = 0.7756905$ cm, $t_3 = 0.25$ cm and $R_5 = R_6 = 1.20$ cm, $t_5 = 0.45$ cm. As usual, the object is at infinity. We will take the half field of view angle to be 20° or 0.349066 radians. With these inputs, the focal length of the system is 8.7985 cm.

To optimize, all of the curvatures and both of the lens separations are made variables. The weights for all of the third-order aberrations and the two paraxial chromatic aberrations are set to 1.0. The target focal length is set to 10.0 cm and its weight is also set to 1.0.

Eight iterations gets the aberration function down to a reasonably small value. The optimized lens prescription is given by $c_1 = 0.231097\ \text{cm}^{-1}$, $c_2 = 0.004161\ \text{cm}^{-1}$, $c_3 = -0.257009\ \text{cm}^{-1}$, $c_4 = 0.271478\ \text{cm}^{-1}$, $c_5 = 0.097439\ \text{cm}^{-1}$, $c_6 = -0.306939\ \text{cm}^{-1}$, $t_a = 1.512611$ cm, and $t_b = 0.748231$ cm. The focal length of this system is almost exactly 10.0 cm. The size of the aperture stop has been adjusted to make sure the system is $f/4.5$.

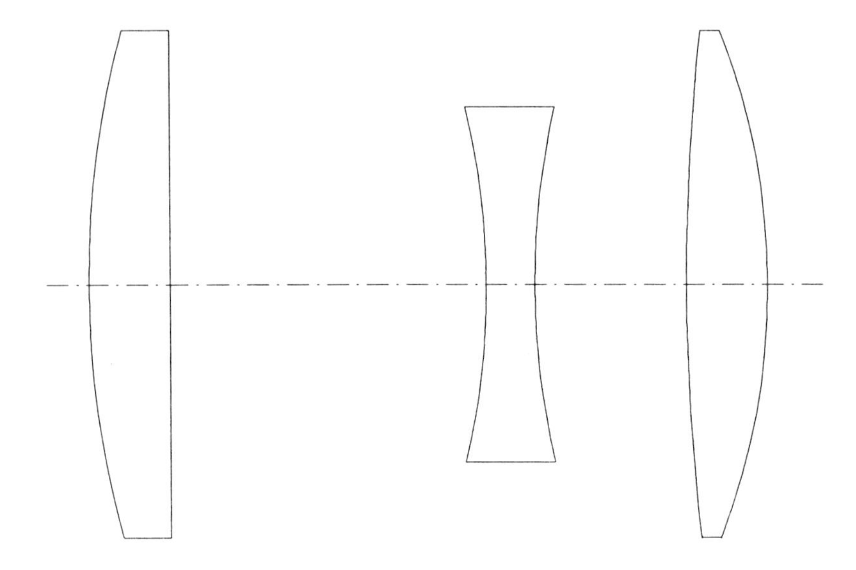

Figure 10.11 The Cooke triplet lens.

Figure 10.11 shows a drawing of the Cooke triplet after optimization. Note that the optimized lens is shorter than our predesign. This is because we should have adjusted the separations between the lenses to account for the principal plane locations. Fortunately, the optimization software can handle this discrepancy.

Figure 10.12 shows the longitudinal spherical aberration, field curvature, and distortion. Note the axes on these graphs indicate that this is a pretty darn good lens.

■

SUMMARY

A lens design problem, like any design problem, is fundamentally different from the academic types of problems you are used to working. A typical textbook problem has well-defined inputs and one correct answer. A lens design problem, on the other hand, is usually poorly defined and may have many, very different solutions. All of this makes lens design a very personal endeavor.

The first step in designing a lens is to select a starting system. In the examples presented in this chapter, the starting systems were deduced from considerations of the third-order aberrations. Lens designers have been using similar approaches for formulating starting systems for many years. They have come up with a whole host of good ideas. It is prudent for a lens designer, novice, or expert to make use of the experience that has already been gained. To take advantage of existing knowledge,

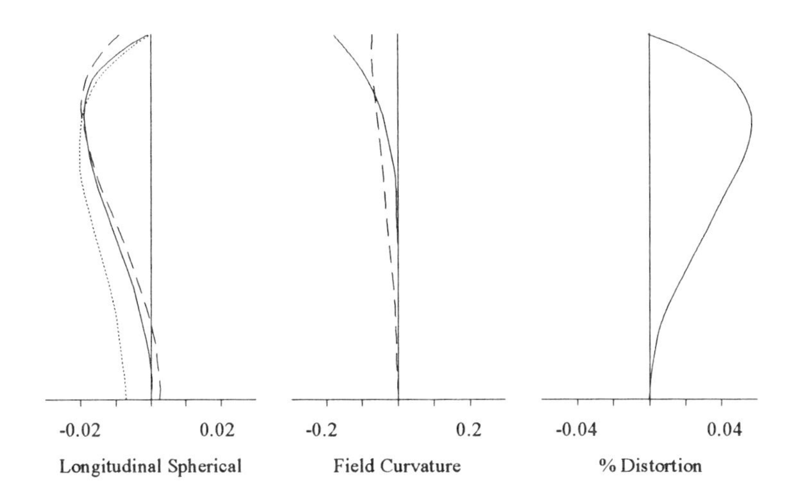

Figure 10.12 Plots of longitudinal spherical aberration, field curvature, and distortion for the Cooke triplet lens. In the LSA plot, the solid line is the *d* light curve, the dashed line is *C* light, and the dotted line is *F* light. In the field curvature plot, the tangential curve is solid and the sagittal curve is dashed.

lens designers often begin a new design by searching the literature on existing designs. There are even several books (Cox, Laiken, and Smith) designed for just this purpose. There is also a considerable amount of patent literature available. Another source of information is the scientific literature in optics journals. The wise lens designer makes use of all of these sources of information for ideas to start a design.

The major consideration in selecting a starting system is that the lens have enough of the right kinds of free parameters to satisfy the system design goals. There must be enough free parameters because to satisfy a particular design goal, such as zero marginal spherical aberration, there must be at least one free parameter available. Yet, having too many free parameters makes the system unnecessarily complex and expensive. Also, changing the free parameters that are available must be effective in satisfying the design goals. It does no good to be able to change the lens thickness, for example, if this change does not significantly affect our aberrations.

Once a starting system is selected, an aberration function must be specified. Clearly, this aberration function must relate directly to the design goals. Including aberrations or constraints in our aberration function which do not address the design goals just makes it more difficult to achieve those goals.

Automatic optimization software takes care of the next step by changing the free parameters to minimize the specified aberration function. Unfortunately, this step is generally not perfectly automatic. The designer must understand what the software is doing and when it is not behaving properly. Don't accept an optimized system uncritically. Look for things that might have gone wrong with the optimization such as thickness or edge violations, or a change in the stop location.

Clearly if something goes wrong at this stage, the previous steps must be reexamined. But even if everything has worked, it is often worthwhile to perform a design through a series of iterations. It is often useful to select only certain parameters and certain aberrations to optimize. Then freeze these parameters and vary others to optimize on other aberrations.

In conclusion, you have only begun your long journey toward becoming a lens designer, but you *have* begun!

REFERENCES

R. Kingslake, *Lens Design Fundamentals* (Academic Press, Orlando, FL, 1978).

A. Cox, *A System of Optical Design* (Focal Press, London, 1964).

M. Laikin, *Lens Design* (Marcel Dekker, New York, 1991).

W. Smith, *Modern Lens Design* (McGraw-Hill, New York, 1992).

PROBLEMS

10.1 Describe the major steps required to design a lens. Which of the steps requires knowledge on the part of the designer and which can be performed automatically with software?

10.2 What considerations determine the focal length of a telescope objective?

10.3 What are the two most important aberrations for a telescope objective?

10.4 What are some ways you could gain an additional degree of freedom in the design of a telescope objective so that coma might be corrected?

10.5 For all of the lens design examples given in this chapter, we started by trying to make the chromatic and third-order aberrations zero. However, none of the final systems had zero chromatic or third-order aberrations. Why did we bother?

10.6 What aberrations are important in the design of a landscape lens? How is spherical aberration controlled in this type of lens?

10.7 In the landscape lens design, why do we correct for tangential field curvature rather than sagittal field curvature?

10.8 If we make a small change in just the second curvature of the Cooke triplet, which aberrations do you think would be affected? Is your answer different if we change the fourth curvature instead? *Hint*: Look carefully at the eight equations used in designing the Cooke triplet.

10.9 What is the maximum half field of view for an astrophotographic camera with focal length 1000 mm? Use a standard 35-mm frame as your field stop.

10.10 Determine the focal length of an objective lens for use with a CCD camera. The active area of the CCD array is 8.6 mm by 6.5 mm. The field of view should be approximately 11.2' by 14.8'.

10.11 Complete the design of the lens described in the previous problem. If the pixel size is 23 μm by 27 μm, then the spot size should be about 23 μm across.

10.12 Make plots of S_t versus c_1 for spherically corrected achromats similar to those given in the text, but put the high-index, low-Abbe-value glass first.

10.13 Make plots of C_t versus c_1 for spherically corrected achromats. Where do these plots have zero values compared with the peaks of the corresponding S_t curves?

10.14 Find a pair of glasses for a spherically corrected achromat so that the peak of the S_t curve is at zero.

10.15 Try designing landscape lenses with the same target parameters given in the text, but use different glasses. Try to make generalizations regarding the best choices for index of refraction and Abbe value.

10.16 Write out the full forms of the eight equations used in the Cooke triplet design.

10.17 Use a computer algebra program to solve the eight Cooke triplet equations for a system using SK15 glass for the first and third lenses and F3 glass for the middle lens.

10.18 Experiment with various glass combinations for the Cooke triplet. Can you reach any generalizations regarding glass choice for this type of lens?

10.19 Derive the equations for the vergence for the second and third lenses in the Cooke triplet.

APPENDIX A

GLASS CATALOG AND MAP[1]

	Name	Abbe Value	Index
FK	FK 3	65.77	1.46450
	FK 5	70.41	1.48749
	FK 51	84.47	1.48656
	FK 52	81.80	1.48605
	FK 54	90.70	1.43700
PK	PK 1	66.92	1.50378
	PK 2	65.05	1.51821
	PK 3	64.66	1.52542
	PK 50	69.71	1.52054
	PK 51A	76.98	1.52855
PSK	PSK 2	63.08	1.56873
	PSK 3	63.46	1.55232
	PSK 50	67.28	1.55753
	PSK 52	65.41	1.60310
	PSK 53A	63.48	1.62014
	PSK 54	64.60	1.58599
BK	BK 1	63.46	1.51009
	BK 3	65.06	1.49831
	BK 6	62.15	1.53113
	BK 7	64.17	1.51680
	U BK 7	64.29	1.51680
	BK 8	63.69	1.52015
	BK 10	66.95	1.49782
BaLK	BaLK N3	60.26	1.51849
K	K 3	58.98	1.51823
	K 4	57.40	1.51895
	K 5	59.48	1.52249
	K 7	60.41	1.51112
	K 10	56.41	1.50137
	K 11	61.44	1.50013
	K 50	60.18	1.52257
	U K 50	60.38	1.52257
ZK	ZK 1	57.98	1.53315
	ZK N7	61.19	1.50847
BaK	BaK 1	57.55	1.57250
	BaK 2	59.71	1.53996
	BaK 4	56.13	1.56883
	BaK 5	58.65	1.55671
	BaK 50	57.99	1.56774

[1]Glass catalog and map are reproduced with the permission of Schott Glass Technologies, Inc.

	Name	Abbe Value	Index
SK	SK 1	56.71	1.61025
	SK 2	56.65	1.60738
	SK 3	58.92	1.60881
	SK 4	58.63	1.61272
	SK 5	61.27	1.58913
	SK 6	56.40	1.61375
	SK 7	59.46	1.60729
	SK 8	55.92	1.61117
	SK 10	56.90	1.62280
	SK 11	60.80	1.56384
	SK 12	59.45	1.58313
	SK 13	58.30	1.59181
	SK 14	60.60	1.60311
	SK 15	58.06	1.62299
	SK 16	60.32	1.62041
	SK 18A	55.42	1.63854
	SK 51	60.31	1.62090
	SK 55	60.12	1.62041
KF	KF 3	54.70	1.51454
	KF 6	52.20	1.51742
	KF 9	51.49	1.52341
BaLF	BaLF 4	53.71	1.57957
	BaLF 5	53.63	1.54739
	BaLF 50	51.37	1.58893
SSK	SSK 1	53.91	1.61720
	SSK 2	53.15	1.62230
	SSK 3	51.16	1.61484
	SSK 4A	55.14	1.61765
	SSK N5	50.88	1.65844
	SSK N8	49.77	1.61772
	SSK 50	52.61	1.61795
	SSK 51	53.63	1.60361

	Name	Abbe Value	Index
LaK	LaK N6	57.96	1.64250
	LaK N7	58.52	1.65160
	LaK 8	53.83	1.71300
	LaK 9	54.71	1.69100
	LaK 10	50.41	1.72000
	LaK 11	57.26	1.65830
	LaK N12	55.20	1.67790
	LaK L12	54.92	1.67790
	LaK N13	53.33	1.69350
	LaK N14	55.41	1.69680
	LaK 16A	51.78	1.73350
	LaK 21	60.10	1.64049
	LaK L21	59.75	1.64048
	LaK N22	55.89	1.65113
	LaK 23	57.38	1.66882
	LaK 28	50.77	1.74429
	LaK 31	56.42	1.69673
	LaK 33	52.43	1.75398
LLF	LLF 1	45.75	1.54814
	LLF 2	47.17	1.54072
	LLF 6	48.76	1.53172
	LLF 7	45.41	1.54883
BaF	BaF 3	46.47	1.58267
	BaF 4	43.93	1.60562
	BaF N6	48.45	1.58900
	BaF 8	47.00	1.62374
	BaF 9	47.96	1.64328
	BaF N10	47.11	1.67003
	BaF N11	48.42	1.66672
	BaF 13	44.96	1.66892
	BaF 50	44.50	1.68273
	BaF 51	44.93	1.65224
	BaF 52	46.44	1.60859
LF	LF 5	40.85	1.58144
	LF 7	41.49	1.57501
	LF 8	43.75	1.56444

	Name	Abbe Value	Index
F	F 1	35.70	1.62588
	F 2	36.37	1.62004
	F 3	37.04	1.61293
	F 4	36.63	1.61659
	F 5	38.03	1.60342
	F 6	35.34	1.63636
	F 7	35.56	1.62536
	F 8	39.18	1.59551
	F 9	38.08	1.62045
	FN 11	36.18	1.62096
	F 13	36.04	1.62237
	F 14	38.23	1.60140
	F 15	37.83	1.60565
BaSF	BaSF 1	38.96	1.62606
	BaSF 2	35.83	1.66446
	BaSF 6	41.93	1.66755
	BaSF 10	39.15	1.65016
	BaSF 12	39.20	1.66998
	BaSF 13	38.57	1.69761
	BaSF 51	38.11	1.72373
	BaSF 52	41.01	1.70181
	BaSF 54	32.15	1.73627
	BaSF 56	36.74	1.65715
	BaSF 57	41.90	1.65147
	BaSF 64A	39.38	1.70400
LaF	LaF 2	44.72	1.74400
	LaF 3	47.96	1.71700
	LaF N7	34.95	1.74950
	LaF N8	41.59	1.73520
	LaF 9	28.39	1.79504
	LaF N10	43.95	1.78443
	LaF 11A	31.70	1.75693
	LaF 13	37.84	1.77551
	LaF 20	48.20	1.68248
	LaF N21	47.47	1.78831
	LaF 22A	37.20	1.78179
	LaF N23	49.71	1.68900
	LaF N24	47.81	1.75719
	LaF N28	49.57	1.77314

	Name	Abbe Value	Index
LaSF	LaSF 3	40.61	1.80801
	LaSF N9	32.17	1.85025
	LaSF N15	38.07	1.87800
	LaSF 18A	32.36	1.91348
	LaSF N30	46.38	1.80318
	LaSF N31	41.01	1.88067
	LaSF 32	30.40	1.80349
	LaSF 33	34.24	1.80596
	LaSF 35	29.06	2.02204
	LaSF 36A	35.08	1.79712
SF	SF 1	29.51	1.71736
	SF 2	33.85	1.64769
	SF 3	28.20	1.74000
	SF 4	27.58	1.75520
	SF L4A	27.40	1.75520
	SF 5	32.21	1.67270
	SF 6	25.43	1.80518
	SF L6	25.39	1.80518
	SF 8	31.18	1.68893
	SF 9	33.65	1.65446
	SF 10	28.41	1.72825
	SF 11	25.76	1.78472
	SF 12	33.84	1.64831
	SF 13	27.60	1.74077
	SF 14	26.53	1.76182
	SF 15	30.07	1.69895
	SF 16	34.05	1.64611
	SF 18	29.25	1.72151
	SF 19	33.01	1.66680
	SF 53	28.69	1.72830
	SF 54	28.09	1.74080
	SF 55	26.95	1.76180
	SF 56A	26.08	1.78470
	SF L56	26.08	1.78470
	SF 57	23.83	1.84666
	SF L57	23.62	1.84666
	SF 58	21.51	1.91761
	SF 59	20.36	1.95250
	SF 63	27.71	1.74840
	SF 64A	30.30	1.70585
TiF	TiF 3	42.20	1.54765
	TiF N5	35.51	1.59355
	TiF 6	30.97	1.61650

	Name	Abbe Value	Index
KzF	KzF N1	49.64	1.55115
	KzF N2	51.63	1.52944
KzFS	KzFS 1	44.34	1.61310
	KzFS N2	54.16	1.55836
	KzFS N4	44.29	1.61340
	KzFS N5	39.63	1.65412
	KzFS 6	48.51	1.59196
	KzFS 7A	37.39	1.68064
	KzFS 8	34.61	1.72047
	KzFS N9	46.90	1.59856

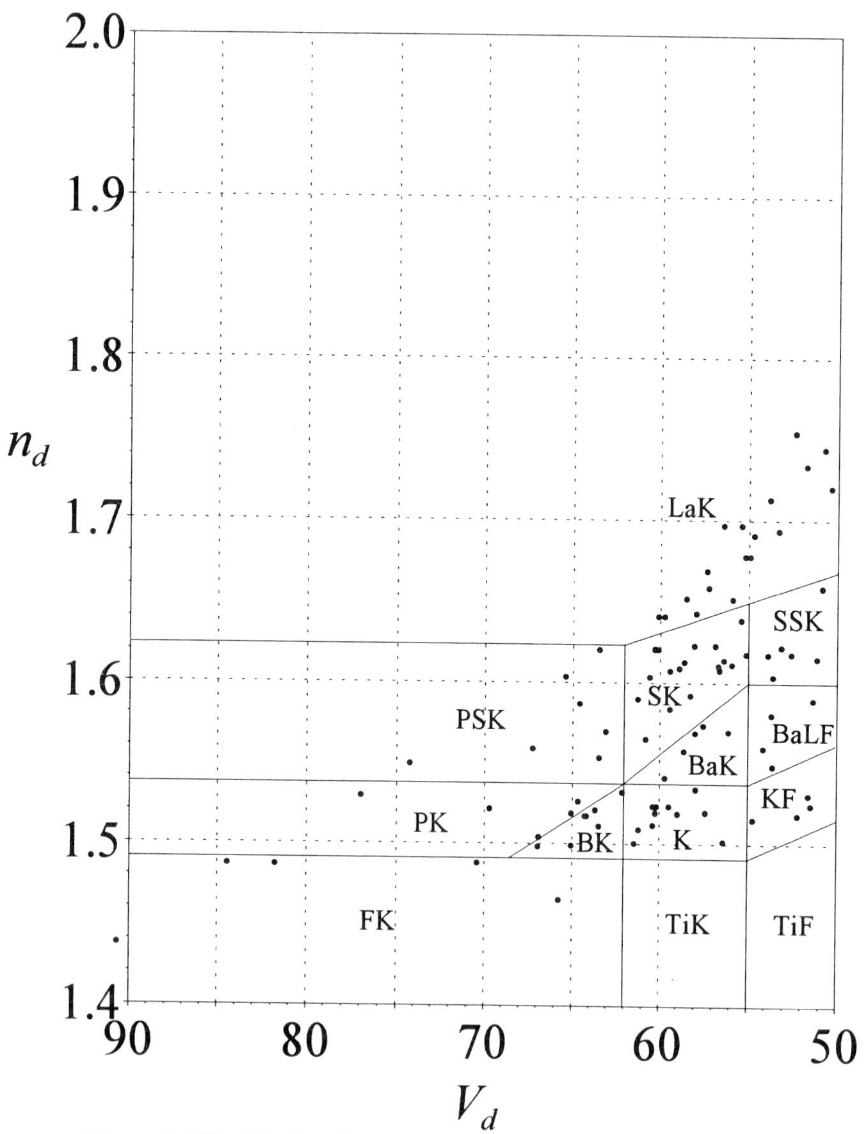

Figure A.1 The left side of the glass map showing mostly crown glasses.

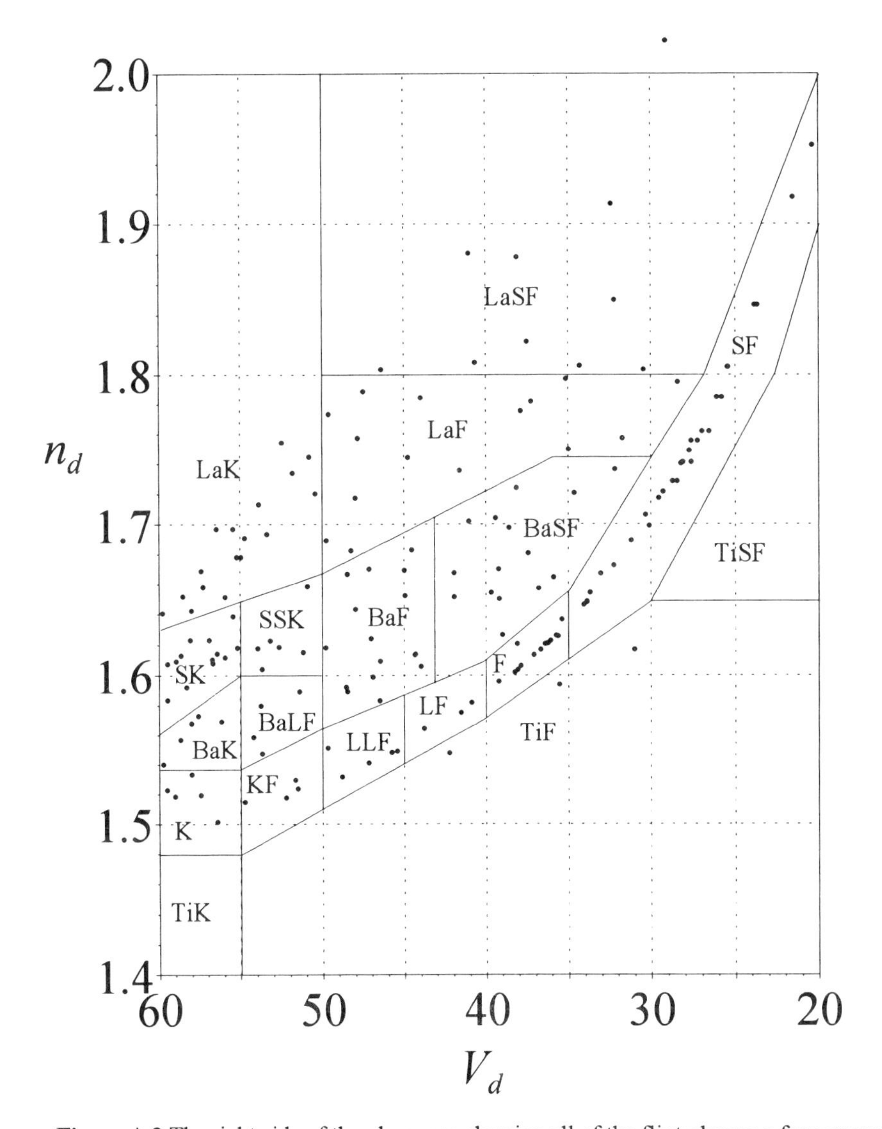

Figure A.2 The right side of the glass map showing all of the flint glasses a few crowns.

APPENDIX B

SIGN CONVENTION AND NOTATION

In general, quantities which relate to the surfaces of an optical system are given a subscript. The subscript refers to a particular surface of the system at or from which the quantity is measured. Values before the surface are unprimed, while values after the surface are primed. Primed and unprimed quantities are also used to denote object space or image space respectively. The variable subscript j refers to a general surface. The object is given the subscript 0. The last surface of the optical system is indicated with a k. Therefore, the image surface, the surface after the last surface of the system, is referred to by a $k + 1$ subscript.

All distances are positive to the right of the surface from which they are measured. Transverse distances are positive upward away from the optical axis and into the plane of the paper. Angles are positive when measured counterclockwise.

Variable	Description	Page
a	a single weighted aberration	307
$\vec{a}$	the aberration vector	326
A	apex angle of a prism	23
A	the third-order astigmatism coefficient	256
AC	axial color coefficient	149
$\overleftrightarrow{B}$	a matrix which takes a ray from the object to the last surface of an optical system, the elements of this matrix—b_{11}, b_{12}, b_{21}, and b_{22}—are useful for calculating the image location and magnification	174
c	speed of light in a vacuum (299,792,458 m/s)	9
c	curvature of a surface (inverse of radius)	85

Variable	Description	Page
c	the net curvature of a thin lens	247
C	the third-order coma coefficient	251
$\overleftrightarrow{C}$	conjugate matrix	170
D	deviation angle, angle between the direction of an incoming ray and the corresponding outgoing ray	22
D	the third-order distortion coefficient	261
$D_{\min}$	minimum deviation angle	24
D^{*}	supplement of the deviation angle	25
E	energy of a photon	14
f	object space focal length measured from the effective optical system to the object space focal point	50
f_{eff}	effective focal length of an optical system	60
f'	image space focal length measured from the effective optical system to the image space focal point	50
$f/\#$	f- number, the ratio of the focal length to the diameter of the entrance pupil	134
F	an intermediate variable used in exact ray tracing	206
$\vec{g}$	gradient vector of the aberration function	325
G	an intermediate variable used in exact ray tracing	206
$\overleftrightarrow{G}$	Hessian matrix of the aberration function	325
h	Planck's constant (6.62×10^{-34} J · m)	14
h	paraxial object size	40
h'	paraxial image size	40
HH'	distance separating the object space principal plane from the image space principal plane measured along the optical axis	109

Variable	Description	Page
HN	distance separating the object space principal plane from the object space nodal point measured along the optical axis	112
$H'N'$	distance separating the image space principal plane from the image space nodal point measured along the optical axis	112
$\overleftrightarrow{H}$	the Hessian matrix of a particular weighted aberration	327
i	angle of incidence measured from the normal of a surface to the incoming ray	2
i_{cr}	critical angle, for any incidence angle larger than the critical angle total internal reflection will occur	20
i'	angle of refraction or refraction measured from the normal of a surface to the outgoing ray	3
$\overleftrightarrow{I}$	the identity matrix	327
$\overleftrightarrow{J}$	the Jacobian matrix of the weighted aberrations	326
l_E	location of the entrance pupil measured from the first surface of a system	131
$l_{E'}$	location of the exit pupil measured from the first surface of a system	132
l_F	front focal length measured from the first surface of an optical system to the object space focal point	98
$l_{F'}$	back focal length measured from the last surface of an optical system to the image space focal point	97
l_H	object space principal plane location measured from the first surface of the optical system	99
$l_{H'}$	image space principal plane location measured from the last surface of the optical system	98
l_N	object space nodal point location measured from the first surface of the optical system	104
$l_{N'}$	image space nodal point location measured from the last surface of the optical system	103

Variable	Description	Page
l_W	location of the entrance window measured from the first surface of the optical system	135
$l_{W'}$	location of the exit window measured from the last surface of the optical system	135
L	direction cosine of an exact ray measured relative to the x axis	204
LC	lateral color coefficient	151
m	transverse magnification	40
M	angular magnification	68
M	direction cosine of an exact ray measured relative to the y axis	204
$\overleftrightarrow{M}$	a matrix which takes a ray from the object to an arbitrary surface in the system; the elements for this matrix are useful for determining the aperture stop	182
n	index of refraction before the surface	3
n'	index of refraction after the optical surface	3
N	direction cosine of an exact ray measured relative to the y axis	204
p	a general system parameter	307
P	the third-order Petzval coefficient	259
P_{gF}	partial dispersion ratio	32
$\overleftrightarrow{P}$	a matrix which takes a ray from the first surface of an optical system to the aperture stop	185
$\overleftrightarrow{P}^{-1}$	the inverse of the $\overleftrightarrow{P}$ matrix which effectively takes a ray from the aperture stop back to the first surface of the system; the elements of this matrix are used for finding the entrance pupil size and location	185
$\overleftrightarrow{P}'$	a matrix which takes rays from the aperture stop to the last surface of a system; the elements of this matrix are used to determine the exit pupil size and location	186

Variable	Description	Page
r	radius of curvature of a surface	50
R	clear radius of a surface measured from the optical axis to the edge of a stop or lens rim	125
R_A	size of the aperture stop	132
R_E	entrance pupil size	132
$R_{E'}$	exit pupil size	132
R_W	entrance window size	135
$R_{W'}$	exit window size	135
$\overleftrightarrow{R}$	refraction or reflection matrix	169
s	object distance measured from the effective optical system to the object	38
s'	image distance measured from the effective optical system to the image	38
S	the third-order spherical coefficient	243
$\overleftrightarrow{S}$	system matrix which takes a ray from the first surface of a system to the last, the elements of this matrix a, b, c, and d are useful for determining the optical power of the system and the locations of the cardinal points	176
t	separation between two optical surfaces measured along the optical axis	46
t_e	edge thickness	126
T	length of an exact ray from one optical surface to the vertex surface of the next optical surface measured along the ray	205
$\overleftrightarrow{T}$	translation matrix	168
u	slope angle of the incident ray measured from the optical axis to the ray	49

Variable	Description	Page
u'	slope angle of the refracted or reflected ray measured from the optical axis to the ray	49
v	velocity of a wave in a medium	9
v	object space vergence	65
v'	image space vergence	65
V	Abbe value	16
w	an aberration weight	307
W	an intermediate variable used in the calculation of the coefficients for astigmatism, coma, and spherical aberration	245
$\bar{W}$	an intermediate variable used in the calculation of third-order distortion	263
X	the x coordinate of the intersection point between an exact ray and an optical surface	204
y	height of a ray at a surface	44
$\vec{y}$	column vector based on ray parameters immediately before a surface	168
$\vec{y}'$	column vector based on ray parameters immediately after a surface	168
Y	the y coordinate of the intersection point between an exact ray and an optical surface	204
z	object distance measured from the object space focal point to the object	63
z	sag of a surface	126
z'	image distance measured from the image space focal point to the image	63
Z	the z coordinate of the intersection point between an exact ray and an optical surface	204

Variable	Description	Page
α	reduced slope angle	86
α	damping factor	321
β	angle measured from the optical axis to the normal to the surface	49
γ	an intermediate variable used in exact ray tracing	211
δn	dispersion, the difference in index of refraction for two wavelengths (typically F and C light)	16
$\delta z'_{AC}$	longitudinal component of the axial chromatic aberration	146
$\delta y'_{LC}$	transverse component of the lateral chromatic aberration	146
Δ	the length of an exact ray from the vertex plane to the actual optical surface measured along the ray	206
ϵ_x'	the x component of the aberration distance	239
ϵ_y'	the y component of the aberration distance	239
η	relative field height	215
θ	angle of a point in the entrance pupil measured from the y axis	215
λ	wavelength of a wave	9
Λ	optical invariant	114
μ	the ratio of the index of refraction before a surface to the index of refraction after the surface	209
ξ	the y coordinate of the intersection point of an exact ray with the vertex plane	205
ρ	distance from the optical axis relative to the entrance pupil radius	215
ρ	a particular value for a general system parameter	310
τ	reduced separation	86

Variable	Description	Page
ϕ	optical power or the change in the vergence	65
χ	the x coordinate of the intersection point of an exact ray with the vertex plane	205
ψ	the weighted aberration function	307

APPENDIX C

ROSE.EXE USER'S MANUAL

ROSE.EXE is a computer program designed specifically for students learning Geometrical Optics. The purpose of the program is to relieve the tedium involved in analyzing and designing simple imaging systems. It is not intended as a professional lens design program. The program uses the tools and techniques, such as *y-nu* ray tracing, discussed in the text. Therefore, it can be used to check your work on homework assignments that are to be done "by hand." Also, the process of learning to use ROSE.EXE will make it easier for the student to learn to use commercial software (i.e., CODE V).

The program is available for downloading on the Wiley ftp server, which can be accessed either through a web browser or ftp program. To access the file containing the ROSE.EXE program, connect online at ftp://ftp.wiley.com/public/sci_tech_med/ and go to the geometrical_optics directory. There you will find direction on how to download and install the program files.

ROSE.EXE is written and compiled in Borland's Turbo Pascal 4.0. It should not be considered a public domain program; rather, it is made available to purchasers of Modern Geometrical Optics. In addition to the compiled program, the source code is included. If you own a copy of the text Modern Geometrical Optics, you have permission to modify the source code, but you cannot sell or give away copies of the program or any modifications to the program.

ROSE.EXE is a menu-driven program. The main menu lists submenus which eventually do the desired calculations. This makes the program easy to learn, but not necessarily efficient to use. ROSE.EXE allows you to analyze optical systems with up to 25 axially symmetric spherical surfaces (reflective or refractive). Thin lens systems can also be studied. System data are entered and modified using a menu style data editor. System data can also be stored on and loaded from disk. All first-order properties of a system may be determined. An arbitrary *y-nu* ray can be traced starting at any surface and ending in object space at one end and image space at the other. Third-order aberration coefficients can be calculated. Exact, skew rays may be traced. The program can also be used for some simple optimization.

The manual which follows consists of a tutorial on how to use the various features of ROSE.EXE.

C.1 SCREEN LAYOUT AND THE MAIN MENU

ROSE.EXE is a very carefully structured program. A consistent user interface should make it easier to learn to use the software.

As shown in Table C.1, each screen in ROSE.EXE consists of four areas. The left-hand side of the first line on each screen is a screen title which tells you where you are in the program. The main area of the screen consists of a menu listing or data display. At the bottom of the screen is the prompt for user input. Any error messages will be displayed just above this prompt.

As mentioned earlier, ROSE.EXE is menu driven. The highlighted menu item will be executed if you press the return key. Highlighting may be accomplished by using the up and down arrow keys. Alternatively, you may directly execute a menu item by pressing the key representing the capital letter of the desired menu item. You may use either a capital or lowercase letter for your selection. In the remainder of this tutorial you will be asked to select particular menu items. It is understood that you should use either the arrows and the return or the key letter for selection. The last menu item is always an escape choice which returns you to the previous menu or exits the program.

The main program menu is shown in Table C.2. When the program loads, this should be the first screen you see after the copyright notice.

C.2 CREATING A SYSTEM

Selecting *Create a system* from the main menu will allow you to enter the system data necessary to begin a system analysis.

Table C.1 Copyright notice.

```
COPYRIGHT NOTICE }screen title

                  Lens Design for Modern Geometrical Optics

menu or
                  Version 6.0
data display
                  January 23, 1997

                  Copyright (C) 1997

                  by

                  Richard Ditteon



(error message when needed)
Hit any key to continue }input prompt
```

Table C.2 Main menu.

```
MAIN MENU

                    Create a system

                    Modify an existing system

                    File handling

                    Display system data

                    Paraxial optics

                    Third order optics

                    Exact optics

                    Optimization

                    Quit (exit program)


Select Option with key letter or arrows
```

If you already have a system entered, the first thing that *Create a system* does is ask you if you want to create a new system. If you respond with *n* or *N*, you will be returned to the main menu. If you respond with *y* or *Y* and the current system has not been saved, you will be asked if you want to save the system.

Next, you will also be asked for number of surfaces in the system. The number entered here should be the actual number of surfaces in the optical system excluding the object and image surfaces but including any stops or apertures. For thick lens or mirror systems, this number will be the number of physical surfaces. For a thin lens system, each thin lens counts as only one surface. This number can be easily changed later. For the example system given in this tutorial the number of surfaces is 3.

Table C.3 shows the menu which is displayed next.

The first menu item is *Curvature (or thin lens power)* depending on the type of surface you need. This is the inverse of the radius of the spherical surface or the inverse of the thin lens focal length.

The *Separation* is the longitudinal distance from the current surface to the next surface (positive to the right). If a surface is intended to be a mirror, then the sign on the separation should indicate the direction of travel of the light. After the first mirror in a system, the separations should be negative until another mirror is encountered.

The *Radius of surface opening* is the clear radius or transverse distance from the optical axis to the edge of the lens rim or stop.

The remaining parameters are the *Index* of refraction in *d* light (587.5618 nm) and the *Abbe value* (calculated between 656.2725 nm and 486.1327 nm) for the glass which follows the surface. These values may be found in a glass catalog for

Table C.3 Surface data for the object surface.

```
SURFACE 0 DATA

          Curvature (or thin lens power) . .    0.000000

          Separation . . . . . . . . . . . .  100.000000

          Radius of surface opening. . . . .    1.000000

          Index of refraction (588 nm) . . .    1.000000

          Abbe value (486 to 656 nm) . . . .    0.000000

          Quit (return to previous menu)


Select Option with key letter or arrows
```

different types of glass. If a surface is intended to be a mirror then both of these values should change sign after each mirror.

The first surface data screen is for the object surface. The program will display default values for the data for the object surface (Surface 0 Data). You can change this default data by first selecting the item you want to change. For example, to change the separation between surfaces (default = 0.0), select Separation. The program will then prompt you for the new separation value. If you enter a large number for this separation (larger than 1.0×10^{12}), the program will treat the object as being at infinity. The clear radius value for the object surface is the desired object size. For objects at infinity, this becomes the desired field-of-view angle in radians. Typically, only the clear radius and separation need to be changed on the object surface. The object is normally a plane surface so the curvature remains 0.0 and the space after the object contains air so the index is 1.0 and the Abbe value is 0.0. Enter the appropriate data for this surface as shown in Table C.3. When you finish changing data, select Quit to go to the next surface.

After defining each surface, you will select *Quit* on the Surface Data Menu. The program will then go on to the following surface until all surfaces have been prescribed. If you make a mistake, you can't go back and correct it here. There is, however, a system modify option which will be discussed later.

Note that you don't need to enter values for any quantity which is not important for the calculation you want to perform. For instance, you can omit clear radius values except when you plan to find the aperture stop and field stop. For thick lens systems, you must specify the middle wavelength index when it differs from 1.0. The Abbe value is used only in chromatic aberration calculations so it can normally be left at 0.0.

Note that the image surface is initially entered without a value for the clear radius (image size) or separation (between the last surface and the image surface). These values will be determined by the program automatically. Also, typically the index of refraction after the last surface of the system (image space) is 1.0 and the Abbe value is 0.0.

Enter the complete system as shown in Tables C.4 to C.6.

C.3 DISPLAYING SYSTEM DATA

After entering your system data, you can display that data by selecting the *Display system* option from the main menu.

System data are displayed in a table format as shown Table C.7. The first column lists the surface number starting with the object surface (surface 0) and ending on the image surface (one larger than the number of surfaces in the system). Values which have meaning only at the surface such as curvature and clear radius are listed on the row with the surface number. Values which are defined between surfaces such as separation and index of refraction are listed on rows between surface numbers.

The middle index is listed as *nd*. This notation stands for the specific spectral line of helium at 587.5618 nm. The Abbe value is derived from indices at 656.2725 nm (red hydrogen *C* line) and 486.1327 nm (blue hydrogen *F* line). However, in ROSE.EXE, there is nothing special about these designations. In fact, any desired combination of wavelengths could be used. The only restriction is that the middle wavelength is always used for calculating the first order properties of the system. The Cauchy formula is used to derive the indices at the long and short wavelengths from the *d* index and the Abbe value. These indices are used for chromatic aberration calculations.

At the bottom of the system data display screen is a prompt asking if you want to send a copy of this data to the printer. Enter *y* or *Y* if you want a hard copy. Make

Table C.4 First surface data.

```
SURFACE 1 DATA

          Curvature (or thin lens power) . .   0.040200
          Separation . . . . . . . . . . . .   1.350000
          Radius of surface opening. . . . .   8.000000
          Index of refraction (588 nm) . . .   1.720000
          Abbe value (486 to 656 nm) . . . .  50.410000
          Quit (return to previous menu)

Select Option with key letter or arrows
```

Table C.5 Second surface data.

```
SURFACE 2 DATA

                  Curvature (or thin lens power) . .    0.077200
                  Separation . . . . . . . . . . . .    3.750000
                  Radius of surface opening. . . . .    8.000000
                  Index of refraction (588 nm) . . .    1.617200
                  Abbe value (486 to 656 nm) . . . .   53.910000
                  Quit (return to previous menu)


Select Option with key letter or arrows
```

sure the printer is on-line and connected to your computer first. Any other input will return you to the previous menu.

C.4 MODIFYING A SYSTEM

If you make a mistake while entering your initial optical system or if you want to change the system for any reason, you do not have to start over from the beginning.

Table C.6 Third surface data.

```
SURFACE 3 DATA

                  Curvature (or thin lens power) . .   0.002048
                  Separation . . . . . . . . . . . .   0.000000
                  Radius of surface opening. . . . .   8.000000
                  Index of refraction (588 nm) . . .   1.000000
                  Abbe value (486 to 656 nm) . . . .   0.000000
                  Quit (return to previous menu)


Select Option with key letter or arrows
```

Table C.7 System data display.

```
SYSTEM DATA

surf   curv        separation   clrad        nd          V

 0     0.000000                  1.000000
                   100.000000                1.000000     0.000000
 1     0.040200                  8.000000
                     1.350000                1.720000    50.410000
 2     0.077200                  8.000000
                     3.750000                1.617200    53.910000
 3     0.002048                  8.000000
                    96.773019                1.000000     0.000000
 4     0.000000                 -1.008754

Send system data to printer?  Yes  No
```

The *Modify an existing system* option of the main menu will let you change any feature of a system.

Table C.8 shows the menu which is displayed when Modify an existing system is selected from the main menu.

The *Modify data for one surface* option asks you to identify which surface needs to be modified and then lists the current data for that surface. You can then change any of the surface properties.

Surfaces can be added to (up to a total of 25) or deleted from (down to 1) the system with the appropriate options. You will be prompted to confirm any additions or deletions.

All lengths can be scaled by selecting this option. For example, if a scale factor of 2.0 were specified, then all separations and clear radii would be doubled. Curvatures or thin lens powers (units of inverse length) would be reduced by half. The index and Abbe values are not affected.

Table C.8 Modify system menu.

```
Modify data for one surface

Add a surface to the system

Delete a surface from the system

Scale entire system

Quit (return to previous menu)
```

If you made any mistakes in entering the tutorial system, correct them using the *Modify system* option.

C.5 FILE HANDLING

A lot of time and effort can be saved by frequently saving your optical system data to a file on a floppy disk.

The File Handling menu is shown in Table C.9. From this menu you can list the files on a disk, save system data to disk, load data from disk, or change the default drive.

To use floppy or hard disk storage effectively, you need to be able to do several functions from within the program. For example, before you load a data file it is helpful to know the names of the files on the data disk. The *Directory of default drive / path* option gives you this information. After selecting this option, the program will ask you for a file specification string. Entering *.* will list all files on the default drive. Entering *.dat will list only those files with dat extensions. Drives / paths other than the default can also be examined. For example, the string b:\data*.dat will list all dat files on drive b: in the directory called "data." You can obtain a hard copy output by responding with *y* or *Y*.

Selecting Save a system to disk will allow you to save data to a disk file. If you try to save data to an existing file, the program will ask you if you want to overwrite the existing file. Do this only if you are sure you no longer need the old data. When saving data you may want to include the extension ".dat." This is not done automatically for you.

To retrieve data, select *Load a system from disk.* You will be asked for the name of the file to retrieve. This input must match the file name exactly or you will get a file not found error or the wrong data.

The last option allows you to change the default drive and default directory path. Changing the defaults can speed up using the other options in file handling. Data or directories can be accessed with the other options by including this information in the file specification string.

Table C.9 File handling menu.

```
Directory of default drive / path

Save a system to disk

Load a system from disk

Change default drive / path

Quit (return to main menu)
```

C.6 PARAXIAL RAY TRACING

Paraxial ray tracing is used to determine all of the first-order properties of the given optical system (except chromatic aberration) as well as providing for tracing an arbitrary paraxial ray. The program performs these calculations exactly as you would do them by hand using a *y-nu* ray trace table. The paraxial ray tracing menu is shown in Table C.10.

Since this program is intended as a learning aid and to check hand calculations, the *Table set-up* option has been included to display the intermediate calculations that are required for a *y-nu* ray trace. This option calculates the negative power (- Phi) and the reduced separation (Tau) as shown in Table C.11. It is not required to access this option before doing any other paraxial calculations. For the other paraxial calculations, the negative power and reduced separation are calculated but not displayed.

The *Image size and location* option allows you to trace a ray from the base of the object through the system to determine the image location (Table C.12). The magnification of the system is determined by the ratio of the reduced slope angles for the axial ray that was traced. The image size is determined by multiplying the object clear radius by the magnification. These calculations are also done automatically when the system data are displayed.

The *Arbitrary ray trace* option allows you to trace a ray starting at any surface and going to object space and image space (Table C.13). The ray can start at any height and have any slope angle after the surface which the user may specify. The ray height (y), reduced slope angle (α), and actual slope angle (u) are displayed at each surface. Since ray height is defined at the surface, it appears on the line with the surface number. The reduced slope angle and slope angle are defined between surfaces and therefore appear on a line between surface lines. This feature is useful for checking *y-nu* ray traces done by hand.

The arbitrary ray trace option allows you to reproduce all of the ray traces that are necessary to determine the first-order properties of a system by hand. However,

Table C.10 The paraxial ray tracing menu.

```
Table set-up

Image size and location

Arbitrary ray trace

Cardinal points

Stops, pupils and windows

Vignetting

Quit (return to main menu)
```

Table C.11 Ray trace table output.

Surf	-Phi	Tau
0		
		100.000000
1	-0.028944	
		0.784884
2	0.007936	
		2.318823
3	0.001264	

Table C.12 Image location and size output.

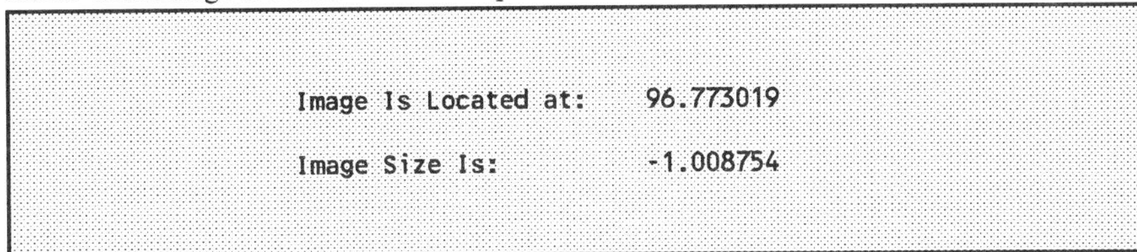

Image Is Located at:	96.773019
Image Size Is:	-1.008754

ROSE.EXE will do these ray traces automatically when the options for determining the first-order properties are selected. With these options, the ray traces are performed but not displayed. Only the final results are shown.

The cardinal points include the focal points, principal points, and nodal points. They are found by tracing a ray from an object at infinity forward into image space and a ray from an image at infinity backward into object space. Output is shown in Table C.14.

The aperture stop, entrance and exit pupils, field stop, and entrance and exit windows are all found using the standard *y-nu* ray tracing technique. *Note that clear radii must be specified when solving for the stops, pupils and windows.* The arbitrary axial ray and chief ray used for finding the stops are scaled to become a marginal

Table C.13 Arbitrary paraxial ray trace output.

Surf	y	alpha	u
0	0.000000		
		0.100000	0.100000
1	10.000000		
		-0.189440	-0.110140
2	9.851312		
		-0.111258	-0.068797
3	9.593323		
		-0.099132	-0.099132
4	0.000000		

Table C.14 Cardinal points output.

```
Overall Lens Power         (Phi) =    0.020015
Effective Focal Length     (F') =   49.962716
Back Focal Length         (LF') =   46.372935
Back Principle Plane      (LH') =   -3.589781
Back Nodal Plane          (LN') =   -3.589781
Image Focal Length          (F) =  -49.962719
Front Focal Length         (LF) =  -50.470856
Front Principle Plane      (LH) =   -0.508140
Front Nodal Plane          (LN) =   -0.508140
```

axial ray and a full-field chief ray. These ray traces are saved for use in the Seidel Aberration option. Output is shown in Table C.15.

The *Vignetting* option calculates the vignetting factor based on the clear radii of the surfaces of the system and the object. Each surface is imaged into object space and then projected onto the entrance pupil as seen from the top of the object. Finally, a numerical integration is performed to find the unvignetted area. This area is divided by area of the entrance pupil to determine the vignetting factor. Output is shown in Table C.16.

The remaining first-order property of a system is its axial and lateral chromatic aberrations. These values are calculated within the Seidel aberration option of the Main Menu.

Table C.15 Stops, pupils, and windows output.

```
Aperture Stop is at Surface #    1
Aperture Stop Size         (RAS) =      8.000000
Entrance Pupil Location     (LE) =      0.000000
Entrance Pupil Size         (RE) =      8.000000
Exit Pupil Location        (LE') =     -3.086757
Exit Pupil Size            (RE') =      7.919456

Field Stop is at Surface #       0
Field Stop Size            (RFS) =      1.000000
Entrance Window Location (LW) =  -100.000000
Entrance Window Size        (RW) =      1.000000
Exit Window Location       (LW') =     96.773019
Exit Window Size           (RW') =     -1.008754
```

Table C.16 Vignetting output.

Surf	radius	displacement
1	8.000000	0.000000
2	8.120746	0.007967
3	8.339133	0.032503

Vignetting factor = 1.000

C.7 THIRD-ORDER ABERRATIONS

The *Third-order optics* option in the Main Menu is used to calculate the third-order aberration coefficients, as well as the axial and lateral color coefficients.

The third-order equations are based on both a marginal axial ray trace and a full-field chief ray trace. If the first-order stops and apertures have not been determined, they will be determined automatically, but the results will not be displayed.

This option first gives the optical invariant based on a marginal axial ray and the full-field chief ray. Then the Seidel aberration coefficients are listed for each surface and the sum over all surfaces is given. Finally, on the last line, the actual aberration contributions are listed. Output is shown in Table C.17.

C.8 EXACT RAY TRACING

Exact ray tracing allows you to determine accurately where any ray entering the system would end up. Exact ray tracing is based on an algorithm for tracing a single

Table C.17 Third-order aberration and chromatic aberration output.

The Optical Invariant is = -0.080000

surf	S	C	A	P	D	AC	LC
1	-0.169318	-0.004216	-0.000105	-0.016828	-0.000005	-0.026679	-0.000664
2	0.108533	0.001339	0.000017	0.002853	0.000000	0.008637	0.000107
3	-0.001405	0.000225	-0.000036	0.000782	0.000005	-0.003455	0.000552
	-0.062190	-0.002652	-0.000124	-0.013193	0.000000	-0.021496	-0.000005
	-0.392089	-0.016722	-0.000784	-0.083179	0.000000	-0.271058	-0.000069

skew ray, but meridional (or tangential) rays may be traced by simply defining the ray to be in the meridional plane. In addition to tracing a single exact ray, the exact ray tracing menu provides five options which are helpful for image evaluation as shown in Table C.18.

The first step in performing an arbitrary ray trace is to define the desired ray. After selecting the *absolute Coordinate ray trace* option, a sub-menu lists default values for the absolute position of the ray in the object plane and the direction cosines of the ray. In addition, the wavelength of the ray to be traced can be specified. The choices are: *short*, *middle* and *long* as specified by your choices of index of refraction and Abbe value. Selecting the appropriate options allows you to change any of these values. Selecting the Perform ray trace option will carry out the ray trace using the currently displayed ray definition.

Selecting the *Relative coordinate ray trace* also traces a certain ray, but now the ray is specified at the object relative to the full-field of view or object size. The direction of the ray is determined by the relative coordinates of the ray at the entrance pupil. As mentioned above, the wavelength can also be specified as shown in Table C.19

The output (Table C.20) for either type of ray trace consists of the intersection point (X, Y, Z) of the ray with each surface and the direction cosines (L, M, N) of the ray after each surface.

In addition to tracing a single exact ray, the exact ray tracing menu provides other options which are helpful in evaluating image quality.

When the *Longitudinal spherical aberration* option is selected from the exact ray trace menu, rays from the on-axis object point are traced through various entrance pupil positions in each of three wavelengths. For each of these different rays, an exact value for the longitudinal spherical aberration is calculated. The data shown in Table C.21 could then be used to draw an LSA plot which shows the effects of spherical aberration, as well as chromatic aberration and spherochromatism.

Table C.18 Exact ray tracing menu.

```
absolute Coordinate ray trace

Relative coordinate ray trace

Longitudinal spherical aberration

ray Intercept plot

Astigmatism plot

Distortion plot

Spot diagram

Quit (return to main menu)
```

Table C.19 Relative exact ray trace menu.

```
Height of ray in entrance pupil .   1.000000

Angle position (degrees). . . . .   0.000000

Object size . . . . . . . . . . .   1.000000

Wavelength (s, m, l). . . . . . .     middle

Perform exact ray trace

Quit (return to exact menu)
```

Table C.20 Exact ray trace output.

Surf	X	Y	Z	L	M	N
0	0.000000	1.000000	0.000000			
				0.000000	0.069829	0.997559
1	0.000000	8.094773	1.353904			
				0.000000	-0.102149	0.994769
2	0.000000	7.825043	2.630653			
				0.000000	-0.063160	0.998003
3	0.000000	7.750312	0.061513			
				0.000000	-0.092327	0.995729
4	0.000000	-1.217104	0.000000			

Table C.21 Longitudinal spherical aberration output.

Rel Aperture	d-LSA	C-LSA	F-LSA
0.0	-0.000000	1.039516	-2.333121
0.1	-0.049297	0.989116	-2.379992
0.2	-0.195448	0.839690	-2.518944
0.3	-0.433110	0.596688	-2.744869
0.4	-0.752926	0.269652	-3.048818
0.5	-1.140740	-0.126988	-3.417246
0.6	-1.576262	-0.572552	-3.830724
0.7	-2.030808	-1.037813	-4.261770
0.8	-2.463459	-1.481088	-4.671167
0.9	-2.814280	-1.841321	-5.001471
1.0	-2.991727	-2.025233	-5.164998

Selecting the *ray Intercept plot* option brings up a new menu as shown in Table C.22. Here you can select a relative field position. You can also choose between a tangential ray plot or a sagittal ray plot. When the *Perform ray intercept* option is selected, a fan of rays is traced from the specified object point in either the tangential or sagittal plane. The positions of the traced rays in the paraxial image plane are output.

Selecting either the *Astigmatism plot* or the *Distortion plot* option results in the corresponding data being displayed. The calculating for these plots are described in Chapter 6.

If the *Spot diagram* option is chosen, another menu appears (Table C.23) which allows you to select how many rays will be traced, the relative object position, and the wavelength. As always, the *Perform spot diagram* option will display the relevant data.

C.9 OPTIMIZATION

The *Optimization* option in the main menu allows you to modify an existing system to improve image quality. The optimization routine is based on damped linear least squares as described in Chapter 9. But optimization is not automatic, there is quite a bit of setting up required as shown in Table C.24.

First, *Variable selection* must be performed. This option displays all of the lens parameters one surface at a time. If a particular parameter is marked with a "no," that parameter is not a variable and will not change during optimization. This is the default status. To make a parameter variable, simply select that parameter. A "yes" should appear next to the parameter. If you change your mind, simply select the variable again.

Next, the error function must be specified. The first part of the error function may consist of one or more aberrations. The *Aberration selection* option lists certain aberrations with weights as shown in Table C.25. The default weight is 0.0, but selecting a particular aberration allows you to input a nonzero weight. Typically specifying a weight of 1.0 is sufficient. The aberrations listed are calculated according to the formulae described in Section 7.7.

Typically the desired focal length must be specified. Selecting the *Focal length* option from the optimization menu brings up the menu shown in Table C.26.

Table C.22 Ray intercept plot menu.

```
Object size (relative) . . . . . . .    0.000000

Direction. . . . . . . . . . . . . .   tangential

Perform exact ray trace

Quit (return to exact menu)
```

Table C.23 Spot diagram menu.

```
Number of rings of rays. . . . . .          4
Object size (relative) . . . . . .   0.000000
Wavelength (s,m,l) . . . . . . . .     middle
Perform spot diagram
Quit (return to exact menu)
```

Selecting *Target focal* length allows you to enter a value for the desired focal length. To enter a weight for the focal length, select *focal length Weight*. Depending on the system being optimized, you may need to adjust the magnitude of the focal length weight relative to the aberration weights to achieve the desired results.

Output selection controls how much or how little information is sent to the printer during an optimization. The optimization output which is sent to the screen consists only of the iteration number, the value of the aberration function, the amount of change in the aberration function from the previous iteration, and the current value of the damping factor. This is all the information you really need. However, this program was designed for students so more information is made available through the output selection menu (Table C.27).

The following information can be printed out by selecting the corresponding option. *Error function* outputs the sum of the squares of the weighted aberrations. Each individual weighted aberration can be viewed by selecting the *Aberration vector* option. As small changes are made in the variable parameters, new aberrations are calculated. The new aberrations can be seen by selecting *Modified aberration vectors*. The *Jacobian matrix* contains all of the first-order derivatives

Table C.24 Optimization menu.

```
Variable selection
Aberration selection
Focal length
Output selection
optimization Controls
Perform optimization
Quit (return to main menu)
```

Table C.25 Aberration selection menu.

```
spherical Aberration . . . . . . .   1.000000
tangential Coma. . . . . . . . . .   0.000000
Sagittal astigmatism . . . . . . .   0.000000
Tangential astigmatism . . . . . .   0.000000
Distortion . . . . . . . . . . . .   0.000000
aXial color. . . . . . . . . . . .   0.000000
Lateral color. . . . . . . . . . .   0.000000
Marginal chromatic . . . . . . . .   1.000000
Quit (return to previous menu)
```

of the aberrations with respect to the free parameters. As discussed in Chapter 9, the first-order derivatives are approximated by finite differences.

The damped least squares method solves a set of equations which are expressed in matrix notation in Eq. 9.43. These equations are linear with respect to parameter changes. The constant terms in the equations can be viewed by selecting the *Constant vector* option. The coefficients are printed out with the *coeFficient matrix* option. The solutions to the set of equations are parameter changes which can be seen by selecting the *chanGe vector* option. Finally, the new set of parameter values is output with the *Parameter vector* option.

The optimization menu (Table C.24) also contains an option for *optimization Controls*. Selecting this option brings up another menu which allows you to select either automatic iteration (the default) or manual iteration. If automatic iteration is set to "yes," then the iteration proceeds until the change in the error function between iterations is smaller than the value set with *Error function change limit*. The default value is 1.0×10^{-8}. If this limit is not reached before the number of iterations set with the *Maximum number of iterations* option is reached, then optimization is

Table C.26 Focal length menu.

```
Target focal length. . . . . . . .   1.000000
focal length Weight. . . . . . . .   0.000000
Quit (return to previous menu)
```

Table C.27 Printer output selection menu.

```
Error function . . . . . . . . . .          yes
Aberration vector. . . . . . . . .           no
Modified aberration vectors. . . .           no
Jacobian matrix. . . . . . . . . .           no
Constant vector. . . . . . . . . .           no
coeFficient matrix . . . . . . . .           no
chanGe vector. . . . . . . . . . .           no
Parameter vector . . . . . . . . .          yes
Quit (return to previous menu)
```

halted. The default value for the maximum number of iterations is 25. If automatic iteration is set to "no," then you will be prompted after each iteration to determine if the optimization should proceed. In this case, the maximum number of iterations and the error function change limit are ignored.

After all of the optimization parameters are set to your liking, choosing *Perform optimization* begins the process.

C.10 THIN LENSES

Just as you can use *y-nu* ray tracing for thin lenses, you can use ROSE.EXE to accomplish the same tasks.

Instead of entering the surface curvatures, you should enter the thin lens power which is just the inverse of the focal length. Enter positive values for positive lenses and negative values for negative lenses. The negative power which is used in the *y-nu* ray trace table is calculated under the paraxial ray trace option. The only other change in data entry is that the index of refraction should be the same on either side of the thin lens (usually 1.0). This is the tip off to the program that thin lens calculations are needed. Clear radii and separations should be entered as usual.

Once a thin lens system is entered, it can be changed, saved, and analyzed just as a thick lens was. Obviously, only paraxial calculations will make any sense for this type of system. Thin lenses can be mixed with thick lenses.

APPENDIX D

THIRD-ORDER OPTICS

I do not like to use equations without seeing the derivation of the equations from a starting point that I am already comfortable with. I think there are several reasons for my skepticism. The first is my training as a physicist. I simply expect to see the derivation. Second, I have seen too many typographical errors in equations, both in the preparation of this book and in other books, to trust what is printed. If I have the derivation, then I can check the equation for myself. Third, it is difficult to fully understand an equation without the derivation. The author must be extremely careful to define all of the symbols used in the equation, as well as restrictions on when the equation is applicable. For all of these reasons, I have included the derivation of the third-order aberration polynomial in this appendix. This derivation is based on the book, *Optical Aberration Coefficients*, by H. A. Buchdahl, Dover, 1968.

D.1 DERIVATION OF THE ABERRATION POLYNOMIAL

Our goal is to derive an equation for the final image aberration of a given optical system. The final image aberration is due to contributions from each surface in an optical system. How do we begin? Well, we do not want to try to take the exact ray tracing equations and convert them to third-order because there are so many equations involved. We will first develop an exact version of an equation for the aberration induced by one surface. Then by summing over all the surfaces, we will get the aberration in the final image. The next step is to expand the equation into a polynomial. That is, we will assume that ray trace parameters like y and u will be small, and this equation will end up with terms that are powers of u and y such as u^3 or y^3 or even uy^2. The paraxial rays that we will use to approximate a general skew ray will be a marginal axial ray and a full-field chief ray.

We begin with some definitions. As always, capital letters refer to exact ray trace parameters and lowercase letters are used for paraxial ray tracing. Let $\vec{h}_j^{\prime}$ be the paraxial image height for the jth surface. As shown in Figure D.1, the paraxial object and image for each surface is in the y–z plane. Let $\vec{H}_j^{\prime}$ be the position vector for the point where an exact ray crosses the paraxial image plane for this surface. $\vec{H}_j^{\prime}$ is not necessarily in the y–z plane. The aberration vector is defined as the difference of these two vectors:

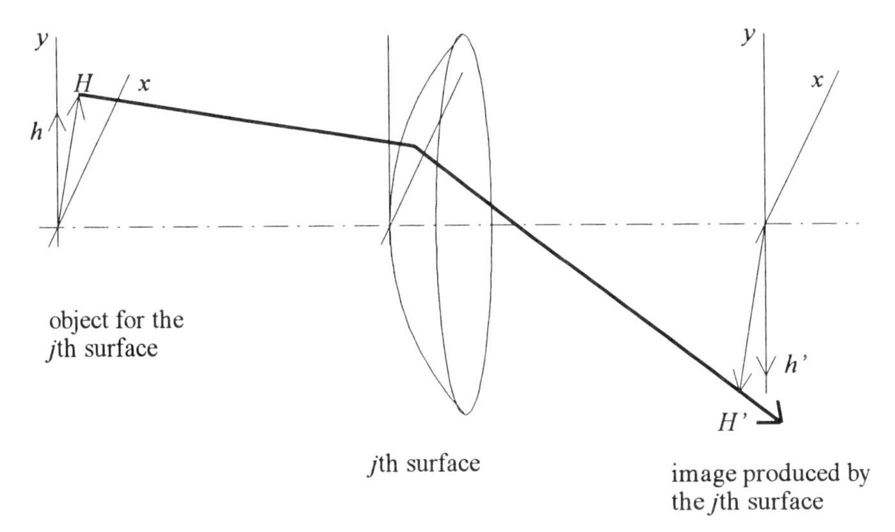

Figure D.1 The aberration of an image is the difference between the paraxial image position and the position of an exact ray.

$$\vec{\epsilon}_j' = \vec{H}_j' - \vec{h}_j' \ . \tag{D.1}$$

Now, as we know, the image produced by one surface becomes the object for the next surface. Therefore, we also need to describe an aberration equation for the object for the jth surface:

$$\vec{\epsilon}_j = \vec{H}_j - \vec{h}_j \ . \tag{D.2}$$

Eventually, we want to find the aberration produced by the last surface since this will be the total aberration of the system. But first we multiply Eq. D.1 by $n_j u_j$ and Eq. D.2 by $n_{j-1}\, u_{j-1}$. The slope angle is for the marginal axial ray. This is a trick that will make some terms cancel out later.

Now, we subtract the modified Eq. D.2 from the modified Eq. D.1:

$$n_j u_j \vec{\epsilon}_j' - n_{j-1}\, u_{j-1}\, \vec{\epsilon}_j =$$

$$n_j u_j \vec{H}_j' - n_{j-1}\, u_{j-1}\, \vec{H}_j + n_{j-1}\, u_{j-1}\, \vec{h}_j - n_j u_j \vec{h}_j' \ . \tag{D.3}$$

From the discussion of optical invariants in Chapter 3 we know that

$$n_j u_j \vec{h}_j' = n_{j-1} u_{j-1} \vec{h}_j , \tag{D.4}$$

and these two terms cancel out. This cancellation makes a lot of sense because paraxial imaging is perfect and should not contribute to the aberrations.

Next, we sum over all surfaces:

$$\sum_{j=1}^{k} \left[n_j u_j \vec{\epsilon}_j' - n_{j-1} u_{j-1} \vec{\epsilon}_j \right] = \sum_{j=1}^{k} \left[n_j u_j \vec{H}_j' - n_{j-1} u_{j-1} \vec{H}_j \right] . \tag{D.5}$$

The left side of this equation looks like

$$n_1 u_1 \vec{\epsilon}_1' - n_0 u_0 \vec{\epsilon}_1 + n_2 u_2 \vec{\epsilon}_2' - n_1 u_1 \vec{\epsilon}_2 + \cdots + n_k u_k \vec{\epsilon}_k' - n_{k-1} u_{k-1} \vec{\epsilon}_k . \tag{D.6}$$

But, by the definition given above, the aberration of the image of one surface is the aberration of the object of the next surface, therefore

$$\vec{\epsilon}_{j-1}' = \vec{\epsilon}_j . \tag{D.7}$$

The terms in Eq. D.6 cancel in pairs, leaving

$$n_k u_k \vec{\epsilon}_k' - n_0 u_0 \vec{\epsilon}_1 . \tag{D.8}$$

Next, we assume that the initial object is a real object. Then, we can set $\vec{\epsilon}_1 = 0$. (If the initial object were actually the image produced by another optical system, then we could apply the formulae we are currently deriving to find a value for $\vec{\epsilon}_1$.) We are left with

$$\vec{\epsilon}_k' = \frac{1}{n_k u_k} \sum_{j=1}^{k} \left[n_j u_j \vec{H}_j' - n_{j-1} u_{j-1} \vec{H}_j \right] . \tag{D.9}$$

The previous equation gives us the contribution of each surface to the final image aberration, however, it is not in a form that is easily converted to a third-order equation. We need some more manipulation. First, for convenience, let

$$\vec{\Gamma}_j = n_j u_j \vec{H}_j' - n_{j-1} u_{j-1} \vec{H}_j . \tag{D.10}$$

Next, define a vector which locates the intersection point of the exact ray with the vertex plane:

$$\vec{\chi}_j = \chi_j \hat{x} + \xi_j \hat{y} . \tag{D.11}$$

This equation is for a point on the incident ray. For the refracted ray we would have

$$\vec{\chi}_j' = \chi_j' \hat{x} + \xi_j' \hat{y} . \tag{D.12}$$

Also, we can define a vector whose components are the ray direction tangents

$$\vec{V}_{j-1} = \frac{L_{j-1}}{N_{j-1}} \hat{x} + \frac{M_{j-1}}{N_{j-1}} \hat{y} . \tag{D.13}$$

The refracted ray tangent vector is given by increasing the subscripts in the previous equation by one

$$\vec{V}_j = \frac{L_j}{N_j} \hat{x} + \frac{M_j}{N_j} \hat{y} . \tag{D.14}$$

As indicated in Figure D.2, simple translation from the object surface to the vertex surface can be written as

$$\vec{\chi}_j = \vec{H}_j - s_j \vec{V}_{j-1} , \tag{D.15}$$

where s_j is the object distance measured from the vertex plane. We continue to use our standard sign convention where distances to the left of the vertex are negative. This equation can be rearranged to give

$$\vec{H}_j = \vec{\chi}_j + s_j \vec{V}_{j-1} . \tag{D.16}$$

A similar equation may be found for the image

$$\vec{H}_j' = \vec{\chi}_j' + s_j' \vec{V}_j . \tag{D.17}$$

Substituting Eqs. D.16 and D.17 into Eq. D.10 gives

$$\vec{\Gamma}_j = n_j u_j \left[\vec{\chi}_j' + s_j' \vec{V}_j \right] - n_{j-1} u_{j-1} \left[\vec{\chi}_j + s_j \vec{V}_{j-1} \right] . \tag{D.18}$$

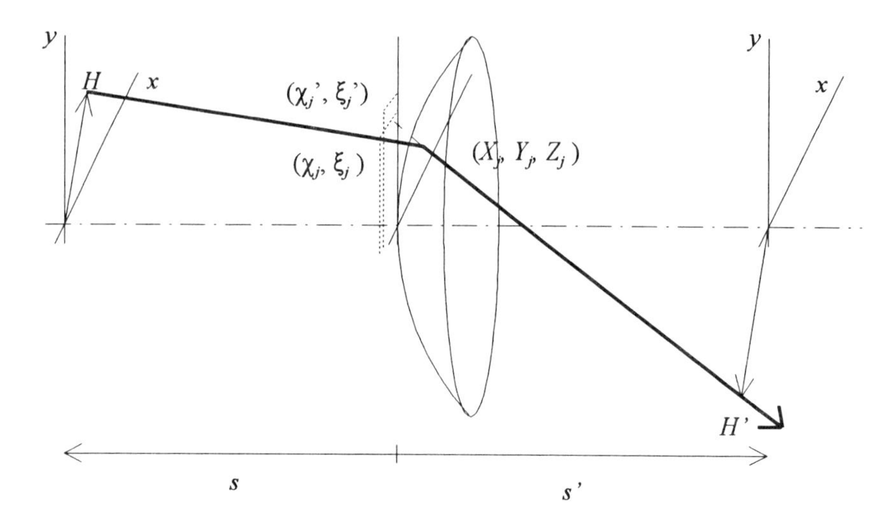

Figure D.2 The relationships between the incoming ray and the ray position in the vertex plane, and the refracted ray and the ray position in the vertex plane are shown. Both rays meet at the intersection point on the spherical surface.

We now begin to play games with this equation, keeping in mind the goal of a polynomial equation in the paraxial ray trace quantities. Note that the paraxial ray height at the jth surface may be expressed as

$$y_j = -s_j u_{j-1} = -s_j' u_j \,. \tag{D.19}$$

Therefore,

$$\vec{\Gamma}_j = n_j\left[u_j \vec{\chi}_j' - y_j \vec{V}_j\right] - n_{j-1}\left[u_{j-1} \vec{\chi}_j - y_j \vec{V}_{j-1}\right]. \tag{D.20}$$

Next, we substitute for the ray height

$$y_j = r_j\left(i_j - u_{j-1}\right) \tag{D.21}$$

to get

$$\vec{\Gamma}_j = n_j\left[u_j \vec{\chi}_j' - r_j\left(i_j' - u_j\right)\vec{V}_j\right] - n_{j-1}\left[u_{j-1} \vec{\chi}_j - r_j\left(i_j - u_{j-1}\right)\vec{V}_{j-1}\right]. \tag{D.22}$$

We can add and subtract terms like $i_j \vec{\chi}_j$ to get

$$\vec{\Gamma}_j = n_j\left\{\left(u_j - i_j'\right)\left[\vec{\chi}_j' + r_j\vec{V}_j\right] + i_j'\vec{\chi}_j'\right\}$$
$$- n_{j-1}\left\{\left(u_{j-1} - i_j\right)\left[\vec{\chi}_j + r_j\vec{V}_{j-1}\right] + i_j\vec{\chi}_j\right\}. \tag{D.23}$$

Recalling that $n_{j-1}i_j = n_j i_j'$ and $i_j - u_{j-1} = i_j' - u_j$, we get

$$\vec{\Gamma}_j = \left(u_{j-1} - i_j\right)\left\{n_j\left[\vec{\chi}_j' + r_j\vec{V}_j\right] - n_{j-1}\left[\vec{\chi}_j + r_j\vec{V}_{j-1}\right]\right\} + n_{j-1}i_j\left[\vec{\chi}_j' - \vec{\chi}_j\right]. \tag{D.24}$$

The terms inside the braces can be simplified by looking at the details of exact ray tracing. Recall that the ray direction is given by

$$\hat{U}_{j-1} = L_{j-1}\hat{x} + M_{j-1}\hat{y} + N_{j-1}\hat{z} \tag{D.25}$$

for the incident ray. For the refracted ray, the ray direction is

$$\hat{U}_j = L_j\hat{x} + M_j\hat{y} + N_j\hat{z}. \tag{D.26}$$

The direction of the normal to the surface is given by

$$\hat{r} = c_jX_j\hat{x} + c_jY_j\hat{y} + \left(c_jZ_j - 1\right)\hat{z}, \tag{D.27}$$

and the law of refraction is given by

$$n_{j-1}\hat{U}_{j-1} = n_j\hat{U}_j + \left(n_j\cos I_j' - n_{j-1}\cos I_j\right)\hat{r}. \tag{D.28}$$

We can simplify this equation by getting rid of the $\hat{r}$ term by taking the cross product of $\hat{r}$ with Eq. D.28. This gives us an alternate, and very simple, form of the law of refraction:

$$n_{j-1}\hat{r} \times \hat{U}_{j-1} = n_j\hat{r} \times \hat{U}_j. \tag{D.29}$$

Equation D.29 must be true for each component. Thus, if we write out just the x component we find

$$n_{j-1}\left[c_jY_jN_{j-1} - \left(c_jZ_j - 1\right)M_{j-1}\right] = n_j\left[c_jY_jN_j - \left(c_jZ_j - 1\right)M_j\right]. \tag{D.30}$$

Multiplying through by r_j and factoring out N_{j-1} or N_j gives

$$n_{j-1} N_{j-1}\left[Y_j - Z_j \frac{M_{j-1}}{N_{j-1}} + r_j \frac{M_{j-1}}{N_{j-1}}\right] = n_j N_j\left[Y_j - Z_j \frac{M_j}{N_j} + r_j \frac{M_j}{N_j}\right]. \tag{D.31}$$

This can be further simplified by remembering that the y component of Eq. D.15 can be expressed as

$$\xi_j = Y_j - Z_j \frac{M_{j-1}}{N_{j-1}}. \tag{D.32}$$

A similar equation, with primes, applies to the refracted ray:

$$\xi_j' = Y_j - Z_j \frac{M_j}{N_j}. \tag{D.33}$$

Upon substitution, we get

$$n_{j-1} N_{j-1}\left[\xi_j + r_j \frac{M_{j-1}}{N_{j-1}}\right] = n_j N_j\left[\xi_j' + r_j \frac{M_j}{N_j}\right]. \tag{D.34}$$

A similar equation may be derived for the x component

$$n_{j-1} N_{j-1}\left[\chi_j + r_j \frac{L_{j-1}}{L_{j-1}}\right] = n_j N_j\left[\chi_j' + r_j \frac{L_j}{N_j}\right]. \tag{D.35}$$

Combining these two equations proves that

$$n_{j-1} N_{j-1}\left[\vec{\chi}_j + r_j \vec{V}_{j-1}\right] = n_j N_j\left[\vec{\chi}_j' + r_j \vec{V}_j\right]. \tag{D.36}$$

We can now factor out a common term from Eq. D.24 to get

$$\vec{\Gamma}_j = (u_{j-1} - i_j) n_{j-1}\left[\vec{\chi}_j + r_j \vec{V}_{j-1}\right]\left[\frac{N_{j-1}}{N_j} - 1\right] + n_{j-1} i_j\left[\vec{\chi}_j' - \vec{\chi}_j\right]. \tag{D.37}$$

Since both the incoming and the refracted rays pass through the intersection point of the incoming ray and the spherical surface, the last term can be rewritten using the ray translation formula

$$\vec{X}_j = \vec{\chi}_j + Z_j \vec{V}_{j-1} = \vec{\chi}_j' + Z_j \vec{V}_j \,, \tag{D.38}$$

which then gives

$$\vec{\Gamma}_j = (u_{j-1} - i_j) n_{j-1} \left[\vec{\chi}_j + r_j \vec{V}_{j-1} \right] \left[\frac{N_{j-1}}{N_j} - 1 \right] - n_{j-1} i_j Z_j \left[\vec{V}_j - \vec{V}_{j-1} \right] . \tag{D.39}$$

This is an exact formula, and it is in a form that can be converted to third-order. All we need to do now is figure out how to write each factor as a series expansion in small quantities.

But before we can begin, we need to discuss just how we will express the exact ray in terms of small quantities. As discussed earlier, the small quantities will be paraxial ray parameters such as ray heights, slope angles, and incident angles. We could figure this out one ray at a time, but it will be easier and clearer to use some combination of the marginal axial ray and the full-field chief ray to approximate each exact ray. We could use the scheme developed in Chapter 8 for the y–$\bar{y}$ diagrams, but convention requires a different system as shown in Figure D.3. Here the full-field chief ray will be in the y–z plane, and the marginal axial ray will be in the x–z plane.

Now, imagine some general exact ray which starts on the object, but not necessarily at the base of the object or the top of the object (see Figure D.3). We will use the parameter

$$\eta = \frac{Y_0}{H} \tag{D.40}$$

to specify the starting position of the ray. If the object is at infinity, we would say that the object is not necessarily on axis or at the full-field angle, and

$$\eta = \frac{U_0}{U_{HFOV}} \,. \tag{D.41}$$

We complete the specification of the ray by indicating where it intersects the entrance pupil. The parameter

$$\rho = \frac{\sqrt{X_E^2 + Y_E^2}}{R_E} \tag{D.42}$$

tells us how far from the optical axis the intersection point is. Finally, θ is the angle that the line from the optical axis to the intersection point makes with respect to the

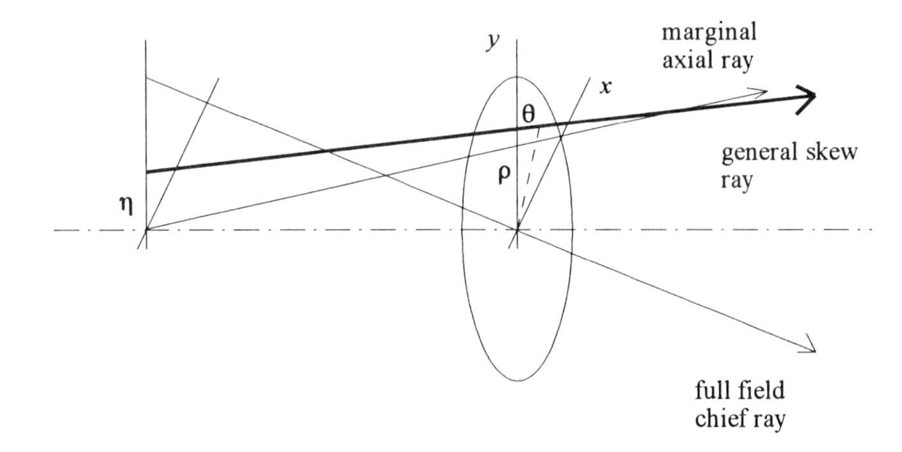

Figure D.3 The definition of relative object height and relative ray position in the entrance pupil.

y axis:

$$\square\square\square\ \theta \quad \frac{X_E}{Y_E}\ . \tag{D.43}$$

Any exact ray may be specified using these three parameters.

At any surface perpendicular to the optical axis, the skew ray coordinates may be expressed as

$$\chi_j = y_j\,\rho \sin\theta \tag{D.44}$$

and

$$\xi_j = \eta\,\bar{y}_j + y_j\,\rho\cos\theta\ . \tag{D.45}$$

Remember that unbarred parameters refer to the marginal axial ray and that barred parameters refer to the full-field chief ray. Angles must behave in the same fashion, so

$$\frac{L_{j-1}}{N_{j-1}} = u_{j-1}\,\rho\sin\theta \tag{D.46}$$

and

$$\frac{M_{j-1}}{N_{j-1}} = \eta\, \bar{u}_{j-1} + u_{j-1}\, \rho \cos\theta \ . \tag{D.47}$$

Similar equations will hold for the outgoing ray with an appropriate change in subscripts.

We could now begin substitution, but we can simplify the application of these equations somewhat by eliminating explicit references to the full-field ray. To do this, we use the optical invariant:

$$\Lambda = n_j\left[u_j \bar{y}_j - \bar{u}_j y_j\right] . \tag{D.48}$$

As we have shown in Problem 3.39, we also can express the optical invariant as

$$\Lambda = n_{j-1}\, r_j \left[u_{j-1}\, \bar{i}_j - \bar{u}_{j-1}\, i_j\right] . \tag{D.49}$$

Rearranging gives

$$\bar{u}_{j-1} = u_{j-1} \frac{\bar{i}_j}{i_j} - \frac{\Lambda}{n_{j-1}\, r_j\, i_j} \ . \tag{D.50}$$

Since we will be using the ratio of the incident angles a lot, let

$$\kappa_j = \frac{\bar{i}_j}{i_j} \ . \tag{D.51}$$

Note that if $i_j = 0$, then $\bar{u}_{j-1}$ is indeterminant. Clearly this is a problem. Eventually, we will get rid of κ_j. It is only used here to simplify the equations a bit.

Next, let

$$\gamma_j = n_{j-1}\, r_j\, i_j \ . \tag{D.52}$$

Therefore,

$$\bar{u}_{j-1} = \kappa_j\, u_{j-1} - \frac{\Lambda}{\gamma_j} \tag{D.53}$$

and

$$\bar{u}_j = \kappa_j u_j - \frac{\Lambda}{\gamma_j} . \tag{D.54}$$

Next, we need to rewrite the equation for ξ_j to get rid of the reference to $\bar{y}_j$. First, substitute

$$\bar{y}_j = r_j \left[\bar{i}_j - \bar{u}_{j-1} \right] . \tag{D.55}$$

Next, substitute $\kappa_j\, i_j$ for the refracted angle and Eq. D.53 for the slope angle to get

$$\xi_j = \eta\, r_j \left[\kappa_j i_j - \kappa_j u_{j-1} + \frac{\Lambda}{\gamma_j} \right] + y_j\, \rho \cos\theta . \tag{D.56}$$

Substituting for $i_j - u_{j-1}$ yields

$$\xi_j = y_j \left[\eta\, \kappa_j + \rho \cos\theta \right] + \frac{\eta\, r_j\, \Lambda}{\gamma_j} . \tag{D.57}$$

Similar substitutions may be made for the angles to give

$$\frac{M_{j-1}}{N_{j-1}} = u_{j-1} \left[\eta\, \kappa_j + \rho \cos\theta \right] - \frac{\eta\, \Lambda}{\gamma_j} \tag{D.58}$$

and

$$\frac{M_j}{N_j} = u_j \left[\eta\, \kappa_j + \rho \cos\theta \right] - \frac{\eta\, \Lambda}{\gamma_j} . \tag{D.59}$$

We are finally ready to begin making substitutions into Eq. D.39, but we will do this one factor at a time. First, we have after some simplification

$$\chi_j + r_j \frac{L_{j-1}}{N_{j-1}} = r_j\, i_j\, \rho \sin\theta \tag{D.60}$$

and

$$\xi_j + r_j \frac{M_{j-1}}{N_{j-1}} = r_j\, i_j \left[\eta\, \kappa_j + \rho \cos\theta \right] . \tag{D.61}$$

Next, the $N_{j-1} / N_j - 1$ term can be rewritten as a power series by recalling that

$$N_{j-1}^2 + M_{j-1}^2 + L_{j-1}^2 = 1 \ . \tag{D.62}$$

A similar equation applies to the ray after the surface. Combining these equations and factoring out N_{j-1} or N_j gives

$$\frac{N_{j-1}}{N_j} = \frac{\sqrt{1 + \left[\frac{M_j}{N_j}\right]^2 + \left[\frac{L_j}{N_j}\right]^2}}{\sqrt{1 + \left[\frac{M_{j-1}}{N_{j-1}}\right]^2 + \left[\frac{L_{j-1}}{N_{j-1}}\right]^2}} \ . \tag{D.63}$$

The binomial expansion may be applied to both the top and the bottom of the fraction on the right side of this equation to get a power series. Keeping only the lowest order terms gives

$$\frac{N_{j-1}}{N_j} - 1 \approx \frac{1}{2}\left[\frac{M_j}{N_j}\right]^2 - \frac{1}{2}\left[\frac{M_{j-1}}{N_{j-1}}\right]^2 + \frac{1}{2}\left[\frac{L_j}{N_j}\right]^2 - \frac{1}{2}\left[\frac{L_{j-1}}{N_{j-1}}\right]^2 . \tag{D.64}$$

Next, from Eqs. D.46 and D.47 we substitute for the direction cosine ratios to get

$$\frac{N_{j-1}}{N_j} - 1 = \frac{1}{2} u_j^2 \rho^2 \sin^2\theta - \frac{1}{2} u_{j-1}^2 \rho^2 \sin^2\theta$$

$$+ \frac{1}{2}\left[u_j(\eta\kappa_j + \rho\cos\theta) - \frac{\eta\Lambda}{\gamma_j}\right]^2 - \frac{1}{2}\left[u_{j-1}(\eta\kappa_j + \rho\cos\theta) - \frac{\eta\Lambda}{\gamma_j}\right]^2 . \tag{D.65}$$

We have dropped the approximately equal sign, but it is understood. After some considerable algebra to combine and cancel terms we get

$$\frac{N_{j-1}}{N_j} - 1 = \frac{1}{2}\left(u_j^2 - u_{j-1}^2\right)\left[\eta^2\kappa_j^2 + 2\eta\kappa_j\rho\cos\theta + \rho^2\right]$$

$$- \left(u_j - u_{j-1}\right)\left[\eta\kappa_j + \rho\cos\theta\right]\frac{\eta\Lambda}{\gamma_j} \ . \tag{D.66}$$

Next we take on the Z_j term. In the paraxial approximation, Z_j goes to zero.

Therefore, we want to write Z_j to second order. Recall that the equation for a sphere whose center is offset along the z axis is

$$Z_j^2 - 2 Z_j r_j + X_j^2 + Y_j^2 = 0 . \tag{D.67}$$

In our small-quantity approximation, $X_j^2 \approx \chi_j^2$, $Y_j^2 \approx \xi_j^2$; and since Z_j is small, Z_j^2 can be ignored. We are left with

$$2 Z_j r_j \approx \chi_j^2 + \xi_j^2 . \tag{D.68}$$

Substituting for χ and ξ gives

$$Z_j = \frac{1}{2 r_j}\left[y_j^2 \rho^2 \sin^2\theta + \left[y_j(\eta \kappa_j + \rho\cos\theta) + \frac{\eta \Lambda r_j}{\gamma_j} \right]^2 \right] . \tag{D.69}$$

Again, after some algebra we get

$$Z_j = \frac{y_j^2}{2 r_j}\left[\eta^2 \kappa_j^2 + 2 \eta \kappa_j \rho \cos\theta + \rho^2 \right]$$

$$+ y_j(\eta \kappa_j + \rho\cos\theta) \frac{\eta \Lambda}{\gamma_j} + \frac{\eta^2 \Lambda^2 r_j}{2 \gamma_j^2} . \tag{D.70}$$

Finally, we are ready to put all of the pieces together. But let's continue with only one component at a time. First, let's look at just the x component. From Eq. D.39 we have

$$\Gamma_{x,j} = (u_{j-1} - i_j) n_{j-1} \left[\chi_j + r_j \frac{L_{j-1}}{N_{j-1}} \right] \left[\frac{N_{j-1}}{N_j} - 1 \right]$$

$$- n_{j-1} i_j Z_j \left[\frac{L_j}{N_j} - \frac{L_{j-1}}{N_{j-1}} \right] . \tag{D.71}$$

Substituting gives

$$\Gamma_{x,j} = (u_{j-1} - i_j)\, n_{n-1} \left[r_j i_j \rho \sin\theta \right]$$

$$\times \frac{1}{2} \left[(u_j^2 - u_{j-1}^2) \left[\eta^2 \kappa_j^2 + 2\eta \kappa_j \rho \cos\theta + \rho^2 \right] \right.$$

$$\left. - 2(u_j - u_{j-1}) \left[\eta \kappa_j + \rho \cos\theta \right] \frac{\eta \Lambda}{\gamma_j} \right]$$

$$- \frac{n_{j-1} i_j}{2} y_j^2 \left[\eta^2 \kappa_j^2 + 2\eta \kappa_j \rho \cos\theta + \rho^2 \right] \times \left[(u_j - u_{j-1}) \rho \sin\theta \right]$$

$$- \frac{n_{j-1}^2 i_j^2 y_j}{2 r_j} (\eta \kappa_j + \rho \cos\theta) \frac{\eta \Lambda r_j}{\gamma_j} \times \left[(u_j - u_{j-1}) \rho \sin\theta \right]$$

$$- \frac{n_{j-1}^2 i_j^2}{4} \frac{\eta^2 \Lambda^2 r_j^2}{\gamma_j^2} \times \left[(u_j - u_{j-1}) \rho \sin\theta \right] . \qquad \text{(D.72)}$$

Some of the terms in this equation do not depend on Λ, some are linear in Λ, and some depend on Λ^2. It can be shown that the linear terms cancel out. In addition, if we substitute $y_j / r_j = i_j - u_{j-1}$ we can factor out many common factors. Thus,

$$\Gamma_{x,j} = -\frac{1}{2} n_{j-1} i_j (u_j - u_{j-1}) \rho \sin\theta$$

$$\times \left[y_j (u_j + i_j) \left[\eta^2 \kappa_j^2 + 2\eta \kappa_j \rho \cos\theta + \rho^2 \right] - \frac{\eta^2 \Lambda^2 r_j}{\gamma_j^2} \right] . \qquad \text{(D.73)}$$

Recall that $u_j - u_{j-1} = i_j' - i_j$. If we factor out an i_j and use the paraxial form of Snell's law we have

$$u_j - u_{j-1} = i_j \left[\frac{n_{j-1}}{n_j} - 1 \right] . \qquad \text{(D.74)}$$

If we now introduce

$$W_j = y_j n_{j-1} \left[\frac{n_{j-1}}{n_j} - 1 \right] (u_j + i_j) , \qquad \text{(D.75)}$$

Eq. D.73 can be simplified to

$$\Gamma_{x,j} = -\frac{1}{2} W_j i_j^2 \left[\eta^2 \kappa_j^2 + 2 \eta \kappa_j \rho \cos\theta + \rho^2 \right] \rho \sin\theta$$

$$- \frac{1}{2} n_{j-1} i_j^2 \left[\frac{n_{j-1}}{n_j} - 1 \right] \frac{\eta^2 \Lambda^2 r_j}{\gamma_j^2} \rho \sin\theta . \qquad \text{(D.76)}$$

Furthermore, recall that $\gamma_j = n_{j-1} r_j i_j$ so that the second term can be changed to yield

$$\Gamma_{x,j} = -\frac{1}{2} W_j i_j^2 \left[\eta^2 \kappa_j^2 + 2 \eta \kappa_j \rho \cos\theta + \rho^2 \right] \rho \sin\theta$$

$$- \frac{1}{2} \left[\frac{n_{j-1} - n_j}{n_{j-1} n_j r_j} \right] \eta^2 \Lambda^2 \rho \sin\theta . \qquad \text{(D.77)}$$

Now, let

$$P_j = \frac{n_{j-1} - n_j}{n_{j-1} n_j r_j} . \qquad \text{(D.78)}$$

Finally, rearrange the formula in powers of ρ and use the trigonometric identity $2 \sin\theta \cos\theta = \sin 2\theta$ to get

$$\Gamma_{x,j} = -\frac{1}{2} W_j i_j^2 \rho^3 \sin\theta - \frac{1}{2} W_j i_j \bar{i}_j \eta \rho \sin 2\theta$$

$$- \frac{1}{2} \left[W_j \bar{i}_j^2 + P_j \Lambda^2 \right] \eta^2 \rho \sin\theta . \qquad \text{(D.79)}$$

Now we can write the x component of the final image aberration by recalling Eq. D.9

$$\epsilon_x' = -\frac{1}{2 n_k u_k} \left[S \rho^3 \sin\theta + C \eta \rho^2 \sin 2\theta \right.$$

$$\left. + (A + P \Lambda^2) \eta^2 \rho \sin\theta \right] , \qquad \text{(D.80)}$$

where

$$
\begin{aligned}
S &= \sum_{j=1}^{k} W_j i_j^2 , \\
C &= \sum_{j=1}^{k} W_j i_j \bar{i}_j , \\
A &= \sum_{j=1}^{k} W_j \bar{i}_j^2 , \\
P &= \sum_{j=1}^{k} \frac{n_{j-1} - n_j}{n_{j-1} n_j r_j} .
\end{aligned}
\tag{D.81}
$$

An almost identical procedure is used to derive the y component so some steps will be omitted. We begin with

$$
\begin{aligned}
\Gamma_{y,j} = {} & n_{j-1}(u_{j-1} - i_j)\left[\xi_j + r_j \frac{M_{j-1}}{N_{j-1}}\right]\left[\frac{N_{j-1}}{N_j} - 1\right] \\
& - n_{j-1} i_j \left[\frac{M_j}{N_j} - \frac{M_{j-1}}{N_{j-1}}\right] Z_j .
\end{aligned}
\tag{D.82}
$$

Then we make the standard substitutions to get

$$
\begin{aligned}
\Gamma_{y,j} = {} & n_{j-1}(u_{j-1} - i_j)\left[r_j i_j (\eta\kappa_j + \rho\cos\theta)\right] \\
& \times \frac{1}{2}\left[(u_j^2 - u_{j-1}^2)\left[\eta^2\kappa_j^2 + 2\eta\kappa_j\rho\cos\theta + \rho^2\right]\right. \\
& \left. - 2(u_j - u_{j-1})\left[\eta\kappa_j + \rho\cos\theta\right]\frac{\eta\Lambda}{\gamma_j}\right] \\
& - n_{j-1} i_j \left[u_j(\eta\kappa_j + \rho\cos\theta) - \frac{\eta\Lambda}{\gamma_j} - u_{j-1}(\eta\kappa_j + \rho\cos\theta) + \frac{\eta\Lambda}{\gamma_j}\right] \\
& \times \frac{1}{2 r_j}\left[y_j^2\left[\eta^2\kappa_j^2 + 2\eta\kappa_j\rho\cos\theta + \rho^2\right]\right. \\
& \left. + 2 y_j(\eta\kappa_j + \rho\cos\theta)\frac{\eta\Lambda}{\gamma_j} + \frac{\eta^2\kappa_j^2 r_j^2}{\gamma_j^2}\right] .
\end{aligned}
\tag{D.83}
$$

As before, the terms linear in Λ cancel out. We can substitute $y_j / r_j = i_j - u_{j-1}$ and combine terms to get

$$\Gamma_{y,j} = -\frac{1}{2} n_{j-1} y_j i_j (\eta \kappa_j + \rho \cos\theta)(u_j - u_{j-1})(u_j + i_j)$$

$$\times (\eta^2 \kappa_j^2 + 2\eta \kappa_j \rho \cos\theta + \rho^2)$$

$$-\frac{1}{2}(u_{j-1} - u_j)(\eta \kappa_j + \rho \cos\theta) \frac{\eta^2 \Lambda^2}{n_{j-1} i_j r_j} . \tag{D.84}$$

Once again, we note that $u_j - u_{j-1} = i_j(n_{j-1} / n_j - 1)$. If we invoke W_j and P_j we can write the previous equation as

$$\Gamma_{y,j} = -\frac{1}{2} W_j i_j^2 (\eta \kappa_j + \rho \cos\theta)(\eta_2 \kappa_j^2 + 2\eta \kappa_j \rho \cos\theta + \rho^2)$$

$$-\frac{1}{2} P_j (\eta \kappa_j + \rho \cos\theta) \Lambda^2 \eta^2 . \tag{D.85}$$

Rearranging in powers of ρ and η gives

$$\Gamma_{y,j} = -\frac{1}{2} W_j i_j^2 \rho^3 \cos\theta - \frac{1}{2} W_j i_j \bar{i}_j \eta \rho^2 (1 + 2\cos^2\theta)$$

$$-\frac{1}{2}(3 W_j \bar{i}_j^2 + P_j \Lambda^2)\eta^2 \rho \cos\theta - \frac{\bar{i}_j}{2 i_j}(W_j \bar{i}_j^2 + P_j \Lambda^2)\eta^3 . \tag{D.86}$$

Next, we use the trigonometry identity $1 + 2\cos^2\theta = 2 + \cos 2\theta$ and sum over all the surfaces to get the y component of the third-order aberration:

$$\epsilon_y^{\prime} = -\frac{1}{2 n_k u_k}\Big[S\rho^3 \cos\theta + C\eta\rho^2 (2 + \cos 2\theta)$$

$$+ (3A + P\Lambda^2)\eta^2 \rho \cos\theta + D\eta^3\Big] , \tag{D.87}$$

where

$$D = \sum_{j=1}^{k} \frac{\bar{i}_j}{i_j}\left[W_j \bar{i}_j^2 + P_j \Lambda^2\right] . \tag{D.88}$$

There is clearly a problem with this formula. As mentioned earlier, it is possible for the incident angle of the marginal axial ray to have a value of zero. Therefore, we must rewrite Eq. D.88 to eliminate i_j. We begin by expanding one of the terms in the sum

$$D_j = \frac{y_j}{i_j} n_{j-1} \left(\frac{n_{j-1}}{n_j} - 1 \right) (u_j + i_j) \bar{i}_j^3 + \bar{i} \left(\frac{n_{j-1}}{n_j} - 1 \right) \frac{\Lambda^2}{n_j r_j i_j} . \tag{D.89}$$

This formula can be rewritten by making use of Eqs. D.49 and D.52

$$D_j = \bar{y}_j n_{j-1} \left(\frac{n_{j-1}}{n_j} - 1 \right) (u_j + i_j) \bar{i}_j^2$$
$$- \frac{\bar{i}_j^2}{i_j} \left(\frac{n_{j-1}}{n_j} - 1 \right) (u_j + i_j) \Lambda + \bar{i}_j \left(\frac{n_{j-1}}{n_j} - 1 \right) \frac{\Lambda^2}{\gamma_j} . \tag{D.90}$$

On our way to making this formula simpler, we first make it more complicated by adding and subtracting $\bar{y}_j n_{j-1} (n_{j-1} / n_j - 1) u_{j-1} \bar{i}_j^2$ and using Eq. D.53 to get

$$D_j = \bar{y}_j n_{j-1} \left(\frac{n_{j-1}}{n_j} - 1 \right) (u_j - u_{j-1}) \bar{i}_j^2$$
$$+ \bar{y}_j n_{j-1} \left(\frac{n_{j-1}}{n_j} - 1 \right) u_{j-1} \bar{i}_j^2 + \bar{y}_j n_{j-1} \left(\frac{n_{j-1}}{n_j} - 1 \right) i_j \bar{i}_j^2$$
$$- \frac{\bar{i}_j^2}{i_j} \left(\frac{n_{j-1}}{n_j} - 1 \right) (u_j + i_j) \Lambda + \frac{\bar{i}_j}{i_j} \left(\frac{n_{j-1}}{n_j} - 1 \right) (u_{j-1} \bar{i}_j - \bar{u}_{j-1} i_j) \Lambda . \tag{D.91}$$

Next, we use the identity from Problem 3.40 to show that

$$D_j = \bar{y}_j n_{j-1} \left(\frac{n_{j-1}}{n_j} - 1 \right) (\bar{u}_j - \bar{u}_{j-1}) i_j \bar{i}_j$$
$$+ \bar{y}_j n_{j-1} \left(\frac{n_{j-1}}{n_j} - 1 \right) u_{j-1} \bar{i}_j^2 + \bar{y}_j n_{j-1} \left(\frac{n_{j-1}}{n_j} - 1 \right) i_j \bar{i}_j^2$$
$$- \frac{\bar{i}_j^2}{i_j} \left(\frac{n_{j-1}}{n_j} - 1 \right) (u_j + i_j) \Lambda + \frac{\bar{i}_j}{i_j} \left(\frac{n_{j-1}}{n_j} - 1 \right) (u_j \bar{i}_j - \bar{u}_j i_j) \Lambda . \tag{D.92}$$

Rearranging and canceling terms yields

$$D_j = \bar{y}_j n_{j-1} \left(\frac{n_{j-1}}{n_j} - 1 \right) (\bar{u}_j + \bar{i}_j) i_j \bar{i}_j$$

$$+ \bar{y}_j n_{j-1} \left(\frac{n_{j-1}}{n_j} - 1 \right) u_{j-1} \bar{i}_j^2 - \bar{y}_j n_{j-1} \left(\frac{n_{j-1}}{n_j} - 1 \right) \bar{u}_{j-1} i_j \bar{i}_j$$

$$- \frac{\bar{i}_j^2}{i_j} \left(\frac{n_{j-1}}{n_j} - 1 \right) i_j \Lambda - \frac{\bar{i}_j}{i_j} \left(\frac{n_{j-1}}{n_j} - 1 \right) \bar{u}_j i_j \Lambda \ . \tag{D.93}$$

The second and third terms can be combined using the optical invariant:

$$D_j = \bar{y}_j n_{j-1} \left(\frac{n_{j-1}}{n_j} - 1 \right) (\bar{u}_j + \bar{i}_j) i_j \bar{i}_j + \bar{y}_j n_{j-1} \left(\frac{n_{j-1}}{n_j} - 1 \right) \bar{i}_j \frac{\Lambda}{r_j}$$

$$- \bar{i}_j^2 \left(\frac{n_{j-1}}{n_j} - 1 \right) \Lambda - \bar{i}_j \left(\frac{n_{j-1}}{n_j} - 1 \right) \bar{u}_j \Lambda \ . \tag{D.94}$$

We next use Eq. D.55 to get

$$D_j = \bar{y}_j n_{j-1} \left(\frac{n_{j-1}}{n_j} - 1 \right) (\bar{u}_j + \bar{i}_j) i_j \bar{i}_j + \left(\frac{n_{j-1}}{n_j} - 1 \right) (\bar{i}_j - \bar{u}_{j-1}) \bar{i}_j \Lambda$$

$$- \bar{i}_j^2 \left(\frac{n_{j-1}}{n_j} - 1 \right) \Lambda - \bar{i}_j \left(\frac{n_{j-1}}{n_j} - 1 \right) \bar{u}_j \Lambda \ . \tag{D.95}$$

Finally, by canceling terms and making use of Eq. D.74 we have

$$D_j = \bar{y}_j n_{j-1} \left(\frac{n_{j-1}}{n_j} - 1 \right) (\bar{u}_j + \bar{i}_j) i_j \bar{i}_j + (\bar{u}_{j-1}^2 - \bar{u}_j^2) \Lambda \ . \tag{D.96}$$

We can now write

$$D = \sum_{j=1}^{k} \left[\bar{W}_j i_j \bar{i}_j + (\bar{u}_{j-1}^2 - \bar{u}_j^2) \Lambda \right] , \tag{D.97}$$

where

$$\bar{W}_j = \bar{y}_j n_{j-1}\left[\frac{n_{j-1}}{n_j} - 1\right](\bar{u}_j + \bar{i}_j) . \tag{D.98}$$

D.2 THIN LENS FORMULAE

For many applications, a thin lens is an adequate first approximation to the final design of a system. We need to derive thin lens versions of the coefficients of the aberration polynomial to help with the predesign of systems. Since a thin lens consists of two surfaces without an effective separation, we can determine the coefficients as shown below.

D.2.1 Spherical Aberration

Spherical aberration is represented by the S term where

$$S = \sum_{j=1}^{k} n_{j-1} y_j i_j^2 \left[\frac{n_{j-1}}{n_j} - 1\right](u_j + i_j) . \tag{D.99}$$

We will represent the thin lens version as S_t. We only need two terms in the sum. In addition, since the lens is thin $y_1 = y_2 = y$, and we are assuming the lens is air so $n_0 = 1$, $n_2 = 1$. Let $n_1 = n$, then

$$S_t = y\left[i_1^2\left(\frac{1-n}{n}\right)(u_1 + i_1) + n i_2^2 (n-1)(u_2 + i_2)\right] . \tag{D.100}$$

Now, we can substitute for all of the angles in the previous equation. First, since u_0 is the slope angle of the marginal axial ray we have

$$u_0 = -\frac{y}{s} = -vy , \tag{D.101}$$

where s is the object distance measured from the thin lens. The vergence v equals $1/s$. For a real object, to the left of the lens, both s and v are negative.

Next, consider the slope angle after the first surface. From the *y-nu* ray trace equations we have

$$n_j u_j = n_{j-1} u_{j-1} - (n_j - n_{j-1}) c_j y_j \ . \tag{D.102}$$

This can be applied to u_1 to yield

$$u_1 = \frac{u_0}{n} - \left(1 - \frac{1}{n} \right) c_1 y \ . \tag{D.103}$$

Substituting for u_0 gives

$$u_1 = y \left[\left(\frac{1 - n}{n} \right) c_1 - \frac{1}{n} v \right] . \tag{D.104}$$

Similarly,

$$u_2 = n u_1 - (1 - n) c_2 y \ . \tag{D.105}$$

We can eliminate the c_2 factor by using $c_2 = c_1 - c$:

$$u_2 = y [(1 - n) c - v] \ . \tag{D.106}$$

Next, we look at the incidence angles. Again, from *y-nu* ray tracing we know that

$$i_j = u_{j-1} + c_j y_j \ . \tag{D.107}$$

Applying the previous equation to i_1 gives

$$i_1 = u_0 + c_1 y \tag{D.108}$$

and

$$i_1 = y [c_1 - v] \ . \tag{D.109}$$

For the second surface we have

$$i_2 = u_1 + c_2 y \tag{D.110}$$

and

$$i_2 = y\left[\frac{1}{n}c_1 - \frac{1}{n}v - c\right]. \tag{D.111}$$

We are now ready to substitute into Eq. D.100:

$$S_t = y^4[c_1 - v]\left[\left(\frac{1-n}{n}\right)c_1 + \left(\frac{n-1}{n}\right)v\right]\left[\frac{1}{n}c_1 - \left(\frac{n+1}{n}\right)v\right]$$
$$+ [c_1 - v - nc]\left[\left(\frac{n-1}{n}\right)c_1 + \left(\frac{1-n}{n}\right)v + (1-n)c\right]$$
$$\times\left[\frac{1}{n}c_1 + \left(\frac{n+1}{n}\right)v - nc\right]. \tag{D.112}$$

The next step is to expand the previous equation. Several terms will cancel out, and the remaining terms can be combined to yield

$$S_t = -y^4(n-1)c\left[\frac{(n+2)}{n}c_1^2 - (2n+1)cc_1 - \frac{4(n+1)}{n}vc_1\right.$$
$$\left. + n^2c^2 + (3n+1)cv + \frac{(3n+2)}{n}v^2\right]. \tag{D.113}$$

D.2.2 Coma

Coma is represented by the C term where

$$C = \sum_{j=1}^{k} n_{j-1}y_j i_j \bar{i}_j\left[\frac{n_{j-1}}{n_j} - 1\right](u_j + i_j). \tag{D.114}$$

We will us the subscript t again to indicate the thin lens version. Making the same substitutions as in the previous section we have

$$C_t = y\left[i_1\bar{i}_1\left(\frac{1-n}{n}\right)(u_1 + i_1) + ni_2\bar{i}_2(n-1)(u_2 + i_2)\right]. \tag{D.115}$$

The angle substitutions can be made as in the derivation of the spherical aberration coefficient, except that now the chief ray incidence angles must be accounted for. In general, the aperture stop will not be at the thin lens as shown in Figure D.4. From this we determine that

$$\bar{u}_0 = \frac{h - \bar{y}}{s} = v(h - \bar{y})\ . \tag{D.116}$$

Substituting into the incidence angle formula gives

$$\bar{i}_1 = vh + (c_1 - v)\bar{y}\ . \tag{D.117}$$

Similar substitutions into the equation for the slope angle after the first surface and the incidence angle at the second surface gives

$$\bar{u}_1 = \frac{1}{n}[hv - \bar{y}(v + (n - 1)c_1]\tag{D.118}$$

and

$$\bar{i}_2 = \frac{1}{n}(h - \bar{y})v + \left(\frac{c_1}{n} - c\right)\bar{y}\ . \tag{D.119}$$

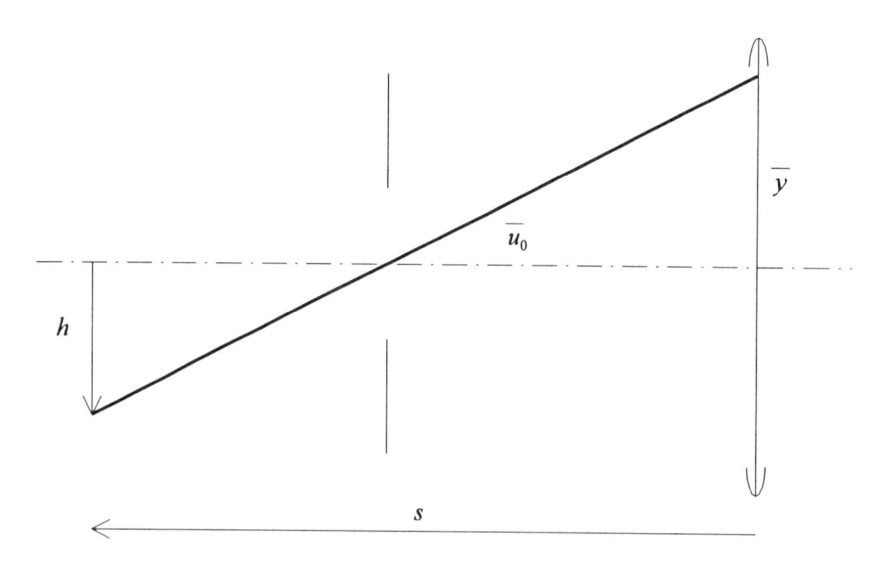

Figure D.4 The relationship between the slope angle of the full-field chief ray and the object height, object position, and ray height at the lens.

The next step is to substitute for the angles in Eq. D.115:

$$C_t = y^2\left[c_1 - v\right]\left[hv + (c_1 - v)y\right]\left[\frac{1-n}{n}\right]$$

$$\times\left[\left[y\left(\frac{1-n}{n}\right)c_1 - \frac{1}{n}v\right] + y[c_1 - v]\right]$$

$$+ ny^2\left[\frac{1}{n}c_1 - \frac{1}{n}v - c\right]\left[\frac{1}{n}hv - \frac{1}{n}v\bar{y} - c\bar{y} + \frac{1}{n}c_1\bar{y}\right]$$

$$\times (n-1)\left[y[(1-n)c - v] + y\left[\frac{1}{n}c_1 - \frac{1}{n}v - c\right]\right]. \qquad \text{(D.120)}$$

Multiplying out all of the factors leads to several terms canceling out. In particular, we can separate terms that contain only powers of y from those with $\bar{y}$, leaving

$$C_t = y^3(n-1)hv\left[nc^2 - \left(\frac{n+1}{n}\right)cc_1 + \left(\frac{2n+1}{n}\right)vc\right]$$

$$- y^3(n-1)\bar{y}\left[n^2c^3 - (2n+1)c^2c_1 + (3n+1)vc^2\right]$$

$$- y^3(n-1)\bar{y}\left[\left(\frac{n+2}{n}\right)cc_1^2\right.$$

$$\left.- 4\left(\frac{n+1}{n}\right)vcc_1 + \left(\frac{3n+2}{n}\right)v^2c\right]. \qquad \text{(D.121)}$$

This can be condensed to

$$C_t = C_t^* + \frac{\bar{y}}{y}S_t, \qquad \text{(D.122)}$$

where

$$C_t^* = y^3(n-1)hv\left[nc^2 - \left(\frac{n+1}{n}\right)cc_1 + \left(\frac{2n+1}{n}\right)vc\right] \qquad \text{(D.123)}$$

is the coma contribution from the surface if the stop is at the surface. The second term in Eq. D.122 is the coma contribution if the stop is shifted away from the surface. This is our first example of a stop shift formula.

It is important to note that if the object is at infinity, then the substitution

$$h v = u_{HFOV} \tag{D.124}$$

should be made.

Notice that Eq. D.122 contains a y in the denominator. It turns out that this is not as big a problem as having an i in the denominator. The marginal axial ray height should only go to zero at the image. This would be a problem only if we had a lens at the image (a field lens). In this case, just don't use the thin lens equations at this lens.

D.2.3 Astigmatism

Astigmatism is given by

$$A = \sum_{j=1}^{k} n_{j-1} y_j \bar{i}_j^{\,2} \left[\frac{n_{j-1}}{n_j} - 1 \right] (u_j + i_j) \ . \tag{D.125}$$

For the thin lens version, we need only two terms in the sum. We also make the standard substitutions for index and ray height:

$$A_t = y \bar{i}_1^{\,2} \left[\frac{1}{n} - 1 \right] (u_1 + i_1) + n y \bar{i}_2^{\,2} [n - 1] (u_2 + i_2) \ . \tag{D.126}$$

Next, use the angle substitutions already discussed to get

$$\begin{aligned} A_t = {} & y^2 (n - 1) \left[h v + (c_1 - v) \bar{y} \right]^2 \left[\left(\frac{n + 1}{n^2} \right) v - \frac{1}{n^2} c_1 \right] \\ & + y^2 (n - 1) n \left[\frac{hv}{n} + \left(\frac{c_1}{n} - \frac{v}{n} - c \right) \bar{y} \right]^2 \\ & \times \left[\frac{c_1}{n} - \left(\frac{n + 1}{n} \right) v - n c \right] . \end{aligned} \tag{D.127}$$

The next step is to do the expansion, looking for terms to cancel. Also, we can factor terms containing $(h v)^2$, $\bar{y} h v$, and $\bar{y}^2$:

$$A_t = -y^2 (h\,v)^2 (n - 1) c$$
$$+ 2y^2 \bar{y} h\,v (n - 1) \left[n c^2 + \left(\frac{2n + 1}{n} \right) v c - \left(\frac{n + 1}{n} \right) c c_1 \right]$$
$$- y^2 \bar{y}^2 (n - 1) \left[n^2 c^3 - (2n + 1) c^2 c_1 + (2n + 1) c^2 v \right] \tag{D.128}$$
$$- y^2 \bar{y}^2 (n - 1) \left[\left(\frac{n + 2}{n} \right) c c_1^2 \right.$$
$$\left. - r \left(\frac{n + 1}{n} \right) v c c_1 + \left(\frac{3n + 2}{n} \right) v^2 c \right] .$$

Once again, this can be shortened to give

$$A_t = A_t^* + 2 \frac{\bar{y}}{y} C_t^* - \left(\frac{\bar{y}}{y} \right)^2 S_t \,, \tag{D.129}$$

where

$$A_t^* = - \frac{(y\,h\,v)^2}{f'} \,. \tag{D.130}$$

Remember for an object at infinity that $h\,v = u_{HFOV}$.

D.2.4 Petzval Curvature

The Petzval curvature is different in that it does not contain any paraxial ray angles

$$P = \sum_{j=1}^{k} \frac{n_{j-1} - n_j}{n_{j-1} n_j} c_j \,. \tag{D.131}$$

Making our standard substitutions for the index of refraction gives

$$P_t = \frac{1 - n}{n} c_1 + \frac{n - 1}{n} c_2 \,. \tag{D.132}$$

Combining this equation with the lens maker's equation for thin lenses gives

$$P_t = -\frac{1}{nf'} . \tag{D.133}$$

D.2.5 Distortion

Since field curvature was so easy to derive, we can expect just the opposite from distortion. We begin by recalling that distortion is given by

$$D = \sum_{j=1}^{k} \kappa_j \left[W_j \bar{i}_j^2 + P_j \Lambda^2 \right] . \tag{D.134}$$

For the thin lens version we can make use of the fact that $A_j = W_j \bar{i}_j^2$ to write

$$D_t = \kappa_1 \left[A_1 + P_1 \Lambda^2 \right] + \kappa_2 \left[A_2 + P_2 \Lambda^2 \right] . \tag{D.135}$$

Now all we need to do is identify all of the pieces.

First, for κ_1 we have

$$\kappa_1 = \frac{\bar{i}_1}{i_1} = \frac{hv}{(c_1 - v)y} + \frac{\bar{y}}{y} . \tag{D.136}$$

Similarly,

$$\kappa_2 = \frac{hv}{(c_1 - v - nc)y} + \frac{\bar{y}}{y} . \tag{D.137}$$

With these substitutions we can combine the terms containing $\bar{y}$ to get

$$D_t = \frac{hv}{(c_1 - v)y}\left(A_1 + P_1 \Lambda^2\right) + \frac{hv}{(c_1 - v - nc)y}\left(A_2 + P_2 \Lambda^2\right) + \frac{\bar{y}}{y}\left(A_t + P_t \Lambda^2\right) . \tag{D.138}$$

To simplify this equation, we will once again make use of the optical invariant. Since Λ is the same at all of the surfaces, we can evaluate it at any one that is convenient. At the object we have

$$\Lambda = -hvy \, . \tag{D.139}$$

We will eventually make this substitution wherever necessary, but first we expand just certain terms for convenience:

$$A_1 + P_1\Lambda^2 = y\bar{i}_1^2\left[\frac{1-n}{n}\right](u_1 + i_1) + \left[\frac{1-n}{n}\right]c_1 h^2 v^2 y^2 \, . \tag{D.140}$$

We make the normal angle substitutions and expand this equation. Some terms will cancel and the remaining terms all have the common factor $y(c_1 - v)$. We are left with

$$\frac{hv}{(c_1 - v)y}\left(A_1 + P_1\Lambda^2\right) = hvy(n-1)$$
$$\times\left\{-\left(\frac{n+1}{n^2}\right)h^2v^2 + 2\left[\left(\frac{n+1}{n^2}\right)v - \frac{c_1}{n^2}\right]h\bar{y}v\right.$$
$$\left. + (c_1 - v)\bar{y}^2\left[\left(\frac{n+1}{n^2}\right)v - \frac{c_1}{n^2}\right]\right\} . \tag{D.141}$$

An identical procedure for the second term from Eq. D.124 yields

$$\frac{hv}{(c_1 - v - nc)y}\left(A_2 + P_2\Lambda^2\right) = hvy\left(\frac{n-1}{n}\right)$$
$$\left\{\left(\frac{n+1}{n}\right)h^2v^2\right.$$
$$\left. + \left[\frac{2}{n}hv\bar{y} + \left(\frac{c_1}{n} - \frac{v}{n} - c\right)\bar{y}^2\right]\left[c_1 - (n+1)v - n^2c\right]\right\} . \tag{D.142}$$

Combining both of these terms yields

$$hvy(n-1)\left\{\bar{y}^2(c_1 - v)\left[\left(\frac{n+1}{n^2}\right)v - \frac{c_1}{n^2}\right] - 2h\bar{y}vc\right.$$
$$\left. + \bar{y}^2\left[\frac{c_1}{n} - \frac{v}{n} - c\right]\left[\frac{c_1}{n} - \left(\frac{n+1}{n}\right)v - nc\right]\right\} . \tag{D.143}$$

This can be simplified by recognizing the thin lens coma and astigmatism equations:

$$2\frac{\bar{y}}{y}A_t^* + \frac{\bar{y}^2}{y^2}C_t^* . \tag{D.144}$$

And finally,

$$D_t = \left(3A_t^* + P_t\Lambda^2\right)\frac{\bar{y}}{y} + 3\,C_t^*\frac{\bar{y}^2}{y^2} + S_t\frac{\bar{y}^3}{y^3} . \tag{D.145}$$

D.2.6 Axial Color

A thin lens equation for axial color has already been developed, but it is presented again here for completeness. Recall that

$$AC = \sum_{j=1}^{k} n_{j-1}y_j i_j\left[\frac{\delta n_{j-1}}{n_{j-1}} - \frac{\delta n_j}{n_j}\right] . \tag{D.146}$$

The thin lens version is

$$AC_t = y\left(\frac{\delta n}{n}\right)\left[-i_1 + n\,i_2\right] . \tag{D.147}$$

Making the usual angle substitutions leads to

$$AC_t = -y^2\,\delta n\,(c_1 - c_2) . \tag{D.148}$$

Multiplying by $n - 1$ over $n - 1$ leads to

$$AC_t = -\frac{y^2}{f'\,V} . \tag{D.149}$$

D.2.7 Lateral Color

The equation for lateral color is

$$LC = \sum_{j=1}^{k} n_{j-1} y_j \bar{i}_j \left[\frac{\delta n_{j-1}}{n_{j-1}} - \frac{\delta n_j}{n_j} \right]. \quad \text{(D.150)}$$

The thin lens version is

$$LC_t = y \left(\frac{\delta n}{n} \right) \left[-\bar{i}_1 + n \bar{i}_2 \right]. \quad \text{(D.151)}$$

After substituting for the angles we are left with

$$LC_t = -y \bar{y} \delta n \, c \,. \quad \text{(D.152)}$$

Again multiplying the top and bottom by $n - 1$ leads to

$$LC_t = -\frac{y \bar{y}}{f' V}. \quad \text{(D.153)}$$

APPENDIX E

ANSWERS TO SELECTED PROBLEMS

CHAPTER 1

1.3 3.00×10^8 m/s, 2.25×10^8 m/s, 1.24×10^8 m/s

1.9 LF5, 1.58144, 40.85

1.11 27.58, SF4, 755276

1.19 192°

1.24 29.77°

1.26 2

CHAPTER 2

2.10 2.00 mm

2.13 −3.33 cm

2.15 2.66 cm, 0.666 cm

2.17 $f' = 4.00$ cm, $f = -6.00$ cm, −8.00 cm, +3.00

2.19 −3.00 cm

2.21 26 cm, −25×

2.23 76.0 cm, 50.0 cm, −37.5 cm

2.27 (a) −20, −10, −1
(b) −10, −40, −20
(c) 10, −10, 2
(d) infinity, −3, 0
(e) −12, +48, −32
(f) −24, −16, −8
(g) 4, 12, 0.333

2.29 (a) −40, −30, 20, 3
(b) −80, −20, 30, −0.333
(c) 1.25, 36, −30, 6
(d) 8, 2, 6, −0.333
(e) 1.5, −36, −3, −0.333
(f) −12, −36, −24, 36

2.32 40.0 cm, −1.0×

2.33 −40×, −2.89 cm, −0.55 cm (toward objective)

2.34 −100×, 0.524°, −52.4°, 3.66 cm

CHAPTER 3

3.14 4.86 cm, −0.783 cm, real, inverted, reduced

3.15 −6.52 cm, 7.27 cm, virtual, upright, enlarged

3.16 −4.86 cm, −0.782, real, inverted, reduced

3.17 6.52 cm, 7.27 cm, virtual, upright, enlarged

3.19 180 cm, −7.99 cm, real, inverted, enlarged

3.21 $f' = 60.1$ cm, $l_{F'} = 59.9$ cm, $l_{H'} = -0.145$ cm, $l_F = -60.1$ cm, $l_H = 0.145$ cm

3.23 180 cm, −7.99 cm, real, inverted, enlarged

3.25 −0.970 cm, 14.6 cm

3.27 0.200 cm

3.28 5.67 cm, −0.380 cm

3.29 $f' = 4.48$ cm, $l_{F'} = 3.96$ cm, $l_{H'} = -0.52$ cm, $l_F = -4.20$ cm, $l_H = 0.28$ cm

3.34 $f' = 1.016$ cm, $l_{F'} = 0.688$ cm, $l_{H'} = -0.328$ cm, $l_F = -688$ cm, $l_H = 0.328$ cm, $s' = 1.255$ cm, $h' = -0.353$ cm

3.35 $f' = 1.005$ cm, $l_{F'} = 0.528$ cm, $l_{H'} = -0.476$ cm, $l_F = -1.540$ cm, $l_H = -0.535$ cm, telephoto

CHAPTER 4

4.5 (a) 5 mm, (b) 5 mm, (c) 1 mm, (d) 1 mm

4.7 surface 1, $l_E = 0.0$, $R_E = 2.00$ cm, $l_{E'} = -4.00$ cm, $R_{E'} = 4.00$ cm

4.9 surface 2, $l_E = 2.00$ cm, $R_E = 4.00$ cm, $l_{E'} = 0.0$, $R_{E'} = 4.00$ cm, $t_0 = 2.00$ cm

4.11 surface 2, $l_E = 4.00$ cm, $R_E = 1.125$ cm, $l_{E'} = -0.556$ cm, $R_{E'} = 0.3125$ cm, $f/\# = 8.89$ cm

4.13 aperture stop is the objective, eyepiece is field stop, $l_E = 0.0$, $R_E = 0.300$ cm, $l_{E'} = 1.06$ cm, $R_{E'} = -0.018$ cm, $l_W = -0.783$ cm, $R_W = -0.011$ cm, $l_{W'} = 0.0$, $R_{W'} = 0.25$ cm

4.15 $f_1' = 47.8$ cm, $f_2' = -132$ cm, $r_1 = 44.4$ cm, $r_2 = -44.4$ cm, $r_3 = -73.8$ cm

4.17 -0.0293 cm^{-1}

4.19 AC = −0.0132, LC = 0.000031

4.23 object is field stop, $l_E = 4.0$, $R_E = 4.0$

CHAPTER 5

5.7 $a = 0.9196$, $b = 0.308$, $c = -0.4532$, $d = 0.9356$

5.9 $f' = 10.0$, $m = -2.0$, $s = -15.0$, $s' = +30.0$

5.11 $f' = 25.0$, $l_{F'} = 20.0$, $l_{H'} = -5.0$, $l_F = -30.0$, $l_H = -5.0$

5.13 $t_k = 61.67$, $h' = -3.33$

5.17 $f' = 0.88$, $l_{F'} = 0.085$, $l_{H'} = -0.795$, $l_F = -0.757$, $l_H = 0.123$

5.19 $l_E = 6.0$, $R_E = 3.0$, $l_{E'} = -5.0$, $R_{E'} = 2.5$

CHAPTER 6

6.7 for all rays: $X_0 = 0$, $Z_0 = 0$, $L_0 = 0$, $Y_0 = 0$ unless stated otherwise
ray 1: $Y_0 = 0$, $M_0 = 0.196116$, $N_0 = 0.980581$
ray 2: $M_0 = -0.099504$, $N_0 = 0.995037$
ray 3: $M_0 = 0.099504$, $N_0 = 0.995037$
ray 4: $L_0 = 0.195180$, $M_0 = -0.975900$, $N_0 = 0.975900$
ray 5: $M_0 = -0.287348$, $N_0 = 0.957826$

6.9 Unless stated otherwise:
$X_0 = 0$, $Z_0 = 0$, $L_0 = 0$, $M_0 = 0.247404$, $N_0 = 0.968912$
ray 1: $Y_0 = 2.0$, $L_0 = 0$, $M_0 = 0$, $N_0 = 1.0$
ray 2: $Y_0 = 0$
ray 3: $Y_0 = 2.0$
ray 4: $X_0 = 2.0$, $Y_0 = 0$
ray 5: $Y_0 = -2.0$

6.13 97.148, -0.21765, -2.4829

6.15 -150, 0.003130, -10.02456

CHAPTER 7

7.18 $S = -0.003071$, $C = -0.003213$, $A = -0.003034$, $P = -0.014970$, $D = 0.000058$

7.23 $S_t = -0.00307$, $C_t = 0.00184$, $A_t = -0.00100$, $P_t = -0.01497$, $D_t = 0$

7.28 0.032256

7.30 0.032256

CHAPTER 8

8.5 25.0

8.9 8.889, 2.0, -8.889

8.11 640, 560, -80

8.13 -0.02

8.25 270, -1.5

8.27 40, 2

8.33 $R_E = 6.0$, $l_E = 0$, $R_{E'} = 3.0$, $l_{E'} = -30$

8.35 $R_E = 3.0$, $l_E = -180$, $R_{E'} = -7.0$, $l_{E'} = 0$

8.37 $l_H = -100$, $l_F = -200$, $l_{H'} = -60$, $l_{F'} = 40$, $f' = 100$

8.39 $l_H = -60$, $l_F = -20$, $l_{H'} = 160$, $l_{F'} = 120$, $f' = -40$

8.41 5.0

8.43 6.0, 4.0

8.51 $t_0 = 18.0$, $f_1' = 1.333$, $t_1 = 4.0$, $f_2' = 2.370$

CHAPTER 9

9.3 3.0, $\psi(3.0) = 1.0$

9.9 5.0, 1.0, $\psi(5.0) = 576$, $\psi(1.0) = 64$

9.15 $p_1 = 2.0$, $p_2 = -3.0$

9.18 $p_1 = 1.0$, $p_2 = 1.5$, $p_3 = 5.0$

9.25 1.412013

INDEX

A

B

C

D

L

M

N

O

P

Q

R

S

T

U

V